ISBN 978-0-260-08423-1
PIBN 10926739

TABLES

INTENDED TO FACILITATE THE OPERATIONS OF

NAVIGATION

AND

NAUTICAL ASTRONOMY,

AS AN ACCOMPANIMENT TO THE

AVIGATION AND NAUTICAL ASTRONOMY,

VOLS. 99 AND 100 OF THE RUDIMENTARY SERIES.

BY J. R. YOUNG,

FORMERLY PROFESSOR OF MATHEMATICS IN BELFAST COLLEGE.

LONDON:

JOHN WEALE, 59, HIGH HOLBORN.

1859.

LONDON:
BRADBURY AND EVANS, PRINTERS, WHITEFRIARS.

TABLES

FOR

NAVIGATION AND NAUTICAL ASTRONOMY.

EXPLANATION OF THE TABLES.

The following Tables are intended to be used conjointly with the "Mathematical Tables"—published in the series of Rudimentary Treatises—in the various computations of Navigation and Nautical Astronomy. They are eleven in number, and the purposes they serve will be readily understood from the following brief explanation:

Tables 1 and 2.

Sines, Cosines, &c., to every Quarter Point of the Compass.

The first of these Tables exhibits the *Natural* sines, cosines, &c., of *Courses* to every quarter-point of the compass, and the second furnishes the *Logarithmic* sines, cosines, &c. of the same angles. Here are two examples of their use:

1. A ship from latitude 37° 3′ N., sails S.W. by S. ½ S. a distance of 148 miles; required the latitude in and the departure made?

BY TABLE 1.

For the diff. lat.

Diff. lat. = cos course × dist.

cos course, $3\frac{1}{2}$ points =	·773
distance =	148
	6184
	3092
	773
diff. lat. S. =	114·404
	= 1° 54′ S.
lat. left .	37 3 N.
lat. in . .	35 9 N.

For the departure.

Dep. = sin course × dist.

sin course, $3\frac{1}{2}$ points =	·6343
distance =	148
	50744
	25372
	6343
departure W. =	93·8764

Hence the departure is 93·9 miles W.

BY TABLE 2.

For the diff. lat.

Diff. lat. = cos course × dist.

		Logs.
cos course, $3\frac{1}{2}$ points	=	9·88818
distance = 148	.	2·17026
diff. lat. S. = 114	.	2·05844

or diff. lat. = 1° 54′ S.

For the departure.

Dep. = sin course × dist.

		Logs.
sin course, $3\frac{1}{2}$ points	=	9·80236
distance = 148	.	2·17026
departure = 93·9	.	1·97262

∴ the departure is 93·9 m. W.

2. A ship sails from latitude 15° 55′ S. on a S.E. $\frac{1}{2}$ E. course till she finds herself in latitude 18° 49′ S.: required the departure made?

BY TABLE 1.

For the departure.

Dep. = tan course × diff. lat.

tan course, $4\frac{1}{2}$ points =	1·2185
diff. lat. =	174
	48740
	85295
	12185
departure E. =	212·019

BY TABLE 2.

For the departure.

Dep. = tan course × diff. lat.

		Logs.
tan course, $4\frac{1}{2}$ points	=	10·08583
diff. lat. = 174	.	2·24055
departure = 212 E.		2·32638

Hence the departure is 212 miles E.

Table 3.

Natural Sines, Cosines, Tangents, and Cotangents.

This is a table of natural sines, cosines, tangents, and cotangents to every degree and minute of the quadrant. It will be found useful in finding the direct course from the beginning to the end of a traverse, and in many other computations of Navigation and Trigonometry in which it may be inconvenient to employ logarithms. Ample illustration of the advantage of this table is given in the Navigation and Nautical Astronomy which accompanies the present volume.

Tables 4 and 5.

Difference of Latitude and Departure, or Traverse Table for Points and for Degrees.

On account of the use of these Tables in working a traverse, they are frequently called *Traverse Tables.* The difference of latitude and departure due to any course and distance are found from one or other of these Tables by *inspection*. The course stands at the head of the page, or if more than 45°, at the foot—expressed in points and quarter-points in Table 4, and in degrees in Table 5—and the difference of latitude and departure which the ship makes in running any distance on that course, from 1 mile to 300 miles is inserted in the body of the Table, and is found as in the examples following:

1. A ship sails N.W. by N. ½ W. a distance of 78 miles: required the difference of latitude and departure by inspection?

The given course is 3½ points; and referring to Table 4, we find the page devoted to this course to be page 38, in which, against 78 in a column headed *Dist.*, stands 60·3 under the head *Lat.*, and 49·5 under the head *Dep.* We conclude therefore that, for the given course and distance, the difference of latitude is 60·3 miles, and the departure 49·5 miles.

2. Suppose the course to be 5½ points, and the distance sailed 78 miles as before.

Then, as the course here exceeds four points, we look for it at the foot of the page (p. 84), and against 78 in the *distance* column, we find 68·8 in the adjacent *departure* column, and 36·8 in the *difference of latitude* column, so that the difference of latitude made is 36·8 miles, and the departure 68·8 miles.

3. Suppose the course to be 30°, and the distance sailed on that course 78 miles.

Then, turning to the page headed 30° in Table 5 (p. 70), we find against 78 in the *Dist.* column, the number 67·5 in the adjacent *Lat.* column, and 39 in the *Dep.* column: we conclude therefore that the difference of latitude made is 67·5 miles, and the corresponding departure 39 miles.

4. But if the course exceed 45°, if, for instance, it be 58°, it will be found at the bottom of the page (p. 72), and against the *distance* 78, there appears 66·1 for *departure*, and 41·3 for *difference of latitude.*

If the distance sailed on any course be greater than 300 miles, since the limits of the Table will then be exceeded, to render the Table still available we must take the half, or the third, or the fourth, &c., of the given distance, so that the part taken may be a distance within the limits of the Table: the diff. lat. and dep. corresponding to this aliquot part of the given distance, being each multiplied by 2, or by 3, or by 4, &c., according to the part taken, will give the diff. lat. and dep. due to the entire distance.

These Tables are employed not only in plane sailing, but also in parallel and mid-latitude sailings, as is sufficiently exemplified in the treatise on Navigation and Nautical Astronomy, to which the present collection is adapted. And in all computations of the parts of a right-angled triangle, provided the angles are expressed in degrees and minutes—seconds being disregarded, Table 5 may be used to save the trouble of arithmetical calculation.

TABLE 6.

Natural Cosines to Degrees, Minutes, and Seconds.

This Table is employed in the Author's method of clearing the lunar distance for the purpose of finding the longitude at sea. The several columns of cosines are headed by the degrees, the accompanying minutes being inserted in the first column on the left of the page: this is equally a column of the seconds, and is accordingly headed by the marks for minutes and seconds. As in the ordinary trigonometrical tables, the cosine of an arc or angle belonging to any number of degrees and minutes is found in the column of cosines, under the degrees, and in a horizontal line with the minutes found in the first column.

Suppose this cosine to have been extracted from the table; then, if there are seconds also in the arc or angle, we again refer to the same first column for these, and in the same horizontal line with them, and in the column headed "parts for ″" which immediately follows the column from which our cosine has been extracted, we shall find the correction for that cosine: this correction is always to be *subtracted*. The remainder will be the cosine of the given degrees, minutes, and seconds. But in taking out a cosine to degrees, minutes, and seconds, it will in general be better to enter the marginal column first with the *seconds*, to write the "parts" for these on a slip of paper, and then, entering the same column with the minutes, instead of extracting the corresponding cosine, to place the slip under it, and subtract the correction written thereon. The Table extends from 0° to 90° only, so that it does not give immediately the cosines of obtuse angles: when therefore the angle is obtuse, we must enter the Table with the supplement of that angle, and regard the corresponding cosine as negative. It was thought better that this trifling amount of trouble should be incurred, than that the extent of the Table should be doubled.

There is indeed a way of avoiding this enlargement of the Table, and yet providing for the supplementary arcs; but as the Table would then have to be used in a peculiar manner—disturbing the general principle upon which the extracts from it are made in the other cases—it was thought preferable, after due consideration, to reject it. The plan is this. Suppose a marginal column added to the right of each page, for the minutes and seconds, proceeding upwards from 0 to 60, and that the degrees supplementary to those at the top of the page (one degree in each case being omitted in the supplement) were given at the bottom, as in the ordinary tables of sines and cosines; we should then have to use the table as in the following instance. Required the cosine of 115° 41′ 34″? Referring to page 96, we should find 115° at the bottom of the column headed 64°, from which column opposite the 41′ on the *right*, we should take the cosine 43397, then referring to the *left* for the 34″, we should extract the "parts" 149, which we should have to *add* to 43397: we should thus get 433546 for the required cosine. As the Table at present stands, however, we enter it with 64° 18′ 26″, the supplement of 115° 41′ 34″, and for 64° 18′ we find 433659, while the "parts" for 26″ are found in a similar manner to be 113, which subtracted from 433659 gives 433546 for the required cosine.

It has not been thought necessary to insert the decimal point before each cosine; indeed, in the operation for which this Table is specially prepared, the numbers may always be regarded as integers. (See p. 227, Naut. Ast.)

Ex. 1. Suppose the natural cosine of 37° 21′ 33″ were required:

Turning to the page containing 37° at the top (page 92), we find the "parts" against 33″ to be 98, and the cosine against 21′ to be 794944, subtracting the 98 from this, we write down 794846 for the cosine required.

2. Again, suppose we wanted the cosine of 118° 16′ 43″: Subtracting this from 180°, the angle in the Table is 61° 43′ 17″. Under 61°, and against 17″, the "parts" are 72, and against 43′, the cosine is 473832: subtracting the 72 from this, we find 473760 for the required cosine, which is negative because the proposed angle is obtuse.

3. Required the angle whose cosine is 452801?

By the Table	452954 = cos 63° 4′	
Given cosine	452801	
Parts for the secs.	153	35″

Hence the angle is 63° 4′ 35″.

If given the cosine had been —452801, then the supplement of the angle thus determined, namely 116° 55′ 25″, would have been the angle to which that cosine belongs.

TABLE 7.

Proportional Logarithms.

These are a peculiar kind of logarithms, first constructed by Dr. Maskelyne to facilitate the operation of finding the Greenwich time, at which a lunar distance taken at sea has place. They are also useful in many other inquiries, in which difference of time varies, as difference of angular measurement. When difference of time is required the Table is to be entered with difference of angular measurement, and when difference of angular measurement is required it is to be entered with the corresponding difference of time. Sufficient illustration of the mode of employing the Table is given in the Nautical Astronomy, pages 236-7.

The last four figures in the Table are decimals, and the greater part of the Table consists of these decimals alone; the decimal point however is suppressed, as well here as in the Nautical Almanac, since in finding the Greenwich time of a lunar distance the logarithms may be always regarded as whole numbers.

1. Suppose the proportional logarithm of $2^h\ 8^m\ 16^s$ is required, or the proportional logarithm of $2°\ 8'\ 16''$: turning to the proper page of the Table (p. 112) we find that for each of these arguments the P. L. is ·1472.

2. Suppose it be required to find the difference of time corresponding to the P. L. ·2954: turning to page 109 we see that this P. L. answers to the difference of time, $1^h\ 31^m\ 10^s$. It also answers to the difference $1°\ 31'\ 10''$ of angular measurement. The Table extends from 0^s to 3^h or 10800^s, or from $0''$ to $3°$: the proportional logarithm of the extreme number of seconds, namely 10800, being 0, the formal insertion of it has not been thought necessary. For the

theory of proportional logarithms reference may be made to the Nautical Astronomy, page 235.

Table 8.

For determining the distance of an object seen in the horizon.

This Table shows the utmost distance at which an object on the surface of the sea can be seen by an eye elevated above it; the elevation of the eye being estimated in feet, and the distance in nautical miles, allowance being made for atmospherical refraction. If the object itself be elevated above the surface, and its summit be just visible in the remote distance, then, if the height of the distant object thus lost to sight all but its top, be previously known, the Table will enable us to find its distance, the height of the spectator's eye being known.

Ex. 1. From the mast head, 130 feet high, a boat was observed as a remote speck, just appearing in the horizon: required its distance?

In the Table opposite 130 feet is found 13·1 miles, the distance required.

2. From the same height the top of a lighthouse known to be 300 feet above the level of the sea, was discerned in the horizon: required the distance from the lighthouse?

For the height 130 the distance is 13·1 miles.
,, 300 ,, 19·9

∴ the distance of the lighthouse is 33 miles.

TABLE 9.

For finding the mean time (nearly) of the Meridian Transits of the Principal Fixed Stars.

In this Table is recorded the mean time at which each of the 100 stars there selected passes the meridian of the ship. The times of transit are given only for every tenth day; but as the stars come to the meridian earlier every day, by a uniform difference of time—about *four* minutes—it is easy to find the time of transit on any intermediate day: we have only to multiply the number of days after the day of transit recorded in the Table by 4, and to subtract the number of minutes in the product from the time of transit on the day given in the Table. Or we may multiply the number of days *before* the next tabular day by 4, and add the resulting minutes to the time of that advanced day's transit: for example,

Suppose it were required to find the mean time of transit of α^2 Centauri on the fifth of November:

By the Table the time of transit Nov. 1, is	$23^h\ 43^m$
And for four days afterwards, we subtract $4 \times 4 =$	16
∴ time of transit Nov. 5th is	23 27

Or, the time of transit Nov. 11, being by the Table $23^h\ 4^m$, by adding $6 \times 4 = 24$ minutes, we have for the time on Nov. 5th, $23^h\ 28^m$.

It will of course be understood that the times of transit furnished by aid of this Table, are only the times *nearly*; but in no case will the time differ from the truth by more than about two or three minutes, and the Table is therefore as accurate as necessary for the purpose intended by it, which purpose is to apprise the mariner about what time he may expect certain well known stars to appear on the meridian whenever the weather permits his taking a star-altitude for his latitude. Should the observer not be sufficiently acquainted with the stars to avail himself readily of this information, he is recommended to procure Mr. Jeans's "Hand Book for the Stars."

But the right star may generally be detected when we know, within about half a degree or so, what altitude it ought to have when on the meridian, and this approximate altitude may be found by help of the star's declination, and the latitude by account; thus:—

1. For a meridian altitude above the Pole.

In this case the star passes from the eastward towards the westward, and *ascends* to the meridian.

When the latitude by account and the declination have the same name. Add 90° to the declination, and subtract the latitude by account; or, which is the same thing, add the colatitude to the declination, the result will be the approximate altitude, measured from the S. in N. lat. and from the N. in S. lat.

When the latitude by account and the declination have different names. Add together the latitude by account, and the declination, and subtract the sum from 90°, or, which is the same thing, subtract the declination from the colatitude, the result will be the approximate altitude from the S. in N. lat. and from the N. in S. lat. If the sum of lat. and dec. exceed 90° the star cannot appear above the horizon.

2. For a meridian altitude below the Pole.

In this case the star passes from the westward towards the eastward and *descends* to the meridian. It can be visible below the Pole only when the latitude and declination have the same name.

From the sum of the latitude and declination subtract 90°: the remainder will be the approximate altitude reckoned from the N. in N. lat. and from the S. in S. lat. If the sum of lat. and dec. be less than 90° the star will pass the meridian below the Pole under the horizon.

To assist in thus getting an approximate meridian altitude, the stars' declinations—each to the nearest degree and minute—are given in the marginal column of the Table. It is scarcely possible to mistake the star, because no other will have nearly the same meridian altitude at the time.

The approximate altitude being found in this way, and the index of the instrument set to it—the sight being directed to the proper point of the horizon—the true meridian altitude, and thence the latitude of the ship, may be readily determined.

It is to be observed that if the mean time at ship be A.M., we must add 12^h to that time for the corresponding time in the Table, from the preceding noon, when the star opposite that time will pass the meridian above the Pole. If the mean time at ship be P.M. the star opposite that time in the Table, will be on the meridian above the Pole; and if 12^h be added to the time, the star opposite the result, will be on the meridian below the Pole.

TABLE 10.

Best Time for taking the Altitude of a Celestial Object, with the view of determining the TIME *at Sea.*

When the time at the ship is to be deduced from an altitude, it is desirable that the object observed should be in such a position that a small error in the observation may have the least possible influence on the magnitude of the hour-angle. And this position is attained when the object is on the prime vertical. If, however, the declination be of a contrary name to the latitude of the place of observation, the object will cross the prime vertical before it rises and after it sets, so that it cannot be *observed* on this circle at all; the observation should then be taken as soon after it rises (or before it sets) as that the altitude of it is sufficient to secure it from the fluctuating effects of the horizontal refraction. The altitude should not be less than 6 or 7 degrees.

The present Table points out, with accuracy enough for the purpose, the time when the prime vertical is crossed *above* the horizon, that is, when the declination is of the same name as the latitude.

If the object observed be the sun, the table is to be entered with that degree of declination which approaches nearest to the sun's declination at the time, and which is found at the top of the table; underneath this declination, and opposite to the latitude found in the first column, is the time before or after *noon* when it will be most advantageous to take the altitude for TIME.*

If, however, the object be a star, we must first ascertain the time when it passes over the meridian; the preceding Table will supply this information. Then, by aid of the present Table, entering it with the star's declination and the latitude of the place, we take out the corresponding time, or hour-angle between the meridian and star, when the latter is on the prime vertical. If the observation is to

* As the Table gives the hour-angle from the meridian in Time, if the object be the sun, the time will be *apparent*; the correction for Equation of Time being applied, will convert it into mean time. For a star, no such correction is requisite.

be made *before* the meridian transit, we must *subtract* the latter time from the former; if the observation is to be made *after* the meridian transit, we must *add* the two times together.

It is only when the latitude of the place is greater than the declination of the object (both being of the same name), that the object actually crosses the prime vertical; when the two are equal, the object is on the prime vertical when it is on the meridian; when the latitude is less than the declination, the Table shows the time of nearest approach to the prime vertical, which will be the best for the altitude to be observed.

We may further observe that, when the latitude of the ship is pretty nearly a mean between two consecutive latitudes in the Table, the time will be obtained more accurately by taking the mean of the times corresponding to those two latitudes.

Ex. 1. At what time will the star α Leonis or Regulus bear due East on the 6th of February, in latitude 47° N.?

	h	m	
Mean time of transit, Feb. 1 —20m (Table 9)	12	53	
Time corresponding to dec. 12° N. and lat. 47°	—5	14	
Mean time of star's bearing due East	7	39	P.M.

2. At what time will the star α Arietis bear nearest to the West on Nov. 21, in latitude 17° 32′ N.?

	h	m	
Time corresponding to dec. 22° and lat. 17°	2	43	
,, ,, ,, lat. 18°	2	26	
	2) 5	9	
,, ,, ,, lat. 17° 30′,	+2	35	
Mean time of star's transit (Table 9)	9	56	
,, ,, bearing nearest to W.	12	31	P.M.
or	0	31	A.M.

that is, at about 31m past midnight.

NOTE.—The mean time of the meridian passage of each of the planets, and also of the moon, is given in the Nautical Almanac for every day in the year. (See Table at p. 120 of the Nautical Astronomy.)

If a planet be favorable for observation, and there be any doubt as to which of the planets it is, the doubt may be removed by noticing what known star is nearest to it, referring to Table 9 for the time of the meridian passage of that star, and then finding, from the Nautical Almanac, which of the planets it is that passes the meridian nearest to that time.

TABLE 11.

For Finding the Altitude of a Celestial Object, most suitable for ascertaining the Time at Sea.

This Table is intended to show, what altitude nearly an object of given declination must have in a given latitude to be most suitable for deducing the time from that altitude; that is to say, it points out approximately the altitude which the object has when on the prime vertical, or when it makes the nearest approach to it. When the object is a star, and the time most suitable for taking its altitude is found from the last Table, the approximate altitude of it at that time, as given by the present Table, will enable the observer readily to discover it, even should he be but little familiar with the constellations.

To these Tables is added a list of the Proper Names of certain of the principal fixed stars, and to this is subjoined the names and sounds of the letters of the Greek Alphabet.

*** At the commencement of the foregoing Explanation, reference is made to Law's "Mathematical Tables," as an accompaniment to the present volume. The first eighteen of these Tables, together with the eleven in the present collection, will be found to contain all that is indispensably necessary in the several operations of Navigation and Nautical Astronomy.

CONTENTS.

TABLE I.

TABLE II.

TABLE III.

TABLE IV.

TABLE V.

TABLE VI.

TABLE VII.

TABLE VIII.

TABLE IX.

TABLE X.

TABLE XI.

TABLE 1.

NATURAL SINES AND TANGENTS TO EVERY QUARTER POINT OF THE COMPASS.

POINTS.	SINE.	COSINE.	TANGENT.	COTANGENT.	
0	0·00000	1·00000	0·00000	Infinite	8
¼	·04907	·99880	·04913	20·35560	7¾
½	·09802	·99518	·09849	10·15319	7½
¾	·14730	·98918	·14834	6 74146	7¼
1	·19509	·98079	·19891	5·02734	7
1¼	·24298	·97003	·25049	3·99222	6¾
1½	·29028	·95694	·30335	3·29483	6½
1¾	·33689	·94154	·35781	2 79481	6¼
2	·38268	·92388	·41421	2·41421	6
2¼	·42756	·90399	·47296	2·11432	5¾
2½	·47140	·88192	·53451	1·87087	5½
2¾	·51410	·85773	·59938	1·66840	5¼
3	·55556	·83147	·66818	1·49661	5
3¼	·59570	·80321	·74165	1·34834	4¾
3½	·63439	·77301	·82068	1·21850	4½
3¾	·67156	·74095	·90635	1·10333	4¼
4	·70711	·70711	1·00000	1·00000	4
	COSINE.	SINE.	COTANGENT.	TANGENT.	POINTS.

TABLE 2.

LOGARITHMIC SINES AND TANGENTS TO EVERY QUARTER POINT OF THE COMPASS.

POINTS.	SINE.	COSINE.	TANGENT.	COTANGENT.	
0	0·00000	10·00000	0·00000	Infinite	8
¼	8·69079	9·99948	8·69132	11·30868	7¾
½	8·99130	9·99790	8.99340	11·00660	7½
¾	9·16652	9·99527	9·17125	10·82875	7¼
1	9·29024	9·99157	9·29866	10·70134	7
1¼	9·38557	9·98679	9·39878	10·60121	6¾
1½	9·46282	9·98088	9·48194	10·51806	6½
1¾	9·52749	9·97384	9 55365	10·44635	6¼
2	9 58284	9·96562	9·61722	10·38278	6
2¼	9·63099	9·95616	9·67483	10·32517	5¾
2½	9 67339	9·94543	9·72796	10·27204	5½
2¾	9·71105	9 93335	9·77770	10·22230	5¼
3	9·74474	9·91985	9·82489	10·17511	5
3¼	9·77503	9·90483	9·87020	10·12980	4¾
3½	9·80236	9·88818	9·91417	10·08583	4½
3¾	9·82708	9·86979	9·95729	10·04270	4¼
4	9·84949	9·84949	10·00000	10·00000	4
	COSINE.	SINE.	COTANGENT.	TANGENT.	POINTS.

′	0°	1°	2°	3°	4°	5°	6°	7°	′
0	00 00	017 45	034 90	052 34	069 76	087 16	104 53	121 87	60
1	29	74	035 19	63	070 05	45	82	122 16	59
2	58	018 03	48	92	34	74	105 11	45	58
3	87	32	77	053 21	63	088 03	40	74	57
4	001 16	62	036 06	50	92	31	69	123 02	56
5	45	91	35	79	071 21	60	97	31	55
6	75	019 20	64	054 08	50	89	106 26	60	54
7	002 04	49	93	37	79	089 18	55	89	53
8	33	78	037 23	66	072 08	47	84	124 18	52
9	62	020 07	52	95	37	76	107 13	47	51
10	91	36	81	055 24	66	090 05	42	76	50
11	003 20	65	038 10	53	95	34	71	125 04	49
12	49	94	39	82	073 24	63	108 00	33	48
13	78	021 23	68	056 11	53	92	29	62	47
14	004 07	52	97	40	82	091 21	58	91	46
15	36	81	039 26	69	074 11	50	87	126 20	45
16	65	022 11	55	98	40	79	109 16	49	44
17	95	40	84	057 27	69	092 08	45	78	43
18	005 24	69	040 13	56	98	37	73	127 06	42
19	53	98	42	85	075 27	66	110 02	35	41
20	82	023 27	71	058 14	56	95	31	64	40
21	006 11	56	041 00	44	85	093 24	60	93	39
22	40	85	29	73	076 14	53	89	128 22	38
23	69	024 14	59	059 02	43	82	111 18	51	37
24	98	43	88	31	72	094 11	47	80	36
25	007 27	72	042 17	60	077 01	40	76	129 08	35
26	56	025 01	46	89	30	69	112 05	37	34
27	85	30	75	060 18	59	98	34	66	33
28	008 14	60	043 04	47	88	095 27	63	95	32
29	44	89	33	76	078 17	56	91	130 24	31
30	73	026 18	62	061 05	46	85	113 20	53	30
31	009 02	47	91	34	75	096 14	49	81	29
32	31	76	044 20	63	079 04	42	78	131 10	28
33	60	027 05	49	92	33	71	114 07	39	27
34	90	34	78	062 21	62	097 00	36	68	26
35	010 18	63	045 07	50	91	29	65	97	25
36	47	92	36	79	080 20	58	94	132 26	24
37	76	028 21	65	063 08	49	87	115 23	54	23
38	011 05	50	94	37	78	098 16	52	83	22
39	34	79	046 23	66	081 07	45	80	133 12	21
40	64	029 08	53	95	36	74	116 09	41	20
41	93	38	82	064 24	65	099 03	38	70	19
42	012 22	67	047 11	53	94	32	67	99	18
43	51	96	40	82	082 23	61	96	134 27	17
44	80	030 25	69	065 11	52	90	117 25	56	16
45	013 09	54	98	40	81	100 19	54	85	15
46	38	83	048 27	69	083 10	48	83	135 14	14
47	67	031 12	56	98	39	77	118 12	43	13
48	96	41	85	066 27	68	101 06	40	72	12
49	014 25	70	049 14	56	97	35	69	136 00	11
50	54	99	43	85	084 26	64	98		10
51	83	032 28	72	067 14	55	92	119 27		9
52	015 13	57	050 01	43	84	102 21	56		8
53	42	86	30	73	085 13	50	85	137 29	7
54	71	033 16	59	068 02	42	79	120 14	58	6
55	016 00	45	88	31	71	103 08	43	73	5
56	29	74	051 17	60	086 00	37	71	138 02	4
57	58	034 03	46	89	29	66	121 00	31	3
58	87	32	76	069 18	58	95	29	60	2
59	017 16	61	052 05	47	87	104 24	58	89	1
60	45	90	34	76	087 16	53	87	139 17	0
′	89°	88°	87°	86°	85°	84°	83°	82°	′

NATURAL COSINE.

′	0°	1°	2°	3°	4°	5°	6°	7°	′
0	000 00	017 46	034 92	052 41	069 93	087 49	105 10	122 78	60
1	29	75	035 21	70	070 22	78	40	123 08	59
2	58	018 04	50	053 00	51	088 07	69	38	58
3	87	33	79	28	80	37	99	67	57
4	001 16	62	036 09	57	071 10	66	106 28	97	56
5	45	91	38	87	39	95	58	124 26	55
6	75	019 20	67	054 16	68	089 25	87	56	54
7	002 04	49	96	45	97	54	107 16	85	53
8	33	78	037 25	74	072 27	83	46	125 15	52
9	62	020 07	54	055 03	56	090 13	75	44	51
10	91	37	83	33	85	42	108 05	74	50
11	003 20	66	038 12	62	073 14	71	34	126 03	49
12	49	95	42	91	44	091 01	63	33	48
13	78	021 24	71	056 20	73	30	93	62	47
14	004 07	53	039 00	49	074 02	59	109 22	92	46
15	36	82	29	78	31	89	52	127 22	45
16	65	022 11	58	057 08	61	092 18	81	51	44
17	95	40	87	37	90	47	110 11	81	43
18	005 24	69	040 16	66	075 19	77	40	128 10	42
19	53	98	46	95	48	093 06	70	40	41
20	82	023 28	75	058 24	78	35	99	69	40
21	006 11	57	041 04	54	076 07	65	111 28	99	39
22	40	86	33	83	36	94	58	129 29	38
23	69	024 15	62	059 12	65	094 23	87	58	37
24	98	44	91	41	95	53	112 17	88	36
25	007 27	73	042 20	70	077 24	82	46	130 17	35
26	56	025 02	50	99	53	095 11	76	47	34
27	85	31	79	060 29	82	41	113 05	76	33
28	008 15	60	043 08	58	078 12	70	35	131 06	32
29	44	89	37	87	41	096 00	64	36	31
30	73	026 19	66	061 16	70	29	94	65	30
31	009 02	48	95	45	99	58	114 23	95	29
32	31	77	044 24	75	079 29	88	53	132 24	28
33	60	027 06	54	062 04	58	097 17	82	54	27
34	89	35	83	33	87	46	115 11	84	26
35	010 18	64	045 12	62	080 17	76	41	133 13	25
36	47	93	41	91	46	098 05	70	42	24
37	76	028 22	70	063 21	75	34	116 00	72	23
38	011 05	51	99	50	081 04	64	29	134 02	22
39	35	81	046 28	79	34	93	59	32	21
40	64	029 10	58	064 08	63	099 23	88	61	20
41	93	39	87	38	92	52	117 18	91	19
42	012 22	68	047 16	67	082 22	81	47	135 21	18
43	51	97	45	96	51	100 11	77	50	17
44	80	030 26	74	065 25	80	40	118 06	80	16
45	013 09	55	048 03	54	083 09	69	36	136 09	15
46	38	84	33	84	39	99	65	39	14
47	67	031 14	62	066 13	68	101 28	95	69	13
48	96	43	91	42	97	58	119 24	98	12
49	014 25	72	049 20	71	084 27	87	54	137 28	11
50	55	032 01	49	067 00	56	102 16	83	58	10
51	84	30	78	30	85	46	120 13	87	9
52	015 13	59	050 07	59	085 14	75	42	138 17	8
53	42	88	37	88	44	103 05	72	47	7
54	71	033 17	66	068 17	73	34	121 01	76	6
55	016 00	46	95	47	086 02	63	31	139 06	5
56	29	76	051 24	76	32	93	60	35	4
57	58	034 05	53	069 05	61	104 22	90	65	3
58	87	34	82	34	90	52	122 19	95	2
59	017 16	63	052 12	63	087 20	81	49	140 24	1
60	46	92	41	93	49	105 10	78	54	0
′	89°	88°	87°	86°	85°	84°	83°	82°	′

NATURAL COTANGENT.

′	8°	9°	10°	11°	12°	13°	14°	15°	′
0	139 17	156 43	173 65	190 81	207 91	224 95	241 92	258 82	60
1	46	72	93	191 09	208 20	225 23	242 20	259 10	59
2	75	157 01	174 22	38	48	52	49	38	58
3	140 04	30	51	67	77	80	77	66	57
4	33	58	79	95	209 05	226 08	243 05	94	56
5	61	87	175 08	192 24	33	37	33	260 22	55
6	90	158 16	37	52	62	65	62	50	54
7	141 19	45	65	81	90	93	90	79	53
8	48	73	94	193 09	210 19	227 22	244 18	261 07	52
9	77	159 02	176 23	38	47	50	46	35	51
10	142 05	31	51	66	76	78	74	63	50
11	34	59	80	95	211 04	228 07	245 03	91	49
12	63	88	177 08	194 23	32	35	31	262 19	48
13	92	160 17	37	52	61	63	59	47	47
14	143 20	46	66	81	89	92	87	75	46
15	49	74	94	195 09	212 18	229 20	246 15	263 03	45
16	78	161 03	178 23	38	46	48	44	31	44
17	144 07	32	52	66	75	77	72	59	43
18	36	60	80	95	213 03	230 05	247 00	87	42
19	64	89	179 09	196 23	31	33	28	264 15	41
20	93	162 18	37	52	60	62	56	43	40
21	145 22	47	66	80	88	90	84	71	39
22	51	75	95	197 09	214 17	231 18	248 13	265 00	38
23	80	163 04	180 23	37	45	46	41	28	37
24	146 08	33	52	66	74	75	69	56	36
25	37	61	81	94	215 02	232 03	97	84	35
26	66	90	181 09	198 23	30	31	249 25	266 12	34
27	95	164 19	38	51	59	60	54	40	33
28	147 23	47	66	80	87	88	82	68	32
29	52	76	95	199 08	216 16	233 16	250 10	96	31
30	81	165 05	182 24	37	44	45	38	267 24	30
31	148 10	33	52	65	72	73	66	52	29
32	38	62	81	94	217 01	234 01	94	80	28
33	67	91	183 09	200 22	29	29	251 22	268 08	27
34	96	166 20	38	51	58	58	51	36	26
35	149 25	48	67	79	86	86	79	64	25
36	54	77	95	201 08	218 14	235 14	252 07	92	24
37	82	167 06	184 24	36	43	42	35	269 20	23
38	150 11	34	52	65	71	71	63	48	22
39	40	63	81	93	99	99	91	76	21
40	69	92	185 09	202 22	219 28	236 27	253 20	270 04	20
41	97	168 20	38	50	56	56	48	32	19
42	151 26	49	67	79	85	84	76	60	18
43	55	78	95	203 07	220 13	237 12	254 04	88	17
44	84	169 06	186 24	36	41	40	32	271 16	16
45	152 12	35	52	64	70	69	60	44	15
46	41	64	81	93	98	97	88	72	14
47	70	92	187 10	204 21	221 26	238 25	255 16	272 00	13
48	99	170 21	38	50	55	53	45	28	12
49	153 27	50	67	78	83	82	73	56	11
50	56	78	95	205 07	222 12	239 10	256 01	84	10
51	85	171 07	188 24	35	40	38	29	273 12	9
52	154 14	36	52	63	68	66	57	40	8
53	42	64	81	92	97	95	85	68	7
54	71	93	189 10	206 20	223 25	240 23	257 13	96	6
55	155 00	172 22	38	49	53	51	41	274 24	5
56	29	50	67	77	82	79	70	52	4
57	57	79	95	207 06	224 10	241 08	98	80	3
58	86	173 08	190 24	34	38	36	258 26	275 08	2
59	156 15	36	52	63	67	64	54	36	1
60	43	65	81	91	95		82	64	0
′	81°	80°	79°	78°	77°	76°	75°	74°	′

NATURAL COSINE.

′	8°	9°	10°	11°	12°	13°	14°	15°	′
0	140 54	158 38	176 33	194 38	212 56	230 87	249 33	267 95	60
1	84	68	63	68	86	231 17	64	268 26	59
2	141 13	98	93	98	213 16	48	95	57	58
3	43	159 28	177 23	195 29	47	79	250 26	88	57
4	73	58	53	59	77	232 09	56	269 20	56
5	142 02	88	83	89	214 08	40	87	51	55
6	32	160 17	178 13	196 19	38	71	251 18	82	54
7	62	47	43	49	69	233 01	49	270 13	53
8	91	77	73	80	99	32	80	44	52
9	143 21	161 07	179 03	197 10	215 29	63	252 11	76	51
10	51	37	33	40	60	93	42	271 07	50
11	81	67	63	70	90	234 24	73	38	49
12	144 10	96	93	198 01	216 21	55	253 04	69	48
13	40	162 26	180 23	31	51	85	35	272 01	47
14	70	56	53	61	82	235 16	66	32	46
15	145 00	86	83	91	217 12	47	97	63	45
16	29	163 16	181 13	199 21	43	78	254 28	94	44
17	59	46	43	52	73	236 08	59	273 26	43
18	88	76	73	82	218 04	39	90	57	42
19	146 18	164 06	182 03	200 12	34	70	255 21	88	41
20	48	35	33	42	64	237 00	52	274 19	40
21	78	65	63	73	95	31	83	51	39
22	147 07	95	93	201 03	219 25	62	256 14	82	38
23	37	165 25	183 23	33	56	93	45	275 13	37
24	67	55	53	64	86	238 23	76	45	36
25	96	85	84	94	220 17	54	257 07	76	35
26	148 26	166 15	184 14	202 24	47	85	38	276 07	34
27	56	45	44	54	78	239 16	69	39	33
28	86	74	74	85	221 08	46	258 00	70	32
29	149 15	167 04	185 04	203 15	39	77	31	277 01	31
30	45	34	34	45	69	240 08	62	32	30
31	75	64	64	76	222 00	39	93	64	29
32	150 05	94	94	204 06	31	69	259 24	95	28
33	34	168 24	186 24	36	61	241 00	55	278 26	27
34	64	54	54	66	92	31	86	58	26
35	94	84	84	97	223 22	62	260 17	89	25
36	151 24	169 14	187 14	205 27	53	93	48	279 21	24
37	53	44	45	57	83	242 23	79	52	23
38	83	74	75	88	224 14	54	261 10	83	22
39	152 13	170 04	188 05	206 18	44	85	41	280 15	21
40	43	33	35	48	75	243 16	72	46	20
41	72	63	65	79	225 06	47	262 03	77	19
42	153 02	93	95	207 09	36	77	35	281 09	18
43	32	171 23	189 25	39	67	244 08	66	40	17
44	62	53	55	70	97	39	97	72	16
45	91	83	86	208 00	226 28	70	263 28	282 03	15
46	154 21	172 13	190 16	30	58	245 01	59	34	14
47	51	43	46	61	89	32	90	66	13
48	81	73	76	91	227 19	62	264 21	97	12
49	155 11	173 03	191 06	209 21	50	93	52	283 29	11
50	40	33	36	52	81	246 24	83	60	10
51	70	63	66	82	228 11	55	265 15	91	9
52	156 00	93	97	210 13	42	86	46	284 23	8
53	30	174 23	192 27	43	72	247 17	77	54	7
54	60	53	57	73	229 03	48	266 08	86	6
55	89	83	87	211 04	34	78	39	285 17	5
56	157 19	175 13	193 17	34	64	248 09	70	49	4
57	49	43	47	64	95	40	267 01	80	3
58	79	73	78	95	230 26	71	33	286 12	2
59	158 09	176 03	194 08	212 25	56	249 02	64	43	1
60	38	33	38	56	87	33	95	74	0
′	81°	80°	79°	78°	77°	76°	75°	74°	′

NATURAL COTANGENT.

′	16°	17°	18°	19°	20°	21°	22°	23°	′
0	275 64	292 37	309 02	325 57	342 02	358 37	374 61	390 73	60
1	92	65	29	84	29	64	88	391 00	59
2	276 20	93	57	326 12	57	91	375 15	27	58
3	48	293 21	85	39	84	359 18	42	53	57
4	76	48	310 12	67	343 11	45	69	80	56
5	277 04	76	40	94	39	73	95	392 07	55
6	31	294 04	68	327 22	66	360 00	376 22	34	54
7	59	32	95	49	93	27	49	60	53
8	87	60	311 23	77	344 21	54	76	87	52
9	278 15	87	51	328 04	48	81	377 03	393 14	51
10	43	295 15	78	32	75	361 08	30	41	50
11	71	43	312 06	59	345 03	35	57	67	49
12	99	71	33	87	30	62	84	94	48
13	279 27	99	61	329 14	57	90	378 11	394 21	47
14	55	296 26	89	42	84	362 17	38	48	46
15	83	54	313 16	69	346 12	44	65	74	45
16	280 11	82	44	97	39	71	92	395 01	44
17	39	297 10	72	330 24	66	98	379 19	28	43
18	67	37	99	51	94	363 25	46	55	42
19	95	65	314 27	79	347 21	52	73	81	41
20	281 23	93	54	331 06	48	79	99	396 08	40
21	50	298 21	82	34	75	364 06	380 26	35	39
22	78	49	315 10	61	348 03	34	53	61	38
23	282 06	76	37	89	30	61	80	88	37
24	34	299 04	65	332 16	57	88	381 07	397 15	36
25	62	32	93	44	84	365 15	34	41	35
26	90	60	316 20	71	349 12	42	61	68	34
27	283 18	87	48	98	39	69	88	95	33
28	46	300 15	75	333 26	66	96	382 15	398 22	32
29	74	43	317 03	53	93	366 23	41	48	31
30	284 02	71	30	81	350 21	50	68	75	30
31	29	98	58	334 08	48	77	95	399 02	29
32	57	301 26	86	36	75	367 04	383 22	28	28
33	85	54	318 13	63	351 02	31	49	55	27
34	285 13	82	41	90	30	58	76	82	26
35	41	302 09	68	335 18	57	85	384 03	400 08	25
36	69	37	96	45	84	368 12	30	35	24
37	97	65	319 24	73	352 11	39	56	62	23
38	286 25	92	51	336 00	39	67	83	88	22
39	52	303 20	79	27	66	94	385 10	401 15	21
40	80	48	320 06	55	93	369 21	37	42	20
41	287 08	76	34	82	353 20	48	64	68	19
42	36	304 03	61	337 10	47	75	91	95	18
43	64	31	89	37	75	370 02	386 17	402 21	17
44	92	59	321 16	64	354 02	29	44	48	16
45	288 20	86	44	92	29	56	71	75	15
46	47	305 14	71	338 19	56	83	98	403 01	14
47	75	42	99	46	84	371 10	387 25	28	13
48	289 03	70	322 27	74	355 11	37	52	55	12
49	31	97	54	339 01	38	64	78	81	11
50	59	306 25	82	29	65	91	388 05	404 08	10
51	87	53	323 09	56	92	372 18	32	34	9
52	290 15	80	37	83	356 19	45	59	61	8
53	42	307 08	64	340 11	47	72	86	88	7
54	70	36	92	38	74	99	389 12	405 14	6
55	98	63	324 19	65	357 01	373 26	39	41	5
56	291 26	91	47	93	28	53	66	67	4
57	54	308 19	74	341 20	55	80	93	94	3
58	82	46	325 02	47	82	374 07	390 20	406 21	2
59	292 09	74	29	75	358 10	34	46	47	1
60	37	309 02	57	342 02	37	61	73	74	0
′	73°	72°	71°	70°	69°	68°	67°	66°	′

′	16°	17°	18°	19°	20°	21°	22°	23°	′
0	286 74	305 73	324 92	344 33	363 97	383 86	404 03	424 47	60
1	287 06	306 05	325 24	65	364 30	384 20	36	82	59
2	38	37	56	98	63	53	70	425 16	58
3	69	69	88	345 30	96	87	405 04	51	57
4	288 01	307 00	326 21	63	365 29	385 20	38	85	56
5	32	32	53	96	62	53	72	426 19	55
6	64	64	85	346 28	95	87	406 06	54	54
7	95	96	327 17	61	366 28	386 20	40	88	53
8	289 27	308 28	49	93	61	54	74	427 22	52
9	58	60	82	347 26	94	87	407 07	57	51
10	90	91	328 14	58	367 27	387 21	41	91	50
11	290 21	309 23	46	91	60	54	75	428 26	49
12	53	55	78	348 24	93	87	408 09	60	48
13	84	87	329 11	56	368 26	388 21	43	94	47
14	291 16	310 19	43	89	59	54	77	429 29	46
15	47	51	75	349 22	92	88	409 11	63	45
16	79	83	330 07	54	369 25	389 21	45	98	44
17	292 10	311 15	40	87	58	55	79	430 32	43
18	42	47	72	350 20	91	88	410 13	67	42
19	74	78	331 04	52	370 24	390 22	47	431 01	41
20	293 05	312 10	36	85	57	55	81	36	40
21	37	42	69	351 18	90	89	411 15	70	39
22	68	74	332 01	50	371 23	391 22	49	432 05	38
23	294 00	313 06	33	83	57	56	83	39	37
24	32	38	66	352 16	90	90	412 17	74	36
25	63	70	98	48	372 23	392 23	51	433 08	35
26	95	314 02	333 30	81	56	57	85	43	34
27	295 26	34	63	353 14	89	90	413 19	78	33
28	58	66	95	46	373 22	393 24	53	434 12	32
29	90	98	334 27	79	55	57	87	47	31
30	296 21	315 30	60	354 12	88	91	414 21	81	30
31	53	62	92	45	374 22	394 25	55	435 16	29
32	85	94	335 24	77	55	58	90	50	28
33	297 16	316 26	57	355 10	88	92	415 24	85	27
34	48	58	89	43	375 21	395 26	58	436 20	26
35	80	90	336 21	76	54	59	92	54	25
36	298 11	317 22	54	356 08	88	93	416 26	89	24
37	43	54	86	41	376 21	396 26	60	437 24	23
38	75	86	337 18	74	54	60	94	58	22
39	299 06	318 18	51	357 07	87	94	417 28	93	21
40	38	50	83	40	377 20	397 27	63	438 28	20
41	70	82	338 16	72	54	61	97	62	19
42	300 01	319 14	48	358 05	87	95	418 31	97	18
43	33	46	81	38	378 20	398 29	65	439 32	17
44	65	78	339 13	71	53	62	99	66	16
45	97	320 10	45	359 04	87	96	419 33	440 01	15
46	301 28	42	78	37	379 20	399 30	68	36	14
47	60	74	340 10	69	53	63	420 02	71	13
48	92	321 06	43	360 02	86	97	36	441 05	12
49	302 24	39	75	35	380 20	400 31	70	40	11
50	55	71	341 08	68	53	65	421 05	75	10
51	87	322 03	40	361 01	86	98	39	442 10	9
52	303 19	35	73	34	381 20	401 32	73	44	8
53	51	67	342 05	67	53	66	422 07	79	7
54	82	99	38	99	86	402 00	42	443 14	6
55	304 14	323 31	70	362 32	382 20	34	76	49	5
56	46	63	343 03	65	53	67	423 10	84	4
57	78	96	35	98	86	403 01	45	444 18	3
58	305 09	324 28	68	363 31	383 20	35	79	53	2
59	41	60	344 00	64	53	69	424 13	88	1
60	73	92	33	97	86	404 03	47	445 23	0
′	73°	72°	71°	70°	69°	68°	67°	66°	′

NATURAL COTANGENT.

′	24°	25°	26°	27°	28°	29°	30°	31°	′
0	406 74	422 62	438 37	453 99	469 47	484 81	500 00	515 04	60
1	407 00	88	63	454 25	73	485 06	25	29	59
2	27	423 15	89	51	99	32	50	54	58
3	53	41	439 16	77	470 24	57	76	79	57
4	80	67	42	455 03	50	83	501 01	516 04	
5	408 06	94	68	29	76	486 08	26	28	
6	33	424 20	94	54	471 01	34	51	53	56
7	60	46	440 20	80	27	59	76	78	[illegible]
8	86	73	46	456 06	53	84	502 01	517 03	52
9	409 13	99	72	32	78	487 10	27	28	51
10	39	425 25	98	58	472 04	35	52	53	50
11	66	52	441 24	84		61	77	78	49
12	92	78	51	457 10		86	503 02	518 03	48
13	410 19	426 04	77	36	29	488 11	27	28	47
14	45	31	442 03	62	473 [illegible]	37	52	52	46
15	72	57	29	87		62	77	77	45
16	98	83	55	458 13		88	504 03	519 02	44
17	411 25	427 09	81	39		489 13	28	27	43
18	51	36	443 07	65	474 32	38	53	52	42
19	78	62	33	91	[illegible]	64	79	77	41
20	412 04	88	59	459 17		89	505 03	520 02	40
21	31	428 15	85	42		490 14	28	26	39
22	57	41	444 11	68	475	40	53	51	38
23	84	67	37	94		65	78	76	37
24	413 10	94	64	460 20		90	506 03	521 01	36
25	37	429 20	90	46		491 16	28	26	35
26	63	46	445 16	72	476	41	54	51	34
27	90	72	42	97		66	79	75	33
28	414 16	99	68	461 23	60	92	507 04	522 00	32
29	43	430 25	94	49	[illegible]	492 17	29	25	31
30	69	51	446 20	75	477 16	42	54	50	30
31	96	77	46	462 01	41	68	79	75	29
32	415 22	431 04	72	26	67	93	508 04	523 00	28
33	49	30	98	52	93	493 18	29	24	27
34	75	56	447 24	78	478 18	44	54	49	26
35	416 02	82	50	463 04	44	69	79	74	25
36	28	432 09	76	30	69	94	509 04	99	24
37	55	35	448 02	55	95	494 19	29	524 23	23
38	81	61	28	81	479 20	45	54	48	22
39	417 07	87	54	464 07	46	70	79	73	21
40	34	433 13	80	33	71	95	510 04	98	20
41	60	40	449 06	58	97	495 21	29	525 22	19
42	87	66	32	84	480 22	46	54	47	18
43	418 13	92	58	465 10	48	71	79	72	17
44	40	434 18	84	36	73	96	511 04	97	16
45	66	45	450 10	61	99	496 22	29	526 21	15
46	92	71	36	87	481 24	47	54	46	14
47	419 19	97	62	466 13	50	72	79	71	13
48	45	435 23	88	39	75	97	512 04	96	12
49	72	49	451 14	64	482 01	497 23	29	527 20	11
50	98	75	40	90	26	48	54	45	10
51	420 24	436 02	66	467 16	52	73	79	70	9
52	51	28	92	42	77	98	513 04	94	8
53	77	54	452 18	67	483 03	498 24	29	528 19	7
54	421 04	80	43	93	28	49	54	44	6
55	30	437 06	69	468 19	54	74	79	69	5
56	56	33	95	44	79	99	514 04	93	4
57	83	59	453 21	70	484 05	499 24	29	529 18	3
58	422 09	85	47	96	30	50	54	43	2
59	35	438 11	73	469 21	56	75	79	67	1
60	62	37	99	47	81	500 00	515 04	92	0
′	65°	64°	63°	62°	61°	60°	59°	58°	′

NATURAL COSINE.

′	24°	25°	26°	27°	28°	29°	30°	31°	′
0	445 23	466 31	487 73	509 53	531 71	554 31	577 35	600 86	60
1	58	66	488 09	89	532 08	69	74	601 26	59
2	93	467 02	45	510 26	46	555 07	578 13	65	58
3	446 27	37	81	63	83	45	51	602 05	57
4	62	73	489 17	99	533 20	83	90	45	56
5	97	468 08	53	511 36	58	556 21	579 29	84	55
6	447 32	43	89	73	95	59	68	603 24	54
7	67	79	490 26	512 09	534 32	97	580 07	64	53
8	448 02	469 14	62	46	70	557 36	46	604 03	52
9	37	50	98	83	535 07	74	85	43	51
10	72	85	491 34	513 20	45	558 12	581 24	83	50
11	449 07	470 21	70	56	83	50	62	605 22	49
12	42	56	492 06	93	536 20	88	582 01	62	48
13	77	92	42	514 30	57	559 26	40	606 02	47
14	450 12	471 28	78	67	94	64	79	42	46
15	47	63	493 15	515 03	537 32	560 03	583 18	81	45
16	82	99	51	40	69	41	57	607 21	44
17	451 17	472 34	87	77	538 07	79	96	61	43
18	52	70	494 23	516 14	44	561 17	584 35	608 01	42
19	87	473 05	59	51	82	56	74	41	41
20	452 22	41	95	88	539 20	94	585 13	81	40
21	57	77	495 32	517 24	57	562 32	52	609 21	39
22	92	474 12	68	61	95	70	91	60	38
23	453 27	48	496 04	98	540 32	563 09	586 31	610 00	37
24	62	83	40	518 35	70	47	70	40	36
25	97	475 19	77	72	541 07	85	587 09	80	35
26	454 32	55	497 13	519 09	45	564 24	48	611 20	34
27	67	90	49	46	83	62	87	60	33
28	455 02	476 26	86	83	542 20	565 01	588 26	612 00	32
29	38	62	498 22	520 20	58	39	65	40	31
30	73	98	58	57	96	77	589 05	80	30
31	456 08	477 33	94	94	543 33	566 16	44	613 20	29
32	43	69	499 31	521 31	71	54	83	60	28
33	78	478 05	67	68	544 09	93	590 22	614 00	27
34	457 13	40	500 04	522 05	46	567 31	61	40	26
35	48	76	40	42	84	69	591 01	80	25
36	84	479 12	76	79	545 22	568 08	40	615 20	24
37	458 19	48	501 13	523 16	60	46	79	61	23
38	54	84	49	53	97	85	592 18	616 01	22
39	89	480 19	85	90	546 35	569 23	58	41	21
40	459 24	55	502 22	524 27	73	62	97	81	20
41	60	91	58	64	547 11	570 00	593 36	617 21	19
42	95	481 27	95	525 01	48	39	76	61	18
43	460 30	63	503 31	38	86	78	594 15	618 01	17
44	65	98	68	75	548 24	571 16	54	42	16
45	461 01	482 34	504 04	526 13	62	55	94	82	15
46	36	70	41	50	549 00	93	595 33	619 22	14
47	71	483 06	77	87	38	572 32	73	62	13
48	462 06	42	505 14	527 24	75	71	596 12	620 03	12
49	42	78	50	61	550 13	573 09	51	43	11
50	77	484 14	87	98	51	48	91	83	10
51	463 12	50	506 23	528 36	89	86	597 30	621 24	9
52	48	86	60	73	551 27	574 25	70	64	8
53	83	485 21	96	529 10	65	64	598 09	622 04	7
54	464 18	57	507 33	47	552 03	575 03	49	45	6
55	54	93	69	85	41	41	88	85	5
56	89	486 29	508 06	530 22	79	80	599 28	623 25	4
57	465 25	65	43	59	553 17	576 19	67	66	3
58	60	487 01	79	96	55	57	600 07	624 06	2
59	95	37	509 16	531 34	93	96	46	47	1
60	466 31	73	53	71	554 31	577 35	86	87	0
′	65°	64°	63°	62°	61°	60°	59°	58°	′

NATURAL COTANGENT.

′	32°	33°	34°	35°	36°	37°	38°	39°	′
0	529 92	544 64	559 19	573 58	587 79	601 82	615 66	629 32	60
1	530 17	88	43	81	588 02	602 05	89	55	59
2	41	545 13	68	574 05	26	28	616 12	77	58
3	66	37	92	29	49	51	35	630 00	57
4	91	61	560 16	53	73	74	58	22	56
5	531 15	86	40	77	96	98	81	45	55
6	40	546 10	64	575 01	589 20	603 21	617 04	68	54
7	65	35	88	24	43	44	26	90	53
8	89	59	561 12	48	67	67	49	631 13	52
9	532 14	83	36	72	90	90	72	35	51
10	38	547 08	60	96	590 14	604 14	95	58	50
11	63	32	84	576 19	37	37	618 18	80	49
12	88	56	562 08	43	61	60	41	632 03	48
13	533 12	81	32	67	84	83	64	25	47
14	37	548 05	56	91	591 08	605 06	87	48	46
15	61	29	80	577 15	31	29	619 09	71	45
16	86	54	563 05	38	54	53	32	93	44
17	534 11	78	29	62	78	76	55	633 16	43
18	35	549 02	53	86	592 01	99	78	38	42
19	60	27	77	578 10	25	606 22	620 01	61	41
20	84	51	564 01	33	48	45	24	83	40
21	535 09	75	25	57	72	68	46	634 06	39
22	34	550 00	49	81	95	91	69	28	38
23	58	24	73	579 04	593 18	607 14	92	51	37
24	83	48	97	28	42	38	621 15	73	36
25	536 07	72	565 21	52	65	61	38	96	35
26	32	97	45	76	89	84	60	635 18	34
27	56	551 21	69	99	594 12	608 07	83	40	33
28	81	45	93	580 23	36	30	622 06	63	32
29	537 05	69	566 17	47	59	53	29	85	31
30	30	94	41	70	82	76	51	636 08	30
31	54	552 18	65	94	595 06	99	74	30	29
32	79	42	89	581 18	29	609 22	97	53	28
33	538 04	66	567 13	41	52	45	623 20	75	27
34	28	91	36	65	76	68	42	98	26
35	53	553 15	60	89	99	91	65	637 20	25
36	77	39	84	582 12	596 22	610 15	88	42	24
37	539 02	63	568 08	36	46	38	624 11	65	23
38	26	88	32	60	69	61	33	87	22
39	51	554 12	56	83	93	84	56	638 10	21
40	75	36	80	583 07	597 16	611 07	79	32	20
41	540 00	60	569 04	31	39	30	625 02	54	19
42	24	84	28	54	63	53	24	77	18
43	49	555 09	52	78	86	76	47	99	17
44	73	33	76	584 01	598 09	99	70	639 22	16
45	97	57	570 00	25	32	612 22	92	44	15
46	541 22	81	24	49	56	45	626 15	66	14
47	46	556 05	47	72	79	68	38	89	13
48	71	30	71	96	599 02	91	60	640 11	12
49	95	54	95	585 19	26	613 14	83	33	11
50	542 20	78	571 19	43	49	37	627 06	56	10
51	44	557 02	43	67	72	60	28	78	9
52	69	26	67	90	95	83	51	641 00	8
53	93	50	91	586 14	600 19	614 06	74	23	7
54	543 17	75	572 15	37	42	29	96	45	6
55	42	99	38	61	65	51	628 19	67	5
56	66	558 23	62	84	89	74	42	90	4
57	91	47	86	587 08	601 12	97	64	642 12	3
58	544 15	71	573 10	31	35	615 20	87	34	2
59	40	95	34	55	58	43	629 09	56	1
60	64	559 19	58	79	82	66	32	79	0
′	57°	56°	55°	54°	53°	52°	51°	50°	′

NATURAL COSINE.

′	32°	33°	34°	35°	36°	37°	38°	39°	′
0	624 87	649 41	674 51	700 21	726 54	753 55	781 29	809 78	60
1	625 27	82	93	64	99	754 01	75	810 27	59
2	68	650 24	675 36	701 07	727 43	47	782 22	75	58
3	626 08	65	78	51	88	92	69	811 23	57
4	49	651 06	676 20	94	728 32	755 38	783 16	71	56
5	89	48	63	702 38	77	84	63	812 20	55
6	627 30	89	677 05	81	729 21	756 29	784 10	68	54
7	70	652 31	48	703 25	66	75	57	813 16	53
8	628 11	72	90	68	730 10	757 21	785 04	64	52
9	52	653 14	678 32	704 12	55	67	51	814 13	51
10	92	55	75	55	731 00	758 12	98	61	50
11	629 33	97	679 17	99	44	58	786 45	815 10	49
12	73	654 38	60	705 42	89	759 04	92	58	48
13	630 14	80	680 02	86	732 34	50	787 39	816 06	47
14	55	655 21	45	706 29	78	96	86	55	46
15	95	63	88	73	733 23	760 42	788 34	817 03	45
16	631 36	656 04	681 30	707 17	68	88	81	52	44
17	77	46	73	60	734 13	761 34	789 28	818 00	43
18	632 17	88	682 15	708 04	57	80	75	49	42
19	58	657 29	58	48	735 02	762 26	790 22	98	41
20	99	71	683 01	91	47	72	70	819 46	40
21	633 40	658 13	43	709 35	92	763 18	791 17	95	39
22	80	54	86	79	736 37	64	64	820 44	38
23	634 21	96	684 29	710 23	81	764 10	792 12	92	37
24	62	659 38	71	66	737 26	56	59	821 41	36
25	635 03	80	685 14	711 10	71	765 02	793 06	90	35
26	44	660 21	57	54	738 16	48	54	822 38	34
27	84	63	686 00	98	61	94	794 01	87	33
28	636 25	661 05	42	712 42	739 06	766 40	49	823 36	32
29	66	47	85	85	51	86	96	85	31
30	637 07	89	687 28	713 29	96	767 33	795 44	824 34	30
31	48	662 30	71	73	740 41	79	91	83	29
32	89	72	688 14	714 17	86	768 25	796 39	825 31	28
33	638 30	663 14	57	61	741 31	71	86	80	27
34	71	56	689 00	715 05	76	769 18	797 34	826 29	26
35	639 12	98	42	49	742 21	64	81	78	25
36	53	664 40	85	93	67	770 10	798 29	827 27	24
37	94	82	690 28	716 37	743 12	57	77	76	23
38	640 35	665 24	71	81	57	771 03	799 24	828 25	22
39	76	66	691 14	717 25	744 02	49	72	74	21
40	641 17	666 08	57	69	47	96	800 20	829 23	20
41	58	50	692 00	718 13	92	772 42	67	72	19
42	99	92	43	57	745 38	89	801 15	830 22	18
43	642 40	667 34	86	719 01	83	773 35	63	71	17
44	81	76	693 29	46	746 28	82	802 11	831 20	16
45	643 22	668 18	72	90	74	774 28	58	69	15
46	63	60	694 16	720 34	747 19	75	803 06	832 18	14
47	644 04	669 02	59	78	64	775 21	54	68	13
48	46	44	695 02	721 22	748 10	68	804 02	833 17	12
49	87	86	45	67	55	776 15	50	66	11
50	645 28	670 28	88	722 11	749 00	61	98	834 15	10
51	69	71	696 31	55	46	777 08	805 46	65	9
52	646 10	671 13	75	99	91	54	94	835 14	8
53	52	55	697 18	723 44	750 37	778 01	806 42	64	7
54	93	97	61	88	82	48	90	836 13	6
55	647 34	672 39	698 04	724 32	751 28	95	807 38	62	5
56	75	82	47	77	73	779 41	86	837 12	4
57	648 17	673 24	91	725 21	752 19	88	808 34	61	3
58	58	66	699 34	65	64	780 35	82	838 11	2
59	99	674 09	77	726 10	753 10	82	809 30	60	1
60	649 41	51	700 21	54	55	781 29	78	839 10	0
′	57°	56°	55°	54°	53°	52°	51°	50°	′

NATURAL COTANGENT.

′	40°	41°	42°	43°	44°	45°	46°	47°	′
0	642 79	656 06	669 13	682 00	694 66	707 11	719 34	731 35	60
1	643 01	28	35	21	87	31	54	55	59
2	23	50	56	42	695 08	52	74	75	58
3	46	72	78	64	29	72	95	95	57
4	68	94	99	85	49	93	720 15	732 15	56
5	90	657 16	670 21	683 06	70	708 13	35	34	55
6	644 12	38	43	27	91	34	55	54	54
7	35	59	64	49	696 12	55	75	74	53
8	57	81	86	70	33	75	95	94	52
9	79	658 03	671 07	91	54	96	721 16	733 14	51
10	645 01	25	29	684 12	75	709 16	36	33	50
11	24	47	51	34	96	37	56	53	49
12	46	69	72	55	697 17	57	76	73	48
13	68	91	94	76	37	78	96	93	47
14	90	659 13	672 15	97	58	98	722 16	734 13	46
15	646 12	35	37	685 18	79	710 19	36	32	45
16	35	56	58	39	698 00	39	57	52	44
17	57	78	80	61	21	59	77	72	43
18	79	660 00	673 01	82	42	80	97	91	42
19	647 01	22	23	686 03	62	711 00	723 17	735 11	41
20	23	44	44	24	83	21	37	31	40
21	46	66	66	45	699 04	41	57	51	39
22	68	88	87	66	25	62	77	70	38
23	90	661 10	674 09	88	46	82	97	90	37
24	648 12	31	30	687 09	66	712 03	724 17	736 10	36
25	34	53	52	30	87	23	37	29	35
26	56	75	73	51	700 08	43	57	49	34
27	78	97	95	72	29	64	77	69	33
28	649 01	662 18	675 16	93	49	84	97	88	32
29	23	40	38	688 14	70	713 05	725 17	737 08	31
30	45				91	25	37	28	30
31	67		59		701 12	45	57	47	29
32	89	663	676 [illegible]		32	66	77	67	28
33	650 11				53	86	97	87	27
34	33			689	74	714 07	726 17	738 06	26
35	55				95	27	37	26	25
36	77				702 15	47	57	46	24
37	651 00	664	677		36	68	77	65	23
38	22	62	23	690 85	57	88	97	85	22
39	44	[illegible]	[illegible]	[illegible]	77	715 08	727 17	739 04	21
40	66	80			98	29	37	24	20
41	88	665 01	73		703 19	49	57	44	19
42	652 10	23	678 [illegible]		39	69	77	63	18
43	32	45		691	60	90	97	83	17
44	54	66	37		81	716 10	728 17	740 02	16
45	76	88	59		704 01	30	37	22	15
46	98	666 10	679 01	46	22	50	57	41	14
47	653 20	32	23	[illegible]	43	71	77	61	13
48	42	53	44	692 14	63	91	97	80	12
49	64	75	65	35	84	717 11	729 17	741 00	11
50	86	97	87		705 05	32	37	20	10
51	654 08	667 18	680 08	56	25	52	57	39	9
52	30	40	29	98	46	72	76	59	8
53	52	62	51	693 19	67	92	96	78	7
54	74	83	72	40	87	718 13	730 16	98	6
55	96	668 05	93	61	706 08	33	36	742 17	5
56	655 18	27	681 15	82	28	53	56	37	4
57	40	48	36	694 03	49	73	76	56	3
58	62	70	57	24	70	94	96	76	2
59	84	91	79	45	90	719 14	731 16	95	1
60	656 06	669 13	682 00	66	707 11	34	35	743 14	0
′	49°	48°	47°	46°	45°	44°	43°	42°	′

NATURAL COSINE.

′	40°	41°	42°	43°	44°	45°	46°	47°	′
0	839 10	869 29	900 40	932 52	965 69	1·00 000	1·03 553	1·07 237	60
1	60	80	93	933 06	966 25	058	613	299	59
2	840 09	870 31	901 46	60	81	116	674	362	58
3	59	82	99	934 15	967 38	175	734	425	57
4	841 08	871 33	902 51	69	94	233	794	487	56
5	58	84	903 04	935 24	968 50	291	855	550	55
6	842 08	872 36	57	78	969 07	350	915	613	54
7	58	87	904 10	936 33	63	408	976	676	53
8	843 07	873 38	63	88	970 20	467	1·04 036	738	52
9	57	89	905 16	937 42	76	525	097	801	51
10	844 07	874 41	69	97	971 33	583	158	864	50
11	57	92	906 21	938 52	89	642	218	927	49
12	845 07	875 43	74	939 06	972 46	701	279	990	48
13	56	95	907 27	61	973 02	759	340	1·08 053	47
14	846 06	876 46	81	940 16	59	818	401	116	46
15	56	98	908 34	71	974 16	876	461	179	45
16	847 06	877 49	87	941 25	72	935	522	243	44
17	56	878 01	909 40	80	975 29	994	583	306	43
18	848 06	52	93	942 35	86	1·01 053	644	369	42
19	56	879 04	910 46	90	976 43	112	705	432	41
20	849 06	55	99	943 45	977 00	170	766	496	40
21	56	880 07	911 53	944 00	56	229	827	559	39
22	850 06	59	912 06	55	978 13	288	888	622	38
23	57	881 10	59	945 10	70	347	949	686	37
24	851 07	62	913 13	65	979 27	406	1·05 010	749	36
25	57	882 14	66	946 20	84	465	072	813	35
26	852 07	65	914 19	76	980 41	524	133	876	34
27	57	883 17	73	947 31	98	583	194	940	33
28	853 07	69	915 26	86	981 55	642	255	1·09 003	32
29	58	884 21	80	948 41	982 13	702	317	067	31
30	854 08	73	916 33	96	70	761	378	131	30
31	58	885 24	87	949 52	983 27	820	439	195	29
32	855 09	76	917 40	950 07	84	879	501	258	28
33	59	886 28	94	62	984 41	939	562	322	27
34	856 10	80	918 47	951 18	99	998	624	386	26
35	60	887 32	919 01	73	985 56	1·02 057	685	450	25
36	857 10	84	55	952 29	986 13	117	747	514	24
37	61	888 36	920 08	84	71	176	809	578	23
38	858 11	88	62	953 40	987 28	236	870	642	22
39	62	889 40	921 16	95	86	295	932	706	21
40	859 12	92	70	954 51	988 43	355	994	770	20
41	63	890 45	922 24	955 06	989 01	414	1·06 056	834	19
42	860 14	97	77	62	58	474	117	899	18
43	64	891 49	923 31	956 18	990 16	533	179	963	17
44	861 15	892 01	85	73	73	593	241	1·10 027	16
45	66	53	924 39	957 29	991 31	653	303	091	15
46	862 16	893 06	93	85	89	713	365	156	14
47	67	58	925 47	959 41	992 47	772	427	220	13
48	863 18	894 10	926 01	97	993 04	832	489	285	12
49	68	63	55	959 52	62	892	551	349	11
50	864 19	895 15	927 09	960 08	994 20	952	613	414	10
51	70	67	63	64	78	1·03 012	676	478	9
52	865 21	896 20	928 17	961 20	995 36	072	738	543	8
53	72	72	72	76	94	132	800	608	7
54	866 23	897 25	929 26	962 32	996 52	192	862	672	6
55	74	77	80	88	997 10	252	925	737	5
56	867 25	898 30	930 34	963 44	68	312	987	802	4
57	76	83	88	964 00	998 26	372	1·07 049	867	3
58	868 27	899 35	931 43	57	84	433	112	931	2
59	78	88	97	965 13	999 42	493	174	996	1
60	869 29	900 40	932 52	69	1·00 000	553	237	1·11 061	0
′	49°	48°	47°	46°	45°	44°	43°	42°	′

NATURAL COTANGENT.

′	48°	49°	50°	51°	52°	53°	54°	55°	′
0	743 14	754 71	766 04	777 15	788 01	798 64	809 02	819 15	60
1	34	90	23	33	19	81	19	32	59
2	53	755 09	42	51	37	99	36	49	58
3	73	28	61	69	55	799 16	53	65	57
4	92	47	79	88	73	34	70	82	56
5	744 12	66	98	778 06	91	51	87	99	55
6	31	85	767 17	24	789 08	68	810 04	820 15	54
7	51	756 04	35	43	26	86	21	32	53
8	70	23	54	61	44	800 03	38	48	52
9	89	42	72	79	62	21	55	65	51
10	745 09	61	91	97	80	38	72	82	50
11	28	81	768 10	779 16	98	56	89	98	49
12	48	757 00	28	34	790 16	73	811 06	821 15	48
13	67	19	47	52	33	91	23	32	47
14	86	38	66	70	51	801 08	40	48	46
15	746 06	57	84	88	69	25	57	65	45
16	25	76	769 03	780 07	87	43	74	81	44
17	44	94	21	25	791 05	60	91	98	43
18	64	758 13	40	43	22	78	812 08	822 14	42
19	83	32	59	61	40	95	25	31	41
20	747 03	51	77	79	58	802 12	42	48	40
21	22	70	96	98	76	30	59	64	39
22	41	89	770 14	781 16	93	47	76	81	38
23	60	759 08	33	34	792 11	64	93	97	37
24	80	27	51	52	29	82	813 10	823 14	36
25	99	46	70	70	47	99	27	30	35
26	748 18	65	88	88	64	803 16	44	47	34
27	38	84	771 07	782 06	82	34	61	63	33
28	57	760 03	25	25	793 00	51	78	80	32
29	76	22	44	43	18	68	95	96	31
30	96	41	62	61	35	86	814 12	824 13	
31	749 15	59	81	79	53	804 03	28	29	
32	34	78	99	97	71	20	45	46	
33	53	97	772 18	783 15	88	38	62	62	
34	73	761 16	36	33	794 06	55	79	78	26
35	92	35	55	51	24	72	96	95	25
36	750 11	54	73	69	41	89	815 13	825 11	24
37	30	73	92	87	59	805 07	30	28	23
38	50	92	773 10	784 05	77	24	46	44	22
39	69	762 10	29	24	94	41	63	61	21
40	88	29	47	42	795 12	58	80	77	20
41	751 07	48	66	60	30	76	97	93	19
42	26	67	84	78	47	93	816 14	826 10	18
43	46	86	774 02	96	65	806 10	31	26	17
44	65	763 04	21	785 14	83	27	47	43	16
45	84	23	39	32	796 00	44	64	59	15
46	752 03	42	58	50	18	62	81	75	14
47	22	61	76	68	35	79	98	92	13
48	41	80	94	86	53	96	817 14	827 08	12
49	61	98	775 13	786 04	71	807 13	31	24	11
50	80	764 17	31	22	88	30	48	41	10
51	99	36	50	40	797 06	48	65	57	9
52	753 18	55	68	58	23	65	82	73	8
53	37	73	86	76	41	82	98	90	7
54	56	92	776 05	94	58	99	818 15	828 06	6
55	75	765 11	23	787 11	76	808 16	32	22	5
56	95	30	41	29	93	33	48	39	4
57	754 14	48	60	47	798 11	50	65	55	3
58	33	67	78	65	29	67	82	71	2
59	52	86	96	83	46	85	99	87	1
60	71	766 04	777 15	788 01	64	809 02	819 15	829 04	0
′	41°	40°	39°	38°	37°	36°	35°	34°	′

NATURAL COSINE.

′	48°	49°	50°	51°	52°	53°	54°	55°	′
0	1·11 061	1·15 037	1·19 175	1·23 490	1·27 994	1·32 704	1·37 638	1·42 815	60
1	126	104	246	563	1·28 071	785	722	903	59
2	191	172	316	637	148	865	807	992	58
3	256	240	387	710	225	946	891	1·43 080	57
4	321	308	457	784	302	1·33 026	976	169	56
5	387	375	528	858	379	107	1·38 060	258	55
6	452	443	599	931	456	188	145	347	54
7	517	511	669	1·24 005	533	268	229	436	53
8	582	579	740	079	610	349	314	525	52
9	648	647	811	153	687	430	399	614	51
10	713	715	882	227	764	511	484	703	50
11	778	783	953	301	842	592	568	792	49
12	844	851	1·20 024	375	919	673	653	881	48
13	909	919	095	449	997	754	738	970	47
14	975	987	166	523	1·29 074	835	824	1·44 060	46
15	1·12 041	1·16 056	237	597	152	916	909	149	45
16	106	124	308	672	229	998	994	239	44
17	172	192	379	746	307	1·34 079	1·39 079	329	43
18	238	261	451	820	385	160	165	418	42
19	303	329	522	895	463	242	250	508	41
20	369	398	593	969	541	323	336	598	40
21	435	466	665	1·25 044	619	405	421	688	39
22	501	535	736	118	696	487	507	778	38
23	567	603	808	193	775	568	593	868	37
24	633	672	879	268	853	650	679	958	36
25	699	741	951	343	931	732	764	1·45 049	35
26	765	809	1·21 023	417	1·30 009	814	850	139	34
27	831	878	094	492	087	896	936	229	33
28	897	947	166	567	166	978	1·40 022	320	32
29	963	1·17 016	238	642	244	1·35 060	109	410	31
30	1·13 029	085	310	717	323	142	195	501	30
31	095	154	382	792	401	224	281	592	29
32	162	223	454	867	480	307	367	682	28
33	228	292	526	943	558	389	454	773	27
34	295	361	598	1·26 018	637	472	540	864	26
35	361	430	670	093	716	554	627	955	25
36	428	500	742	169	795	637	714	1·46 046	24
37	494	569	814	244	873	719	800	137	23
38	561	638	887	320	952	802	887	229	22
39	627	708	959	395	1·31 031	885	974	320	21
40	694	777	1·22 031	471	110	968	1·41 061	411	20
41	761	846	104	546	190	1·36 051	148	503	19
42	828	916	176	622	269	134	235	595	18
43	894	986	249	698	348	217	322	686	17
44	961	1·18 055	321	774	427	300	409	778	16
45	1·14 028	125	394	849	507	383	497	870	15
46	095	194	467	925	586	466	584	962	14
47	162	264	539	1·27 001	666	549	672	1·47 054	13
48	229	334	612	077	745	633	759	146	12
49	296	404	685	153	825	716	847	238	11
50	363	474	758	230	904	800	934	330	10
51	430	544	831	306	984	883	1·42 022	422	9
52	498	614	904	382	1·32 064	967	110	514	8
53	565	684	977	458	144	1·37 050	198	607	7
54	632	754	1·23 050	535	224	134	286	699	6
55	699	824	123	611	304	218	374	792	5
56	767	894	196	688	384	302	462	885	4
57	834	964	270	764	464	386	550	977	3
58	902	1·19 035	343	841	544	470	638	1·48 070	2
59	969	105	416	917	624	554	726	163	1
60	1·15 037	175	490	994	704	638	815	256	0
′	41°	40°	39°	38°	37°	36°	35°	34°	′

NATURAL COTANGENT.

′	56°	57°	58°	59°	60°	61°	62°	63°	′
0	829 04	838 67	848 05	857 17	866 03	874 62	882 95	891 01	60
1	20	83	20	32	17	76	883 08	14	59
2	36	99	36	47	32	90	22	27	58
3	53	839 15	51	62	46	875 04	36	40	57
4	69	30	66	77	61	18	49	53	56
5	85	46	82	92	75	32	63	67	55
6	830 01	62	97	858 06	90	46	77	80	54
7	17	78	849 13	21	867 04	61	90	93	53
8	34	94	28	36	19	75	884 04	892 06	52
9	50	840 09	43	51	33	89	17	19	51
10	66	25	59	66	48	876 03	31	32	50
11	82	41	74	81	62	17	45	45	49
12	98	57	89	96	77	31	58	59	48
13	831 15	72	850 05	859 11	91	45	72	72	47
14	31	88	20	26	868 05	59	85	85	46
15	47	841 04	35	41	20	73	99	98	45
16	63	20	51	56	34	87	885 12	893 11	44
17	79	35	66	70	49	877 01	26	24	43
18	95	51	81	85	63	15	39	37	42
19	832 12	67	96	860 00	78	29	53	50	41
20	28	82	851 12	15	92	43	66	63	40
21	44	98	27	30	869 06	56	80	76	39
22	60	842 14	42	45	21	70	93	89	38
23	76	30	57	59	35	84	886 07	894 02	37
24	92	45	73	74	49	98	20	15	36
25	833 08	61	88	89	64	878 12	34	28	35
26	24	77	852 03	861 04	78	26	47	41	34
27	40	92	18	19	93	40	61	54	33
28	56	843 08	34	33	870 07	54	74	67	32
29	73	24	49	48	21	68	88	80	31
30	89	39	64	63	36	82	887 01	93	30
31	834 05	55	79	78	50	96	15	895 06	29
32	21	70	94	92	64	879 09	28	19	28
33	37	86	853 10	862 07	79	23	41	32	27
34	53	844 02	25	22	93	37	55	45	26
35	69	17	40	37	871 07	51	68	58	25
36	85	33	55	51	21	65	82	71	24
37	835 01	48	70	66	36	79	95	84	23
38	17	64	85	81	50	93	888 08	97	22
39	33	80	854 01	95	64	880 06	22	896 10	21
40	49	95	16	863 10	78	20	35	23	20
41	65	845 11	31	25	93	34	48	36	19
42	81	26	46	40	872 07	48	62	49	18
43	97	42	61	54	21	62	75	62	17
44	836 13	57	76	69	35	75	88	74	16
45	29	73	91	84	50	89	889 02	87	15
46	45	88	855 06	98	64	881 03	15	897 00	14
47	61	846 04	21	864 13	78	17	28	13	13
48	76	19	36	27	92	30	42	26	12
49	92	35	51	42	873 06	44	55	39	11
50	837 08	50	67	57	21	58	68	52	10
51	24	66	82	71	35	72	81	64	9
52	40	81	97	86	49	85	95	77	8
53	56	97	856 12	865 01	63	99	890 08	90	7
54	72	847 12	27	15	77	882 13	21	898 03	6
55	88	28	42	30	91	26	35	16	5
56	838 04	43	57	44	874 06	40	48	28	4
57	20	59	72	59	20	54	61	41	3
58	35	74	87	73	34	67	74	54	2
59	51	89	857 02	88	48	81	87	67	1
60	67	848 05	17	866 03	62	95	891 01	79	0
′	33°	32°	31°	30°	29°	28°	27°	26°	′

NATURAL COSINE.

′	56°	57°	58°	59°	60°	61°	62°	63°	′
0	1·48 256	1·53 987	1·60 033	1·66 428	1·73 205	1·80 405	1·88 073	1·96 261	60
1	349	1·54 085	137	538	321	529	205	402	59
2	442	183	241	647	438	653	337	544	58
3	536	281	345	757	555	777	469	685	57
4	629	379	449	867	671	901	602	827	56
5	722	478	553	978	788	1·81 025	734	969	55
6	816	576	657	1·67 088	905	150	867	1·97 111	54
7	909	675	761	198	1·74 022	274	1·89 000	253	53
8	1·49 003	774	865	309	140	399	133	395	52
9	097	873	970	419	257	524	266	538	51
10	190	972	1·61 074	530	375	649	400	681	50
11	284	1·55 071	179	641	492	774	533	823	49
12	378	170	283	752	610	899	667	966	48
13	472	269	388	863	728	1·82 025	801	1·98 110	47
14	566	368	493	974	846	150	935	253	46
15	661	467	598	1·68 085	964	276	1·90 069	396	45
16	755	567	703	196	1·75 082	402	203	540	44
17	849	666	809	308	200	528	337	684	43
18	944	766	914	419	319	654	472	828	42
19	1·50 038	866	1·62 019	531	437	780	607	972	41
20	133	966	125	643	556	906	741	1·99 116	40
21	228	1·56 065	230	754	675	1·83 033	876	261	39
22	322	165	336	866	794	159	1·91 012	406	38
23	417	265	442	979	913	286	147	550	37
24	512	366	548	1·69 091	1·76 032	413	282	695	36
25	607	466	654	203	151	540	418	841	35
26	702	566	760	316	271	667	554	986	34
27	797	667	866	428	390	794	690	2·00 131	33
28	893	767	972	541	510	922	826	277	32
29	988	868	1·63 079	653	630	1·84 049	962	423	31
30	1·51 084	969	185	766	749	177	1·92 098	569	30
31	179	1·57 069	292	879	869	305	235	715	29
32	275	170	398	992	990	433	371	862	28
33	370	271	505	1·70 106	1·77 110	561	508	2·01 008	27
34	466	372	612	219	230	689	645	155	26
35	562	474	719	332	351	818	782	302	25
36	658	575	826	446	471	946	920	449	24
37	754	676	934	560	592	1·85 075	1·93 057	596	23
38	850	778	1·64 041	673	713	204	195	743	22
39	946	879	148	787	834	333	332	891	21
40	1·52 043	981	256	901	955	462	470	2·02 039	20
41	139	1·58 083	363	1·71 015	1·78 077	591	608	187	19
42	235	184	471	129	198	720	746	335	18
43	332	286	579	244	319	850	885	483	17
44	429	388	687	358	441	979	1·94 023	631	16
45	525	490	795	473	563	1·86 109	162	780	15
46	622	593	903	588	685	239	301	929	14
47	719	695	1·65 011	702	807	369	440	2·03 078	13
48	816	797	120	817	929	499	579	227	12
49	913	900	228	932	1·79 051	630	718	376	11
50	1·53 010	1·59 002	337	1·72 047	174	760	858	526	10
51	107	105	445	163	296	891	997	675	9
52	205	208	554	278	419	1·87 021	1·95 137	825	8
53	302	311	663	393	542	152	277	975	7
54	400	414	772	509	665	283	417	2·04 125	6
55	497	517	881	625	788	415	557	276	5
56	595	620	990	741	911	546	698	426	4
57	693	723	1·66 099	857	1·80 034	677	838	577	3
58	791	826	209	973	158	809	979	728	2
59	888	930	318	1·73 089	281	941	1·96 120	879	1
60	987	1·60 033	428	205	405	1·88 073	261	2·05 030	0
′	33°	32°	31°	30°	29°	28°	27°	26°	′

NATURAL COTANGENT.

′	64°	65°	66°	67°	68°	69°	70°	71°	′
0	898 79	906 31	913 55	920 50	927 18	933 58	939 69	945 52	60
1	92	43	66	62	29	68	79	61	59
2	899 05	55	78	73	40	79	89	71	58
3	18	68	90	85	51	89	99	80	57
4	30	80	914 02	96	62	934 00	940 09	90	56
5	43	92	14	921 07	73	10	19	99	55
6	56	907 04	25	19	84	20	29	946 09	54
7	68	17	37	30	94	31	39	18	53
8	81	29	49	41	928 05	41	49	27	52
9	94	41	61	52	16	52	58	37	51
10	900 07	53	72	64	27	62	68	46	50
11	19	66	84	75	38	72	78	56	49
12	32	78	96	86	49	83	88	65	48
13	45	90	915 08	98	59	93	98	74	47
14	57	908 02	19	922 09	70	935 03	941 08	84	46
15	70	14	31	20	81	14	18	93	45
16	82	26	43	31	92	24	27	947 02	44
17	95	39	55	43	929 03	34	37	12	43
18	901 08	51	66	54	13	44	47	21	42
19	20	63	78	65	24	55	57	30	41
20	33	75	90	76	35	65	67	40	40
21	46	87	916 01	87	45	75	76	49	39
22	58	99	13	99	56	85	86	58	38
23	71	909 12	25	923 10	67	96	96	68	37
24	83	24	36	21	78	936 06	942 06	77	36
25	96		48	32	88	16	16	86	35
26	902 08		60	43	99	26	25	95	34
27	21		71	55	930 10	37	35	948 05	33
28	33	36	83	66	20	47	45	14	32
29	46	[illegible]	94	77	31	57	54	23	31
30	59	96	917 06	88	42	67	64	32	30
31	71	910 08	18	99	52	77	74	42	29
32	84	20	29	924 10	63	88	84	51	28
33	96	32	41	21	74	98	93	60	27
34	903 09	44	52	32	84	937 08	943 03	69	26
35	21	56	64	44	95	18	13	78	25
36	34	68	75	55	931 06	28	22	88	24
37	46	80	87	66	16	38	32	97	23
38	58	92	99	77	27	48	42	949 06	22
39	71	911 04	918 10	88	37	59	51	15	21
40	83	16	22	99	48	69	61	24	20
41	96	28	33	925 10	59	79	70	33	19
42	904 08	40	45	21	69	89	80	43	18
43	21	52	56	32	80	99	90	52	17
44	33	64	68	43	90	938 09	99	61	16
45	46	76	79	54	932 01	19	944 09	70	15
46	58	88	91	65	11	29	18	79	14
47	70	912 00	919 02	76	22	39	28	88	13
48	83	12	14	87	32	49	38	97	12
49	95	24	25	98	43	59	47	950 06	11
50	905 07	36	36	926 09	53	69	57	15	10
51	20	48	48	20	64	79	66	24	9
52	32	60	59	31	74	89	76	33	8
53	45	72	71	42	85	99	85	43	7
54	57	83	82	53	95	939 09	95	52	6
5	69	95	94	64	933 06	19	945 04	61	5
56	82	913 07	920 05	75	16	29	14	70	4
57	94	19	16	86	27	39	23	79	3
58	906 06	31	28	97	37	49	33	88	2
59	18	43	39	927 07	48	59	42	97	1
60	31	55	50	18	58	69	52	951 06	0
′	25°	24°	23°	22°	21°	20°	19°	18°	′

NATURAL COSINE.

′	64°	65°	66°	67°	68°	69°	70°	71°	′
0	2·05 030	2·14 451	2·24 604	2·35 585	2·47 509	2·60 509	2·74 748	2·90 421	60
1	182	614	780	776	716	736	997	696	59
2	333	777	956	967	924	963	2·75 246	971	58
3	485	940	2·25 132	2·36 158	2·48 132	2·61 190	496	2·91 246	57
4	637	2·15 104	309	349	340	418	746	523	56
5	790	268	486	541	549	646	996	799	55
6	942	432	663	733	758	874	2·76 247	2·92 076	54
7	2·06 094	596	840	925	967	2·62 103	498	354	53
8	247	760	2·26 018	2·37 118	2·49 177	332	750	632	52
9	400	925	196	311	386	561	2·77 002	910	51
10	553	2·16 090	374	504	597	791	254	2·93 189	50
11	706	255	552	697	807	2·63 021	507	468	49
12	860	420	730	891	2·50 018	252	761	748	48
13	2·07 014	585	909	2·38 084	229	483	2·78 014	2·94 028	47
14	167	751	2·27 088	279	440	714	269	309	46
15	321	917	267	473	652	945	523	591	45
16	476	2·17 083	447	668	864	2·64 177	778	872	44
17	630	249	626	863	2·51 076	410	2·79 033	2·95 155	43
18	785	416	806	2·39 058	289	642	289	437	42
19	939	582	987	253	502	875	545	721	41
20	2·08 094	749	2·28 167	449	715	2·65 109	802	2·96 004	40
21	250	916	348	645	929	342	2·80 059	288	39
22	405	2·18 084	528	841	2·52 142	576	316	573	38
23	560	251	710	2·40 038	357	811	574	858	37
24	716	419	891	235	571	2·66 046	833	2·97 144	36
25	872	587	2·29 073	432	786	281	2·81 091	430	35
26	2·09 028	755	254	629	2·53 001	516	350	717	34
27	184	923	437	827	217	752	610	2·98 004	33
28	341	2·19 092	619	2·41 025	432	989	870	292	32
29	498	261	801	223	648	2·67 225	2·82 130	580	31
30	654	430	984	421	865	462	391	869	30
31	811	599	2·30 167	620	2·54 082	700	653	2·99 158	29
32	969	769	351	819	299	937	914	447	28
33	2·10 126	938	534	2·42 019	516	2·68 175	2·83 176	738	27
34	284	2·20 108	718	218	734	414	439	3·00 028	26
35	442	278	902	418	952	653	702	319	25
36	600	449	2·31 086	618	2·55 170	892	965	611	24
37	758	619	271	819	389	2·69 131	2·84 229	903	23
38	916	790	456	2·43 019	608	371	494	3·01 196	22
39	2·11 075	961	641	220	827	612	758	489	21
40	233	2·21 132	826	422	2·56 046	853	2·85 023	783	20
41	392	304	2·32 012	623	266	2·70 094	289	3·02 077	19
42	552	475	197	825	487	335	555	372	18
43	711	647	383	2·44 027	707	577	822	667	17
44	871	819	570	230	928	819	2·86 089	963	16
45	2·12 030	992	756	433	2·57 150	2·71 062	356	3·03 260	15
46	190	2·22 164	943	636	371	305	624	556	14
47	350	337	2·33 130	839	593	548	892	854	13
48	511	510	317	2·45 043	815	792	2·87 161	3·04 152	12
49	671	683	505	246	2·58 038	2·72 036	430	450	11
50	832	857	693	451	261	281	700	749	10
51	993	2·23 030	881	655	484	526	970	3·05 049	9
52	2·13 154	204	2·34 069	860	708	771	2·88 240	349	8
53	316	376	258	2·46 065	932	2·73 017	511	649	7
54	477	553	447	270	2·59 156	263	783	950	6
55	639	727	636	476	381	509	2·89 055	3·06 252	5
56	801	902	825	682	606	756	327	544	4
57	963	2·24 077	2·35 015	888	831	2·74 004	600	857	3
58	2·14 125	252	205	2·47 095	2·60 057	251	873	3·07 160	2
59	288	428	395	302	283	499	2·90 147	464	1
60	451	604	585	509	509	748	421	768	0
′	25°	24°	23°	22°	21°	20°	19°	18°	′

NATURAL COTANGENT.

′	72°	73°	74°	75°	76°	77°	78°	79°	′
0	951 06	956 30	961 26	965 98	970 30	974 37	978 15	981 63	60
1	15	39	34	966 00	37	44	21	68	59
2	24	47	42	08	44	50	27	74	58
3	33	56	50	15	51	57	33	79	57
4	42	64	58	23	58	63	39	85	56
5	51	73	66	30	65	70	45	90	55
6	59	81	74	38	72	76	51	96	54
7	68	90	82	45	79	83	57	982 01	53
8	77	98	90	53	86	89	63	07	52
9	86	957 07	98	60	93	96	69	12	51
10	95	15	962 06	67	971 00	975 02	75	18	50
11	952 04	24	14	75	06	08	81	23	49
12	13	32	22	82	13	15	87	29	48
13	22	40	30	90	20	21	93	34	47
14	31	49	38	97	27	28	99	40	46
15	40	57	46	967 05	34	34	979 05	45	45
16	48	66	53	12	41	41	10	50	44
17	57	74	61	19	48	47	16	56	43
18	66	82	69	27	55	53	22	61	42
19	75	91	77	34	62	60	28	67	41
20	84	99	85	42	69	66	34	72	40
21	93	958 07	93	49	76	73	40	77	39
22	953 01	16	963 01	56	82	79	46	83	38
23	10	24	08	64	89	85	52	88	37
24	19	32	16	71	96	92	58	94	36
25	28	41	24	78	972 03	98	63	99	35
26	37	49	32	86	10	976 04	69	983 04	34
27	45	57	40	93	17	11	75	10	33
28	54	65	47	968 00	23	17	81	15	32
29	63	74	55	07	30	23	87	20	31
30	72	82	63	15	37	30	92	25	30
31	80	90	71	22	44	36	98	31	29
32	89	98	79	29	51	42	980 04	36	28
33	98	959 07	86	37	57	48	10	41	27
34	954 07	15	94	44	64	55	16	47	26
35	15	23	964 02	51	71	61	21	52	25
36	24	31	10	58	78	67	27	57	24
37	33	40	17	66	84	73	33	62	23
38	41	48	25	73	91	80	39	68	22
39	50	56	33	80	98	86	44	73	21
40	59	64	40	87	973 04	92	50	78	20
41	67	72	48	94	11	98	56	83	19
42	76	81	56	969 02	18	977 05	61	89	18
43	85	89	63	09	25	11	67	94	17
44	93	97	71	16	31	17	73	99	16
45	955 02	960 05	79	23	38	23	79	984 04	15
46	11	13	86	30	45	29	84	09	14
47	19	21	94	37	51	35	90	14	13
48	28	29	965 02	45	58	42	96	20	12
49	36	37	09	52	65	48	981 01	25	11
50	45	46	17	59	71	54	07	30	10
51	54	54	24	66	78	60	12	35	9
52	62	62	32	73	84	66	18	40	8
53	71	70	40	80	91	72	24	45	7
54	79	78	47	87	98	78	29	50	6
55	88	86	55	94	974 04	84	35	55	5
56	96	94	62	970 01	11	91	40	61	4
57	956 05	961 02	70	08	17	97	46	66	3
58	13	10	78	15	24	978 03	52	71	2
59	22	18	85	23	30	09	57	76	1
60	30	26	93	30	37	15	63	81	0
′	17°	16°	15°	14°	13°	12°	11°	10°	′

′	72°	73°	74°	75°	76°	77°	78°	79°	′
0	3·07 768	3·27 085	3·48 741	3·73 205	4·01 078	4·33 148	4·70 463	5·14 455	60
1	3·08 073	426	3·49 125	640	576	723	4·71 137	5·15 256	59
2	379	767	509	3·74 075	4·02 074	4·34 300	813	5·16 058	58
3	685	3·28 109	894	512	574	879	4·72 490	863	57
4	991	452	3·50 279	950	4·03 076	4·35 459	4·73 170	5·17 671	56
5	3·09 298	795	666	3·75 388	578	4·36 040	851	5·18 480	55
6	606	3·29 139	3·51 053	828	4·04 081	623	4·74 534	5·19 293	54
7	914	483	441	3·76 268	586	4·37 207	4·75 219	5·20 107	53
8	3·10 223	829	829	709	4·05 092	793	906	925	52
9	532	3·30 174	3·52 219	3·77 152	599	4·38 381	4·76 595	5·21 744	51
10	842	521	609	595	4·06 107	969	4·77 286	5·22 566	50
11	3·11 153	868	3·53 001	3·78 040	616	4·39 560	978	5·23 391	49
12	464	3·31 216	393	485	4·07 127	4·40 152	4·78 673	5·24 218	48
13	775	565	785	931	639	745	4·79 370	5·25 048	47
14	3·12 087	914	3·54 179	3·79 378	4·08 152	4·41 340	4·80 068	880	46
15	400	3·32 264	573	827	666	936	769	5·26 715	45
16	713	614	968	3·80 276	4·09 182	4·42 534	4·81 471	5·27 553	44
17	3·13 027	965	3·55 364	726	699	4·43 134	4·82 175	5·28 393	43
18	341	3·33 317	761	3·81 177	4·10 216	735	882	5·29 235	42
19	656	670	3·56 159	630	736	4·44 338	4·83 590	5·30 080	41
20	972	3·34 023	557	3·82 083	4·11 256	942	4·84 300	928	40
21	3·14 288	377	957	537	778	4·45 548	4·85 013	5·31 778	39
22	605	732	3·57 357	992	4·12 301	4·46 155	727	5·32 631	38
23	922	3·35 087	758	3·83 449	825	764	4·86 444	5·33 487	37
24	3·15 240	443	3·58 160	906	4·13 350	4·47 374	4·87 162	5·34 345	36
25	558	800	562	3·84 364	877	986	882	5·35 206	35
26	877	3·36 158	966	824	4·14 405	4·48 600	4·88 605	5·36 070	34
27	3·16 197	516	3·59 370	3·85 284	934	4·49 215	4·89 330	936	33
28	517	875	775	745	4·15 465	832	4·90 056	5·37 805	32
29	838	3·37 234	3·60 181	3·86 208	997	4·50 451	785	5·38 677	31
30	3·17 159	594	588	671	4·16 530	4·51 071	4·91 516	5·39 552	30
31	481	955	996	3·87 136	4·17 064	693	4·92 249	5·40 429	29
32	804	3·38 317	3·61 405	601	600	4·52 316	984	5·41 309	28
33	3·18 127	679	814	3·88 068	4·18 137	941	4·93 721	5·42 192	27
34	451	3·39 042	3·62 224	536	675	4·53 568	4·94 460	5·43 078	26
35	775	406	636	3·89 004	4·19 215	4·54 196	4·95 201	966	25
36	3·19 100	771	3·63 048	474	756	826	945	5·44 857	24
37	426	3·40 136	461	945	4·20 298	4·55 458	4·96 690	5·45 751	23
38	752	502	874	3·90 417	842	4·56 091	4·97 438	5·46 648	22
39	3·20 079	869	3·64 289	890	4·21 387	726	4·98 188	5·47 548	21
40	406	3·41 236	705	3·91 364	933	4·57 363	940	5·48 451	20
41	734	604	3·65 121	839	4·22 481	4·58 001	4·99 695	5·49 356	19
42	3·21 063	973	538	3·92 316	4·23 030	641	5·00 451	5·50 264	18
43	392	3·42 343	957	793	580	4·59 283	5·01 210	5·51 176	17
44	722	713	3·66 376	3·93 271	4·24 132	927	971	5·52 090	16
45	3·22 053	3·43 084	796	751	685	4·60 572	5·02 734	5·53 007	15
46	384	456	3·67 217	3·94 232	4·25 239	4·61 219	5·03 499	927	14
47	715	829	638	713	795	868	5·04 267	5·54 851	13
48	3·23 048	3·44 202	3·68 061	3·95 196	4·26 352	4·62 518	5·05 037	5·55 777	12
49	381	576	485	680	911	4·63 171	809	5·56 706	11
50	714	951	909	3·96 165	4·27 471	825	5·06 584	5·57 638	10
51	3·24 049	3·45 327	3·69 335	651	4·28 032	4·64 480	5·07 360	5·58 573	9
52	383	703	761	3·97 139	595	4·65 138	5·08 139	5·59 511	8
53	719	3·46 080	3·70 188	627	4·29 159	797	921	5·60 452	7
54	3·25 055	458	616	3·98 117	724	4·66 458	5·09 794	5·61 397	6
55	392	837	3·71 046	607	4·30 291	4·67 121	5·10 490	3·62 344	5
56	729	3·47 216	476	3·99 099	860	786	5·11 279	5·63 295	4
57	3·26 067	596	907	592	4·31 430	4·68 452	5·12 069	5·64 248	3
58	406	977	3·72 338	4·00 086	4·32 001	4·69 121	862	5·65 205	2
59	745	3·48 359	771	582	573	791	5·13 658	5·66 165	1
60	3·27 085	741	3·73 205	4·01 078	4·33 148	4·70 463	5·14 455	5·67 128	0
′	17°	16°	15°	14°	13°	12°	11°	10°	′

NATURAL COTANGENT.

′	80°	81°	82°	83°	84°	85°	86°	87°	′
0	9848 1	9876 9	9902 7	9925 5	9945 2	9961 9	9975 6	9986 3	60
1	6	9877 3	9903 1	8	5	9962 2	8	4	59
2	9849 1	8	5	9926 2	8	5	9976 0	6	58
3	6	9878 2	9	5	9946 1	7	2	7	57
4	9850 1	7	9904 3	9	4	9963 0	4	9	56
5	6	9879 1	7	9927 2	7	2	6	9987 0	55
6	9851 1	6	9905 1	6	9947 0	5	8	2	54
7	6	9880 0	5	9	3	7	9977 0	3	53
8	9852 1	5	9	9928 3	6	9	2	5	52
9	6	9	9906 3	6	9	9964 2	4	6	51
10	9853 1	9881 4	7	9929 0	9948 2	4	6	8	50
11	6	8	9907 1	3	5	7	8	9	49
12	9854 1	9882 3	5	7	8	9	9978 0	9988 1	48
13	6	7	9	9930 0	9949 1	9965 2	2	2	47
14	9855 1	9883 2	9908 3	3	4	4	4	3	46
15	6	6	7	7	7	7	6	5	45
16	9856 1	9884 1	9909 1	9931 0	9950 0	9	8	6	44
17	5	5	4	4	3	9966 1	9979 0	8	43
18	9857 0	9	8	7	6	4	2	9	42
19	5	9885 4	9910 2	9932 0	8	6	3	9989 0	41
20	9858 0	8	6	4	9951 1	8	5	2	40
21	5	9886 3	9911 0	7	4	9967 1	7	3	39
22	9859 0	7	4	9933 1	7	3	9	4	38
23	5	9887 1	8	4	9952 0	6	9980 1	6	37
24	9860 0	6	9912 2	7	3	8	3	7	36
25	4	9888 0	5	9934 1	6	9968 0	5	8	35
26	9	4	9	4	8	3	6	9990 0	34
27	9861 4	9	9913 3	7	9953 1	5	8	1	33
28	9	9889 3	7	9935 1	4	7	9981 0	2	32
29	9862 4	7	9914 1	4	7	9	2	4	31
30	9	9890 2	4	7	9954 0	9969 2	3	5	30
31	9863 3	6	8	9936 0	2	4	5	6	29
32	8	9891 0	9915 2	4	5	6	7	7	28
33	9864 3	4	6	7	8	9	9	9	27
34	8	9	9916 0	9937 0	9955 1	9970 1	9982 1	9991 0	26
35	9865 2	9892 3	3	4	3	3	2	1	25
36	7	7	7	7	6	5	4	2	24
37	9866 2	9893 1	9917 1	9938 0	9	8	6	4	23
38	7	6	5	3	9956 2	9971 0	7	5	22
39	9867 1	9894 0	8	6	4	2	9	6	21
40	6	4	9918 2	9939 0	7	4	9983 1	7	20
41	9868 1	8	6	3	9957 0	6	3	8	19
42	6	9895 3	9	6	2	9	4	9	18
43	9869 0	7	9919 3	9	5	9972 1	6	9992 1	17
44	5	9896 1	7	9940 2	8	3	8	2	16
45	9870 0	5	9920 0	6	9958 0	5	9	3	15
46	4	9	4	9	3	7	9984 1	4	14
47	9	9897 3	8	9941 2	6	9	2	5	13
48	9871 4	8	9921 1	5	8	9973 1	4	6	12
49	8	9898 2	5	8	9959 1	4	6	7	11
50	9872 3	6	9	9942 1	4	6	7	9	10
51	8	9899 0	9922 2	4	6	8	9	9993 0	9
52	9873 2	4	6	8	9	9974 0	9985 1	1	8
53	7	8	9923 0	9943 1	9960 2	2	2	2	7
54	9874 1	9900 2	3	4	4	4	4	3	6
55	6	6	7	7	7	6	5	4	5
56	9875 1	9901 1	9924 0	9944 0	9	8	7	5	4
57	5	5	4	3	9961 2	9975 0	8	6	3
58	9876 0	9	8	6	4	2	9986 0	7	2
59	4	9902 3	9925 1	9	7	4	1	8	1
60	9	7	5	9945 2	9	6	3	9	0
′	9°	8°	7°	6°	5°	4°	3°	2°	′

NATURAL COSINE.

′	80°	81°	82°	83°	84°	85°	86°	87°	′
0	5·6 7128	6·3 1375	7·1 1537	8·1 4435	9·5 1436	11· 4301	14· 3007	19· 0811	60
1	8094	2566	3042	6398	4106	4685	3607	1879	59
2	9064	3761	4553	8370	6791	5072	4212	2959	58
3	5·7 0037	4961	6071	8·2 0352	9490	5461	4823	4051	57
4	1013	6165	7594	2344	9·6 2205	5853	5438	5156	56
5	1992	7374	9125	4345	4935	6248	6059	6273	55
6	2974	8587	7·2 0661	6355	7680	6645	6685	7403	54
7	3960	9804	2204	8376	9·7 0441	7045	7317	8546	53
8	4949	6·4 1026	3754	8·3 0406	3217	7448	7954	9702	52
9	5941	2258	5310	2446	6009	7853	8596	20· 0872	51
10	6937	3484	6873	4496	8817	8262	9244	2056	50
11	7936	4720	8442	6555	9·8 1641	8673	9898	3253	49
12	8938	5961	7·3 0018	8625	4482	9087	15· 0557	4465	48
13	9944	7206	1600	8·4 0705	7338	9504	1222	5691	47
14	5·8 0953	8456	3190	2795	9·9 0211	9923	1893	6932	46
15	1966	9710	4786	4896	3101	12· 0346	2571	8188	45
16	2982	6·5 0970	6389	7007	6007	0772	3254	9460	44
17	4001	2234	7999	9128	8931	1201	3943	21· 0747	43
18	5024	3503	9616	8·5 1259	10· 0187	1632	4638	2049	42
19	6051	4777	7·4 1240	3402	0483	2067	5340	3369	41
20	7080	6055	2871	5555	0780	2505	6048	4704	40
21	8114	7339	4509	7718	1080	2946	6762	6056	39
22	9151	8627	6154	9893	1381	3390	7483	7426	38
23	5·9 0191	9921	7806	8·6 2078	1683	3838	8211	8813	37
24	1236	6·6 1219	9465	4275	1988	4288	8945	22· 0217	36
25	2283	2523	7·5 1132	6482	2294	4742	9687	1640	35
26	3335	3831	2806	8701	2602	5199	16· 0435	3081	34
27	4390	5144	4487	8·7 0931	2913	5660	1190	4541	33
28	5448	6463	6176	3172	3224	6124	1952	6020	32
29	6510	7787	7872	5425	3538	6591	2722	7519	31
30	7576	9116	9575	7689	3854	7062	3499	9038	30
31	8646	6·7 0450	7·6 1287	9964	4172	7536	4283	23· 0577	29
32	9720	1789	3005	8·8 2252	4491	8014	5075	2137	28
33	6·0 0797	3133	4732	4551	4813	8496	5874	3718	27
34	1878	4483	6466	6862	5136	8981	6681	5321	26
35	2962	5838	8208	9185	5462	9469	7496	6945	25
36	4051	7199	9957	8·9 1520	5789	9962	8319	8593	24
37	5143	8564	7·7 1715	3867	6118	13· 0458	9150	24· 0263	23
38	6240	9936	3480	6227	6450	0958	9990	1957	22
39	7340	6·8 1312	5254	8598	6783	1461	17· 0837	3675	21
40	8444	2694	7035	9·0 0983	7119	1969	1693	5418	20
41	9552	4082	8825	3379	7457	2480	2558	7185	19
42	6·1 0664	5475	7·8 0622	5789	7797	2996	3432	8978	18
43	1779	6874	2428	8211	8139	3515	4314	25· 0798	17
44	2899	8278	4242	9·1 0646	8483	4039	5205	2644	16
45	4023	9688	6064	3093	8829	4566	6106	4517	15
46	5151	6·9 1104	7895	5554	9178	5098	7015	6418	14
47	6283	2525	9734	8028	9529	5634	7934	8348	13
48	7419	3952	7·9 1582	9·2 0516	9882	6174	8863	26· 0307	12
49	8559	5385	3438	3016	11· 0237	6719	9802	2296	11
50	9703	6823	5302	5530	0594	7267	18· 0750	4316	10
51	6·2 0851	8263	7176	8058	0954	7821	1708	6367	9
52	2003	9718	9058	9·3 0599	1316	8378	2677	8450	8
53	3160	7·0 1174	8·0 0948	3155	1681	8940	3655	27· 0566	7
54	4321	2637	2848	5724	2048	9507	4645	2715	6
55	5486	4105	4756	8307	2417	14· 0079	5645	4899	5
56	6655	5579	6674	9·4 0904	2789	0655	6656	7117	4
57	7829	7059	8600	3515	3163	1235	7678	9372	3
58	9007	8546	8·1 0536	6141	3540	1821	8711	28· 1664	2
59	6·3 0189	7·1 0038	2481	8781	3919	2411	9755	3994	1
60	1375	1537	4435	9·5 1436	4301	3007	19· 0811	6363	0
′	9°	8°	7°	6°	5°	4°	3°	2°	′

NATURAL COTANGENT.

NATURAL SINE.

'	88°	89°	'
0	9993 9	9998 5	60
1	9994 0	5	59
2	1	6	58
3	2	6	57
4	3	7	56
5	4	7	55
6	5	8	54
7	6	8	53
8	7	9	52
9	8	9	51
10	9	9	50
11	9995 0	9999 0	49
12	1	0	48
13	2	1	47
14	2	1	46
15	3	1	45
16	4	2	44
17	5	2	43
18	6	3	42
19	7	3	41
20	8	3	40
21	9	4	39
22	9996 0	4	38
23	0	4	37
24	1	5	36
25	2	5	35
26	3	5	34
27	3	5	33
28	4	6	32
29	5	6	31
30	6	6	30
31	6	6	29
32	7	7	28
33	8	6	27
34	9	7	26
35	9	7	25
36	9997 0	8	24
37	1	8	23
38	2	8	22
39	2	8	21
40	3	8	20
41	4	8	19
42	4	9	18
43	5	9	17
44	6	9	16
45	6	9	15
46	7	9	14
47	7	9	13
48	8	9	12
49	9	9	11
50	9	1·0000 0	10
51	9998 0	0	9
52	0	0	8
53	1	0	7
54	2	0	6
55	2	0	5
56	3	0	4
57	3	0	3
58	4	0	2
59	4	0	1
60	5	0	0
'	1°	0°	'

NATURAL COSINE.

NATURAL TANGENT.

'	88°	89°	'
0	28·6363	57·2900	60
1	8771	58·2612	59
2	29·1220	59·2659	58
3	3711	60·3058	57
4	6245	61·3829	56
5	8823	62·4992	55
6	30·1446	63·6567	54
7	4116	64·8580	53
8	6833	66·1055	52
9	9599	67·4019	51
10	31·2416	68·7501	50
11	5284	70·1533	49
12	8205	71·6151	48
13	32·1181	73·1390	47
14	4213	74·7292	46
15	7303	76·3900	45
16	33·0452	78·1263	44
17	3662	79·9434	43
18	6935	81·8470	42
19	34·0273	83·8435	41
20	3678	85·9398	40
21	7151	88·1436	39
22	35·0695	90·4633	38
23	4313	92·9085	37
24	8006	95·4895	36
25	36·1776	98·2179	35
26	5627	101·1069	34
27	9560	104·1709	33
28	37·3579	107·4265	32
29	7686	110·8921	31
30	38·1885	114·5887	30
31	6177	118·5402	29
32	39·0568	122·7740	28
33	5059	127·3213	27
34	9655	132·2185	26
35	40·4358	137·5075	25
36	9174	143·2371	24
37	41·4106	149·4650	23
38	9158	156·2591	22
39	42·4335	163·7002	21
40	9641	171·8854	20
41	43·5081	180·9322	19
42	44·0661	190·9842	18
43	6386	202·2188	17
44	45·2261	214·8576	16
45	8294	229·1817	15
46	46·4489	245·5520	14
47	47·0853	264·4408	13
48	7395	286·4777	12
49	48·4121	312·5214	11
50	49·1039	343·7737	10
51	8157	381·9710	9
52	50·5485	429·7176	8
53	51·3032	491·1060	7
54	52·0807	572·9572	6
55	8821	687·5489	5
56	53·7086	859·4363	4
	54·5613	1145·9153	3
	55·4415	1718·8732	2
	56·3506	3437·7467	1
60	57·2900	Infinite.	0
'	1°	0°	'

NATURAL COTANGENT.

Table 4.

Dist.	Lat.	Dep.	Dist.	Lat.	Dep.	Dist.	Lat.	Dep	Dist.	Lat.	Dep.	Dist.	Lat.	Dep.
1	01·0	00·0	61	60·9	03·0	121	120·9	05·9	181	180·8	08·9	241	240·7	11·8
2	02·0	00·1	62	61·9	03·0	122	121·9	06·0	182	181·8	08·9	242	241·7	11·9
3	03·0	00·1	63	62·9	03·1	123	122·9	06·0	183	182·8	09·0	243	242·7	11·9
4	04·0	00·2	64	63·9	03·1	124	123·9	06·1	184	183·8	09·0	244	243·7	12·0
5	05·0	00·2	65	64·9	03·2	125	124·9	06·1	185	184·8	09·1	245	244·7	12·0
6	06·0	00·3	66	65·9	03·2	126	125·8	06·2	186	185·8	09·1	246	245·7	12·1
7	07·0	00·3	67	66·9	03·3	127	126·8	06·2	187	186·8	09·2	247	246·7	12·1
8	08·0	00·4	68	67·9	03·3	128	127·8	06·3	188	187·8	09·2	248	247·7	12·2
9	09·0	00·4	69	68·9	03·4	129	128·8	06·3	189	188·8	09·3	249	248·7	12·2
10	10·0	00·5	70	69·9	03·4	130	129·8	06·4	190	189·8	09·3	250	249·7	12·3
11	11·0	00·5	71	70·9	03·5	131	130·8	06·4	191	190·8	09·4	251	250·7	12·3
12	12·0	00·6	72	71·9	03·5	132	131·8	06·5	192	191·8	09·4	252	251·7	12·4
13	13·0	00·6	73	72·9	03·6	133	132·8	06·5	193	192·8	09·5	253	252·7	12·4
14	14·0	00·7	74	73·9	03·6	134	133·8	06·6	194	193·8	09·5	254	253·7	12·5
15	15·0	00·7	75	74·9	03·7	135	134·8	06·6	195	194·8	09·6	255	254·7	12·5
16	16·0	00·8	76	75·9	03·7	136	135·8	06·7	196	195·8	09·6	256	255·7	12·6
17	17·0	00·8	77	76·9	03·8	137	136·8	06·7	197	196·8	09·7	257	256·7	12·6
18	18·0	00·9	78	77·9	03·8	138	137·8	06·8	198	197·8	09·7	258	257·7	12·7
19	19·0	00·9	79	78·9	03·9	139	138·8	06·8	199	198·8	09·8	259	258·7	12·7
20	20·0	01·0	80	79·9	03·9	140	139·8	06·9	200	199·8	09·8	260	259·7	12·8
21	21·0	01·0	81	80·9	04·0	141	140·8	06·9	201	200·8	09·9	261	260·7	12·8
22	22·0	01·1	82	81·9	04·0	142	141·8	07·0	202	201·8	09·9	262	261·7	12·9
23	23·0	01·1	83	82·9	04·1	143	142·8	07·0	203	202·8	10·0	263	262·7	12·9
24	24·0	01·2	84	83·9	04·1	144	143·8	07·1	204	203·8	10·0	264	263·7	13·0
25	25·0	01·2	85	84·9	04·2	145	144·8	07·1	205	204·8	10·1	265	264·7	13·0
26	26·0	01·3	86	85·9	04·2	146	145·8	07·2	206	205·8	10·1	266	265·7	13·1
27	27·0	01·3	87	86·9	04·3	147	146·8	07·2	207	206·8	10·2	267	266·7	13·1
28	28·0	01·4	88	87·9	04·3	148	147·8	07·3	208	207·8	10·2	268	267·7	13·2
29	29·0	01·4	89	88·9	04·4	149	148·8	07·3	209	208·8	10·3	269	268·7	13·2
30	30·0	01·5	90	89·9	04·4	150	149·8	07·4	210	209·8	10·3	270	269·7	13·3
31	31·0	01·5	91	90·9	04·5	151	150·8	07·4	211	210·7	10·4	271	270·7	13·3
32	32·0	01·6	92	91·9	04·5	152	151·8	07·5	212	211·7	10·4	272	271·7	13·3
33	33·0	01·6	93	92·9	04·6	153	152·8	07·5	213	212·7	10·5	273	272·7	13·4
34	34·0	01·7	94	93·9	04·6	154	153·8	07·6	214	213·7	10·5	274	273·7	13·4
35	35·0	01·7	95	94·9	04·7	155	154·8	07·6	215	214·7	10·6	275	274·7	13·5
36	36·0	01·8	96	95·9	04·7	156	155·8	07·7	216	215·7	10·6	276	275·7	13·5
37	37·0	01·8	97	96·9	04·8	157	156·8	07·7	217	216·7	10·7	277	276·7	13·6
38	38·0	01·9	98	97·9	04·8	158	157·8	07·8	218	217·7	10·7	278	277·7	13·6
39	39·0	01·9	99	98·9	04·9	159	158·8	07·8	219	218·7	10·8	279	278·7	13·7
40	40·0	02·0	100	99·9	04·9	160	159·8	07·9	220	219·7	10·8	280	279·7	13·7
41	41·0	02·0	101	100·9	05·0	161	160·8	07·9	221	220·7	10·8	281	280·7	13·8
42	41·9	02·1	102	101·9	05·0	162	161·8	08·0	222	221·7	10·9	282	281·7	13·8
43	42·9	02·1	103	102·9	05·1	163	162·8	08·0	223	222·7	10·9	283	282·7	13·9
44	43·9	02·2	104	103·9	05·1	164	163·8	08·1	224	223·7	11·0	284	283·7	13·9
45	44·9	02·2	105	104·9	05·2	165	164·8	08·1	225	224·7	11·0	285	284·7	14·0
46	45·9	02·3	106	105·9	05·2	166	165·8	08·2	226	225·7	11·1	286	285·7	14·0
47	46·9	02·3	107	106·9	05·3	167	166·8	08·2	227	226·7	11·1	287	286·7	14·1
48	47·9	02·4	108	107·9	05·3	168	167·8	08·2	228	227·7	11·2	288	287·7	14·1
49	48·9	02·4	109	108·9	05·4	169	168·8	08·3	229	228·7	11·2	289	288·7	14·2
50	49·9	02·5	110	109·9	05·4	170	169·8	08·3	230	229·7	11·3	290	289·7	14·2
51	50·9	02·5	111	110·9	05·5	171	170·8	08·4	231	230·7	11·3	291	290·7	14·3
52	51·9	02·6	112	111·9	05·5	172	171·8	08·4	232	231·7	11·4	292	291·7	14·3
53	52·9	02·6	113	112·9	05·5	173	172·8	08·5	233	232·7	11·4	293	292·7	14·4
54	53·9	02·7	114	113·9	05·6	174	173·8	08·5	234	233·7	11·5	294	293·6	14·4
55	54·9	02·7	115	114·9	05·6	175	174·8	08·6	235	234·7	11·5	295	294·6	14·5
56	55·9	02·8	116	115·9	05·7	176	175·8	08·6	236	235·7	11·6	296	295·6	14·5
57	56·9	02·8	117	116·9	05·7	177	176·8	08·7	237	236·7	11·6	297	296·6	14·6
58	57·9	02·9	118	117·9	05·8	178	177·8	08·7	238	237·7	11·7	298	297·6	14·6
59	58·9	02·9	119	118·9	05·8	179	178·8	08·8	239	238·7	11·7	299	298·6	14·7
60	59·9	02·9	120	119·9	05·9	180	179·8	08·8	240	239·7	11·8	300	299·6	14·7
Dist.	Dep.	Lat.	Dist.	Dep.	Lat.	Dist.	Dep.	Lat.	Dist.	Dep.	Lat.	Dist.	Dep.	Lat.

FOR 7¾ POINTS.

Dist.	Lat.	Dep.	Dist.	Lat.	Dep.	Dist.	Lat.	Dep.	Dist.	Lat.	Dep.	Dist.	Lat.	Dep.
1	01·0	00·1	61	60·7	06·0	121	120·4	11·9	181	180·1	17·7	241	239·8	23·6
2	02·0	00·2	62	61·7	06·1	122	121·4	12·0	182	181·1	17·8	242	240·8	23·7
3	03·0	00·3	63	62·7	06·2	123	122·4	12·1	183	182·1	17·9	243	241·8	23·8
4	04·0	00·4	64	63·7	06·3	124	123·4	12·2	184	183·1	18·0	244	242·8	23·9
5	05·0	00·5	65	64·7	06·4	125	124·4	12·3	185	184·1	18·1	245	243·8	24·0
6	06·0	00·6	66	65·7	06·5	126	125·4	12·3	186	185·1	18·2	246	244·8	24·1
7	07·0	00·7	67	66·7	06·6	127	126·4	12·4	187	186·1	18·3	247	245·8	24·2
8	08·0	00·8	68	67·7	06·7	128	127·4	12·5	188	187·1	18·4	248	246·8	24·3
9	09·0	00·9	69	68·7	06·8	129	128·4	12·6	189	188·1	18·5	249	247·8	24·4
10	10·0	01·0	70	69·7	06·9	130	129·4	12·7	190	189·1	18·6	250	248·8	24·5
11	10·9	01·1	71	70·7	07·0	131	130·4	12·8	191	190·1	18·7	251	249·8	24·6
12	11·9	01·2	72	71·7	07·1	132	131·4	12·9	192	191·1	18·8	252	250·8	24·7
13	12·9	01·3	73	72·6	07·2	133	132·4	13·0	193	192·1	18·9	253	251·8	24·8
14	13·9	01·4	74	73·6	07·3	134	133·4	13·1	194	193·1	19·0	254	252·8	24·9
15	14·9	01·5	75	74·6	07·4	135	134·3	13·2	195	194·1	19·1	255	253·8	25·0
16	15·9	01·6	76	75·6	07·4	136	135·3	13·3	196	195·1	19·2	256	254·8	25·1
17	16·9	01·7	77	76·6	07·5	137	136·3	13·4	197	196·1	19·3	257	255·8	25·2
18	17·9	01·8	78	77·6	07·6	138	137·3	13·5	198	197·0	19·4	258	256·8	25·3
19	18·9	01·9	79	78·6	07·7	139	138·3	13·6	199	198·0	19·5	259	257·8	25·4
20	19·9	02·0	80	79·6	07·8	140	139·3	13·7	200	199·0	19·6	260	258·7	25·5
21	20·9	02·1	81	80·6	07·9	141	140·3	13·8	201	200·0	19·7	261	259·7	25·6
22	21·9	02·2	82	81·6	08·0	142	141·3	13·9	202	201·0	19·8	262	260·7	25·7
23	22·9	02·3	83	82·6	08·1	143	142·3	14·0	203	202·0	19·9	263	261·7	25·8
24	23·9	02·4	84	83·6	08·2	144	143·3	14·1	204	203·0	20·0	264	262·7	25·9
25	24·9	02·4	85	84·6	08·3	145	144·3	14·2	205	204·0	20·1	265	263·7	26·0
26	25·9	02·5	86	85·6	08·4	146	145·3	14·3	206	205·0	20·2	266	264·7	26·1
27	26·9	02·6	87	86·6	08·5	147	146·3	14·4	207	206·0	20·3	267	265·7	26·2
28	27·9	02·7	88	87·6	08·6	148	147·3	14·5	208	207·0	20·4	268	266·7	26·3
29	28·9	02·8	89	88·6	08·7	149	148·3	14·6	209	208·0	20·5	269	267·7	26·4
30	29·9	02·9	90	89·6	08·8	150	149·3	14·7	210	209·0	20·6	270	268·7	26·5
31	30·9	03·0	91	90·6	08·9	151	150·3	14·8	211	210·0	20·7	271	269·7	26·6
32	31·8	03·1	92	91·6	09·0	152	151·3	14·9	212	211·0	20·8	272	270·7	26·7
33	32·8	03·2	93	92·6	09·1	153	152·3	15·0	213	212·0	20·9	273	271·7	26·8
34	33·8	03·3	94	93·5	09·2	154	153·3	15·1	214	213·0	21·0	274	272·7	26·9
35	34·8	03·4	95	94·5	09·3	155	154·3	15·2	215	214·0	21·1	275	273·7	27·0
36	35·8	03·5	96	95·5	09·4	156	155·2	15·3	216	215·0	21·2	276	274·7	27·1
37	36·8	03·6	97	96·5	09·5	157	156.2	15·4	217	216·0	21·3	277	275·7	27·2
38	37·8	03·7	98	97·3	09·6	158	157·2	15·5	218	216·9	21·4	278	276·7	27·3
39	38·8	03·8	99	98·5	09·7	159	158·2	15·6	219	217·9	21·5	279	277·7	27·3
40	39·8	03·9	100	99·5	09·8	160	159·2	15·7	220	218·9	21·6	280	278·7	27·4
41	40·8	04·0	101	100·5	09·9	161	160·2	15·8	221	219·9	21·7	281	279·6	27·5
42	41·8	04·1	102	101·5	10·0	162	161·2	15·9	222	220·9	21·8	282	280·6	27·6
43	42·8	04·2	103	102·5	10·1	163	162·2	16·0	223	221·9	21·9	283	281·6	27·7
44	43·8	04·3	104	103·5	10·2	164	163·2	16·1	224	222·9	22·0	284	282·6	27·8
45	44·8	04·4	105	104·5	10·3	165	164·2	16·2	225	223·9	22·1	285	283·6	27·9
46	45·8	04·5	106	105·5	10·4	166	165·2	16·3	226	224·9	22·2	286	284·6	28·0
47	46·8	04·6	107	106·5	10·5	167	166·2	16·4	227	225·9	22·2	287	285·6	28·1
48	47·8	04·7	108	107·5	10·6	168	167·2	16·5	228	226·9	22·3	288	286·6	28·2
49	48·8	04·8	109	108·5	10·7	169	168·2	16·6	229	227·9	22·4	289	287·6	28·3
50	49·8	04·9	110	109·5	10·8	170	169·2	16·7	230	228·9	22·5	290	288·6	28·4
51	50·8	05·0	111	110·5	10·9	171	170·2	16·8	231	229·9	22·6	291	289·6	28·5
52	51·7	05·1	112	111·5	11·0	172	171·2	16·9	232	230·9	22·7	292	290·6	28·6
53	52·7	05·2	113	112·5	11·1	173	172·2	17·0	233	231·9	22·8	293	291·6	28·7
54	53·7	05·3	114	113·5	11·2	174	173·2	17·1	234	232·9	22·9	294	292·6	28·8
55	54·7	05·4	115	114·4	11·3	175	174·2	17·2	235	233·9	23·0	295	293·6	28·9
56	55·7	05·5	116	115·4	11·4	176	175·2	17·3	236	234·9	23·1	296	294·6	29·0
57	56·7	05·6	117	116·4	11·5	177	176·1	17·4	237	235·9	23·2	297	295·6	29·1
58	57·7	05·7	118	117·4	11·6	178	177·1	17·4	238	236·9	23·3	298	296·6	29·2
59	58·7	05·8	119	118·4	11·7	179	178·1	17·5	239	237·8	23·4	299	297·6	29·3
60	59·7	05·9	120	119·4	11·8	180	179·1	17·6	240	238·8	23·5	300	298·6	29·4
Dist.	Dep.	Lat.	Dist.	Dep.	Lat.	Dist.	Dep.	Lat.	Dist.	Dep.	Lat.	Dist.	Dep.	Lat.

FOR 7¾ POINTS.

Dist.	Lat.	Dep.	Dist.	Lat.	Dep.	Dist.	Lat.	Dep.	Dist.	Lat.	Dep.	Dist.	Lat.	Dep.
1	01·0	00·1	61	60·3	09·0	121	119·7	17·8	181	179·0	26·6	241	238·4	35·4
2	02·0	00·3	62	61·3	09·1	122	120·7	17·9	182	180·0	26·7	242	239·4	35·5
3	03·0	00·4	63	62·3	09·2	123	121·7	18·1	183	181·0	26·9	243	240·4	35·7
4	04·0	00·6	64	63·3	09·4	124	122·7	18·2	184	182·0	27·0	244	241·4	35·8
5	04·9	00·7	65	64·3	09·5	125	123·7	18·3	185	183·0	27·2	245	242·4	36·0
6	05·9	00·9	66	65·3	09·7	126	124·6	18·5	186	184·0	27·3	246	243·3	36·1
7	06·9	01·0	67	66·3	09·8	127	125·6	18·6	187	185·0	27·4	247	244·3	36·2
8	07·9	01·2	68	67·3	10·0	128	126·6	18·8	188	186·0	27·6	248	245·3	36·4
9	08·9	01·3	69	68·3	10·1	129	127·6	18·9	189	187·0	27·7	249	246·3	36·5
10	09·9	01·5	70	69·2	10·3	130	128·6	19·1	190	187·9	27·9	250	247·3	36·7
11	10·9	01·6	71	70·2	10·4	131	129·6	19·2	191	188·9	28·0	251	248·3	36·8
12	11·9	01·8	72	71·2	10·6	132	130·6	19·4	192	189·9	28·2	252	249·3	37·0
13	12·9	01·9	73	72·2	10·7	133	131·6	19·5	193	190·9	28·3	253	250·3	37·1
14	13·9	02·1	74	73·2	10·9	134	132·6	19·7	194	191·9	28·5	254	251·3	37·3
15	14·8	02·2	75	74·2	11·0	135	133·5	19·8	195	192·9	28·6	255	252·2	37·4
16	15·8	02·3	76	75·2	11·2	136	134·5	20·0	196	193·9	28·8	256	253·2	37·6
17	16·8	02·5	77	76·2	11·3	137	135·5	20·1	197	194·9	28·9	257	254·2	37·7
18	17·8	02·6	78	77·2	11·4	138	136·5	20·3	198	195·9	29·1	258	255·2	37·9
19	18·8	02·8	79	78·1	11·6	139	137·5	20·4	199	196·8	29·2	259	256·2	38·0
20	19·8	02·9	80	79·1	11·7	140	138·5	20·5	200	197·8	29·4	260	257·2	38·2
21	20·8	03·1	81	80·1	11·9	141	139·5	20·7	201	198·8	29·5	261	258·2	38·3
22	21·8	03·2	82	81·1	12·0	142	140·5	20·8	202	199·8	29·6	262	259·2	38·4
23	22·8	03·4	83	82·1	12·2	143	141·5	21·0	203	200·8	29·8	263	260·2	38·6
24	23·7	03·5	84	83·1	12·3	144	142·4	21·1	204	201·8	29·9	264	261·1	38·7
25	24·7	03·7	85	84·1	12·5	145	143·4	21·3	205	202·8	30·1	265	262·1	38·9
26	25·7	03·8	86	85·1	12·6	146	144·4	21·4	206	203·8	30·2	266	263·1	39·0
27	26·7	04·0	87	86·1	12·8	147	145·4	21·6	207	204·8	30·4	267	264·1	39·2
28	27·7	04·1	88	87·1	12·9	148	146·4	21·7	208	205·8	30·5	268	265·1	39·3
29	28·7	04·3	89	88·0	13·1	149	147·4	21·9	209	206·7	30·7	269	266·1	39·5
30	29·7	04·4	90	89·0	13·2	150	148·4	22·0	210	207·7	30·8	270	267·1	39·6
31	30·7	04·6	91	90·0	13·4	151	149·4	22·2	211	208·7	31·0	271	268·1	39·8
32	31·7	04·7	92	91·0	13·5	152	150·4	22·3	212	209·7	31·1	272	269·1	39·9
33	32·6	04·8	93	92·0	13·7	153	151·3	22·5	213	210·7	31·3	273	270·0	40·1
34	33·6	05·0	94	93·0	13·8	154	152·3	22·6	214	211·7	31·4	274	271·0	40·2
35	34·6	05·1	95	94·0	13·9	155	153·3	22·7	215	212·7	31·6	275	272·0	40·4
36	35·6	05·3	96	95·0	14·1	156	154·3	22·9	216	213·7	31·7	276	273·0	40·5
37	36·6	05·4	97	96·0	14·2	157	155·3	23·0	217	214·7	31·8	277	274·0	40·6
38	37·6	05·6	98	96·9	14·4	158	156·3	23·2	218	215·6	32·0	278	275·0	40·8
39	38·6	05·7	99	97·9	14·5	159	157·3	23·3	219	216·6	32·1	279	276·0	40·9
40	39·6	05·9	100	98·9	14·7	160	158·3	23·5	220	217·6	32·3	280	277·0	41·1
41	40·6	06·0	101	99·9	14·8	161	159·3	23·6	221	218·6	32·4	281	278·0	41·2
42	41·6	06·2	102	100·9	15·0	162	160·3	23·8	222	219·6	32·6	282	279·0	41·4
43	42·5	06·3	103	101·9	15·1	163	161·2	23·9	223	220·6	32·7	283	279·9	41·5
44	43·5	06·5	104	102·9	15·3	164	162·2	24·1	224	221·6	32·9	284	280·9	41·7
45	44·5	06·6	105	103·9	15·4	165	163·2	24·2	225	222·6	33·0	285	281·9	41·8
46	45·5	06·8	106	104·9	15·6	166	164·2	24·4	226	223·6	33·2	286	282·9	42·0
47	46·5	06·9	107	105·8	15·7	167	165·2	24·5	227	224·5	33·3	287	283·9	42·1
48	47·5	07·0	108	106·8	15·9	168	166·2	24·7	228	225·5	33·5	288	284·9	42·3
49	48·5	07·2	109	107·8	16·0	169	167·2	24·8	229	226·5	33·6	289	285·9	42·4
50	49·5	07·3	110	108·8	16·1	170	168·2	24·9	230	227·5	33·8	290	286·9	42·6
51	50·5	07·5	111	109·8	16·3	171	169·2	25·1	231	228·5	33·9	291	287·9	42·7
52	51·4	07·6	112	110·8	16·4	172	170·1	25·2	232	229·5	34·0	292	288·8	42·9
53	52·4	07·8	113	111·8	16·6	173	171·1	25·4	233	230·5	34·2	293	289·8	43·0
54	53·4	07·9	114	112·8	16·7	174	172·1	25·5	234	231·5	34·3	294	290·8	43·1
55	54·4	08·1	115	113·8	16·9	175	173·1	25·7	235	232·5	34·5	295	291·8	43·3
56	55·4	08·2	116	114·7	17·0	176	174·1	25·8	236	233·4	34·6	296	292·8	43·4
57	56·4	08·4	117	115·7	17·2	177	175·1	26·0	237	234·4	34·8	297	293·8	43·6
58	57·4	08·5	118	116·7	17·3	178	176·1	26·1	238	235·4	34·9	298	294·8	43·7
59	58·4	08·7	119	117·7	17·5	179	177·1	26·3	239	236·4	35·1	299	295·8	43·9
60	59·4	08·8	120	118·7	17·6	180	178·1	26·4	240	237·4	35·2	300	296·8	44·0
Dist.	Dep.	Lat.	Dist.	Dep.	Lat.	Dist.	Dep.	Lat.	Dist.	Dep.	Lat.	Dist.	Dep.	Lat.

FOR 7¼ POINTS.

Dist.	Lat.	Dep.	Dist.	Lat.	Dep.	Dist.	Lat.	Dep.	Dist.	Lat.	Dep.	Dist.	Lat.	Dep.
1	01·0	00·2	61	59·8	11·9	121	118·7	23·6	181	177·5	35·3	241	236·4	47·0
2	02·0	00·4	62	60·8	12·1	122	119·7	23·8	182	178·5	35·5	242	237·4	47·2
3	02·9	00·6	63	61·8	12·3	123	120·6	24·0	183	179·5	35·7	243	238·3	47·4
4	03·9	00·8	64	62·8	12·5	124	121·6	24·2	184	180·5	35·9	244	239·3	47·6
5	04·9	01·0	65	63·8	12·7	125	122·6	24·4	185	181·5	36·1	245	240·3	47·8
6	05·9	01·2	66	64·7	12·9	126	123·6	24·6	186	182·4	36·3	246	241·3	48·0
7	06·9	01·4	67	65·7	13·1	127	124·6	24·8	187	183·4	36·5	247	242·3	48·2
8	07·8	01·6	68	66·7	13·3	128	125·5	25·0	188	184·4	36·7	248	243·2	48·4
9	08·8	01·8	69	67·7	13·5	129	126·5	25·2	189	185·4	36·9	249	244·2	48·6
10	09·8	02·0	70	68·7	13·7	130	127·5	25·4	190	186·4	37·1	250	245·2	48·8
11	10·8	02·2	71	69·6	13·9	131	128·5	25·6	191	187·3	37·3	251	246·2	49·0
12	11·8	02·3	72	70·6	14·0	132	129·5	25·8	192	188·3	37·5	252	247·2	49·2
13	12·8	02·5	73	71·6	14·2	133	130·5	26·0	193	189·3	37·7	253	248·1	49·4
14	13·7	02·7	74	72·6	14·4	134	131·4	26·1	194	190·3	37·8	254	249·1	49·6
15	14·7	02·9	75	73·6	14·6	135	132·4	26·3	195	191·3	38·0	255	250·1	49·7
16	15·7	03·1	76	74·5	14·8	136	133·4	26·5	196	192·2	38·2	256	251·1	49·9
17	16·7	03·3	77	75·5	15·0	137	134·4	26·7	197	193·2	38·4	257	252·1	50·1
18	17·7	03·5	78	76·5	15·2	138	135·4	26·9	198	194·2	38·6	258	253·0	50·3
19	18·6	03·7	79	77·5	15·4	139	136·3	27·1	199	195·2	38·8	259	254·0	50·5
20	19·6	03·9	80	78·5	15·6	140	137·3	27·3	200	196·2	39·0	260	255·0	50·7
21	20·6	04·1	81	79·4	15·8	141	138·3	27·5	201	197·1	39·2	261	256·0	50·9
22	21·6	04·3	82	80·4	16·0	142	139·3	27·7	202	198·1	39·4	262	257·0	51·1
23	22·6	04·5	83	81·4	16·2	143	140·3	27·9	203	199·1	39·6	263	258·0	51·3
24	23·5	04·7	84	82·4	16·4	144	141·2	28·1	204	200·1	39·8	264	258·9	51·5
25	24·5	04·9	85	83·4	16·6	145	142·2	28·3	205	201·1	40·0	265	259·9	51·7
26	25·5	05·1	86	84·4	16·8	146	143·2	28·5	206	202·0	40·2	266	260·9	51·9
27	26·5	05·3	87	85·3	17·0	147	144·2	28·7	207	203·0	40·4	267	261·9	52·1
28	27·5	05·5	88	86·3	17·2	148	145·2	28·9	208	204·0	40·6	268	262·9	52·3
29	28·4	05·7	89	87·3	17·4	149	146·1	29·1	209	205·0	40·8	269	263·8	52·5
30	29·4	05·9	90	88·3	17·6	150	147·1	29·3	210	206·0	41·0	270	264·8	52·7
31	30·4	06·0	91	89·3	17·8	151	148·1	29·5	211	207·0	41·2	271	265·8	52·9
32	31·4	06·2	92	90·2	18·0	152	149·1	29·7	212	207·9	41·4	272	266·8	53·1
33	32·4	06·4	93	91·2	18·1	153	150·1	29·9	213	208·9	41·6	273	267·8	53·3
34	33·4	06·6	94	92·2	18·3	154	151·0	30·0	214	209·9	41·8	274	268·7	53·5
35	34·3	06·8	95	93·2	18·5	155	152·0	30·2	215	210·9	41·9	275	269·7	53·6
36	35·3	07·0	96	94·2	18·7	156	153·0	30·4	216	211·9	42·1	276	270·7	53·8
37	36·3	07·2	97	95·1	18·9	157	154·0	30·6	217	212·8	42·3	277	271·7	54·0
38	37·3	07·4	98	96·1	19·1	158	155·0	30·8	218	213·8	42·5	278	272·7	54·2
39	38·3	07·6	99	97·1	19·3	159	156·0	31·0	219	214·8	42·7	279	273·6	54·4
40	39·2	07·8	100	98·1	19·5	160	156·9	31·2	220	215·8	42·9	280	274·6	54·6
41	40·2	08·0	101	99·1	19·7	161	157·9	31·4	221	216·8	43·1	281	275·6	54·8
42	41·2	08·2	102	100·0	19·9	162	158·9	31·6	222	217·7	43·3	282	276·6	55·0
43	42·2	08·4	103	101·0	20·1	163	159·9	31·8	223	218·7	43·5	283	277·6	55·2
44	43·2	08·6	104	102·0	20·3	164	160·9	32·0	224	219·7	43·7	284	278·5	55·4
45	44·1	08·8	105	103·0	20·5	165	161·8	32·2	225	220·7	43·9	285	279·5	55·6
46	45·1	09·0	106	104·0	20·7	166	162·8	32·4	226	221·7	44·1	286	280·5	55·8
47	46·1	09·2	107	104·9	20·9	167	163·8	32·6	227	222·6	44·3	287	281·5	56·0
48	47·1	09·4	108	105·9	21·1	168	164·8	32·8	228	223·6	44·5	288	282·5	56·2
49	48·1	09·6	109	106·9	21·3	169	165·8	33·0	229	224·6	44·7	289	283·5	56·4
50	49·0	09·8	110	107·9	21·5	170	166·7	33·2	230	225·6	44·9	290	284·4	56·6
51	50·0	10·0	111	108·9	21·7	171	167·7	33·4	231	226·6	45·1	291	285·4	56·8
52	51·0	10·1	112	109·9	21·9	172	168·7	33·6	232	227·5	45·3	292	286·4	57·0
53	52·0	10·3	113	110·8	22·0	173	169·7	33·8	233	228·5	45·5	293	287·4	57·2
54	53·0	10·5	114	111·8	22·2	174	170·7	34·0	234	229·5	45·7	294	288·4	57·4
55	53·9	10·7	115	112·8	22·4	175	171·6	34·1	235	230·5	45·9	295	289·3	57·6
56	54·9	10·9	116	113·8	22·6	176	172·6	34·3	236	231·5	46·0	296	290·3	57·7
57	55·9	11·1	117	114·8	22·8	177	173·6	34·5	237	232·5	46·2	297	291·3	57·9
58	56·9	11·3	118	115·7	23·0	178	174·6	34·7	238	233·4	46·4	298	292·3	58·1
59	57·9	11·5	119	116·7	23·2	179	175·6	34·9	239	234·4	46·6	299	293·3	58·3
60	58·8	11·7	120	117·7	23·4	180	176·5	35·1	240	235·4	46·8	300	294·2	58·5
Dist.	Dep.	Lat.	Dist.	Dep.	Lat.	Dist.	Dep.	Lat.	Dist.	Dep.	Lat.	Dist.	Dep.	Lat.

FOR 7 POINTS.

Dist.	Lat.	Dep.	Dist.	Lat.	Dep.	Dist.	Lat.	Dep.	Dist.	Lat.	Dep.	Dist.	Lat.	Dep.
1	01·0	00·2	61	59·2	14·8	121	117·4	29·4	181	175·6	44·0	241	233·8	58·6
2	01·9	00·5	62	60·1	15·1	122	118·4	29·6	182	176·5	44·2	242	234·8	58·8
3	02·9	00·7	63	61·1	15·3	123	119·3	29·9	183	177·5	44·5	243	235·7	59·0
4	03·9	01·0	64	62·1	15·6	124	120·3	30·1	184	178·5	44·7	244	236·7	59·3
5	04·9	01·2	65	63·1	15·8	125	121·3	30·4	185	179·5	45·0	245	237·7	59·5
6	05·8	01·5	66	64·0	16·0	126	122·2	30·6	186	180·4	45·2	246	238·6	59·8
7	06·8	01·7	67	65·0	16·3	127	123·2	30·9	187	181·4	45·4	247	239·6	60·0
8	07·8	01·9	68	66·0	16·5	128	124·2	31·1	188	182·4	45·7	248	240·6	60·3
9	08·7	02·2	69	66·9	16·8	129	125·1	31·3	189	183·3	45·9	249	241·6	60·5
10	09·7	02·4	70	67·9	17·0	130	126·1	31·6	190	184·3	46·2	250	242·5	60·8
11	10·7	02·7	71	68·9	17·3	131	127·1	31·8	191	185·3	46·4	251	243·5	61·0
12	11·6	02·9	72	69·9	17·5	132	128·1	32·1	192	186·2	46·7	252	244·5	61·2
13	12·6	03·2	73	70·8	17·7	133	129·0	32·3	193	187·2	46·9	253	245·4	61·5
14	13·6	03·4	74	71·8	18·0	134	130·0	32·6	194	188·2	47·1	254	246·4	61·7
15	14·6	03·6	75	72·8	18·2	135	131·0	32·8	195	189·2	47·4	255	247·4	62·0
16	15·5	03·9	76	73·7	18·5	136	131·9	33·1	196	190·1	47·6	256	248·3	62·2
17	16·5	04·1	77	74·7	18·7	137	132·9	33·3	197	191·1	47·9	257	249·3	62·5
18	17·5	04·4	78	75·7	19·0	138	133·9	33·5	198	192·1	48·1	258	250·3	62·7
19	18·4	04·6	79	76·6	19·2	139	134·8	33·8	199	193·0	48·4	259	251·3	62·9
20	19·4	04·9	80	77·6	19·4	140	135·8	34·0	200	194·0	48·6	260	252·2	63·2
21	20·4	05·1	81	78·6	19·7	141	136·8	34·3	201	195·0	48·8	261	253·2	63·4
22	21·3	05·4	82	79·6	19·9	142	137·8	34·5	202	196·0	49·1	262	254·2	63·7
23	22·3	05·6	83	80·5	20·2	143	138·7	34·8	203	196·9	49·3	263	255·1	63·9
24	23·3	05·8	84	81·5	20·4	144	139·7	35·0	204	197·9	49·6	264	256·1	64·2
25	24·3	06·1	85	82·5	20·7	145	140·7	35·2	205	198·9	49·8	265	257·1	64·4
26	25·2	06·3	86	83·4	20·9	146	141·6	35·5	206	199·8	50·1	266	258·0	64·6
27	26·2	06·6	87	84·4	21·1	147	142·6	35·7	207	200·8	50·3	267	259·0	64·9
28	27·2	06·8	88	85·4	21·4	148	143·6	36·0	208	201·8	50·5	268	260·0	65·1
29	28·1	07·1	89	86·3	21·6	149	144·5	36·2	209	202·7	50·8	269	261·0	65·4
30	29·1	07·3	90	87·3	21·9	150	145·5	36·5	210	203·7	51·0	270	261·9	65·6
31	30·1	07·5	91	88·3	22·1	151	146·5	36·7	211	204·7	51·3	271	262·9	65·9
32	31·0	07·8	92	89·3	22·4	152	147·4	36·9	212	205·7	51·5	272	263·9	66·1
33	32·0	08·0	93	90·2	22·6	153	148·4	37·2	213	206·6	51·8	273	264·8	66·3
34	33·0	08·3	94	91·2	22·8	154	149·4	37·4	214	207·6	52·0	274	265·8	66·6
35	34·0	08·5	95	92·2	23·1	155	150·4	37·7	215	208·6	52·2	275	266·8	66·8
36	34·9	08·8	96	93·1	23·3	156	151·3	37·9	216	209·5	52·5	276	267·7	67·1
37	35·9	09·0	97	94·1	23·6	157	152·3	38·2	217	210·5	52·7	277	268·7	67·3
38	36·9	09·2	98	95·1	23·8	158	153·3	38·4	218	211·5	53·0	278	269·7	67·6
39	37·8	09·5	99	96·0	24·1	159	154·2	38·6	219	212·5	53·2	279	270·7	67·8
40	38·8	09·7	100	97·0	24·3	160	155·2	38·9	220	213·4	53·5	280	271·6	68·0
41	39·8	10·0	101	98·0	24·5	161	156·2	39·1	221	214·4	53·7	281	272·6	68·3
42	40·7	10·2	102	99·0	24·8	162	157·2	39·4	222	215·4	53·9	282	273·6	68·5
43	41·7	10·5	103	99·9	25·0	163	158·1	39·6	223	216·3	54·2	283	274·5	68·8
44	42·7	10·7	104	100·9	25·3	164	159·1	39·9	224	217·3	54·4	284	275·5	69·0
45	43·7	10·9	105	101·9	25·5	165	160·1	40·1	225	218·3	54·7	285	276·5	69·3
46	44·6	11·2	106	102·8	25·8	166	161·0	40·3	226	219·2	54·9	286	277·4	69·5
47	45·6	11·4	107	103·8	26·0	167	162·0	40·6	227	220·2	55·2	287	278·4	69·7
48	46·6	11·7	108	104·8	26·2	168	163·0	40·8	228	221·2	55·4	288	279·4	70·0
49	47·5	11·9	109	105·7	26·5	169	163·9	41·1	229	222·2	55·6	289	280·4	70·2
50	48·5	12·2	110	106·7	26·7	170	164·9	41·3	230	223·1	55·9	290	281·3	70·5
51	49·5	12·4	111	107·7	27·0	171	165·9	41·6	231	224·1	56·1	291	282·3	70·7
52	50·4	12·6	112	108·7	27·2	172	166·9	41·8	232	225·1	56·4	292	283·3	71·0
53	51·4	12·9	113	109·6	27·5	173	167·8	42·0	233	226·0	56·6	293	284·2	71·2
54	52·4	13·1	114	110·6	27·7	174	168·8	42·3	234	227·0	56·9	294	285·2	71·4
55	53·4	13·4	115	111·6	27·9	175	169·8	42·5	235	228·0	57·1	295	286·2	71·7
56	54·3	13·6	116	112·5	28·2	176	170·7	42·8	236	228·9	57·3	296	287·1	71·9
57	55·3	13·9	117	113·5	28·4	177	171·7	43·0	237	229·9	57·6	297	288·1	72·2
58	56·3	14·1	118	114·5	28·7	178	172·7	43·3	238	230·9	57·8	298	289·1	72·4
59	57·2	14·3	119	115·4	28·9	179	173·6	43·5	239	231·8	58·1	299	290·1	72·7
60	58·2	14·6	120	116·4	29·2	180	174·6	43·7	240	232·8	58·3	300	291·0	72·9
Dist.	Dep.	Lat.	Dist.	Dep.	Lat.	Dist.	Dep.	Lat.	Dist.	Dep.	Lat.	Dist.	Dep.	Lat.

DIFFERENCE OF LATITUDE AND DEPARTURE FOR 1½ POINT. [TABLE 4.

Dist.	Lat.	Dep.	Dist.	Lat.	Dep.	Dist.	Lat.	Dep.	Dist.	Lat.	Dep.	Dist.	Lat.	Dep.
1	01.0	00.3	61	58.4	17.7	121	115.8	35.1	181	173.2	52.5	241	230.6	70.0
2	01.9	00.6	62	59.3	18.0	122	116.8	35.4	182	174.2	52.8	242	231.6	70.3
3	02.9	00.9	63	60.3	18.3	123	117.7	35.7	183	175.1	53.1	243	232.5	70.5
4	03.8	01.2	64	61.2	18.6	124	118.7	36.0	184	176.1	53.4	244	233.5	70.8
5	04.8	01.5	65	62.2	18.9	125	119.6	36.3	185	177.0	53.7	245	234.5	71.1
6	05.7	01.7	66	63.2	19.2	126	120.6	36.6	186	178.0	54.0	246	235.4	71.4
7	06.7	02.0	67	64.1	19.5	127	121.5	36.9	187	179.0	54.3	247	236.4	71.7
8	07.7	02.3	68	65.1	19.7	128	122.5	37.2	188	179.9	54.6	248	237.3	72.0
9	08.6	02.6	69	66.0	20.0	129	123.5	37.5	189	180.9	54.9	249	238.3	72.3
10	09.6	02.9	70	67.0	20.3	130	124.4	37.7	190	181.8	55.2	250	239.2	72.6
11	10.5	03.2	71	67.9	20.6	131	125.4	38.0	191	182.8	55.4	251	240.2	72.9
12	11.5	03.5	72	68.9	20.9	132	126.3	38.3	192	183.7	55.7	252	241.2	73.2
13	12.4	03.8	73	69.9	21.2	133	127.3	38.6	193	184.7	56.0	253	242.1	73.4
14	13.4	04.1	74	70.8	21.5	134	128.2	38.9	194	185.7	56.3	254	243.1	73.7
15	14.4	04.4	75	71.8	21.8	135	129.2	39.2	195	186.6	56.6	255	244.0	74.0
16	15.3	04.6	76	72.7	22.1	136	130.1	39.5	196	187.6	56.9	256	245.0	74.3
17	16.3	04.9	77	73.7	22.4	137	131.1	39.8	197	188.5	57.2	257	245.9	74.6
18	17.2	05.2	78	74.6	22.6	138	132.1	40.1	198	189.5	57.5	258	246.9	74.9
19	18.2	05.5	79	75.6	22.9	139	133.0	40.4	199	190.4	57.8	259	247.9	75.2
20	19.1	05.8	80	76.6	23.2	140	134.0	40.6	200	191.4	58.1	260	248.8	75.5
21	20.1	06.1	81	77.5	23.5	141	134.9	40.9	201	192.3	58.4	261	249.8	75.8
22	21.1	06.4	82	78.5	23.8	142	135.9	41.2	202	193.3	58.6	262	250.7	76.1
23	22.0	06.7	83	79.4	24.1	143	136.8	41.5	203	194.3	58.9	263	251.7	76.3
24	23.0	07.0	84	80.4	24.4	144	137.8	41.8	204	195.2	59.2	264	252.6	76.6
25	23.9	07.3	85	81.3	24.7	145	138.8	42.1	205	196.2	59.5	265	253.6	76.9
26	24.9	07.6	86	82.3	25.0	146	139.7	42.4	206	197.1	59.8	266	254.6	77.2
27	25.8	07.8	87	83.3	25.3	147	140.7	42.7	207	198.1	60.1	267	255.5	77.5
28	26.8	08.1	88	84.2	25.5	148	141.6	43.0	208	199.0	60.4	268	256.5	77.8
29	27.8	08.4	89	85.2	25.8	149	142.6	43.3	209	200.0	60.7	269	257.4	78.1
30	28.7	08.7	90	86.1	26.1	150	143.5	43.5	210	201.0	61.0	270	258.4	78.4
31	29.7	09.0	91	87.1	26.4	151	144.5	43.8	211	201.9	61.3	271	259.3	78.7
32	30.6	09.3	92	88.0	26.7	152	145.5	44.1	212	202.9	61.5	272	260.3	79.0
33	31.6	09.6	93	89.0	27.0	153	146.4	44.4	213	203.8	61.8	273	261.2	79.3
34	32.5	09.9	94	90.0	27.3	154	147.4	44.7	214	204.8	62.1	274	262.2	79.5
35	33.5	10.2	95	90.9	27.6	155	148.3	45.0	215	205.7	62.4	275	263.2	79.8
36	34.5	10.5	96	91.9	27.9	156	149.3	45.3	216	206.7	62.7	276	264.1	80.1
37	35.4	10.7	97	92.8	28.2	157	150.2	45.6	217	207.7	63.0	277	265.1	80.4
38	36.4	11.0	98	93.8	28.5	158	151.2	45.9	218	208.6	63.3	278	266.0	80.7
39	37.3	11.3	99	94.7	28.7	159	152.2	46.2	219	209.6	63.6	279	267.0	81.0
40	38.3	11.6	100	95.7	29.0	160	153.1	46.4	220	210.5	63.9	280	267.9	81.3
41	39.2	11.9	101	96.7	29.3	161	154.1	46.7	221	211.5	64.2	281	268.9	81.6
42	40.2	12.2	102	97.6	29.6	162	155.0	47.0	222	212.4	64.4	282	269.9	81.9
43	41.2	12.5	103	98.6	29.9	163	156.0	47.3	223	213.4	64.7	283	270.8	82.2
44	42.1	12.8	104	99.5	30.2	164	156.9	47.6	224	214.4	65.0	284	271.8	82.4
45	43.1	13.1	105	100.5	30.5	165	157.9	47.9	225	215.3	65.3	285	272.7	82.7
46	44.0	13.4	106	101.4	30.8	166	158.9	48.2	226	216.3	65.6	286	273.7	83.0
47	45.0	13.6	107	102.4	31.1	167	159.8	48.5	227	217.2	65.9	287	274.6	83.3
48	45.9	13.9	108	103.4	31.4	168	160.8	48.8	228	218.2	66.2	288	275.6	83.6
49	46.9	14.2	109	104.3	31.6	169	161.7	49.1	229	219.1	66.5	289	276.6	83.9
50	47.9	14.5	110	105.3	31.9	170	162.7	49.4	230	220.1	66.8	290	277.5	84.2
51	48.8	14.8	111	106.2	32.2	171	163.6	49.6	231	221.1	67.1	291	278.5	84.5
52	49.8	15.1	112	107.2	32.5	172	164.6	49.9	232	222.0	67.3	292	279.4	84.8
53	50.7	15.4	113	108.1	32.8	173	165.6	50.2	233	223.0	67.6	293	280.4	85.0
54	51.7	15.7	114	109.1	33.1	174	166.5	50.5	234	223.9	67.9	294	281.3	85.3
55	52.6	16.0	115	110.1	33.4	175	167.5	50.8	235	224.9	68.2	295	282.3	85.6
56	53.6	16.3	116	111.0	33.7	176	168.4	51.1	236	225.8	68.5	296	283.3	85.9
57	54.6	16.6	117	112.0	34.0	177	169.4	51.4	237	226.8	68.8	297	284.2	86.2
58	55.5	16.8	118	112.9	34.3	178	170.3	51.7	238	227.8	69.1	298	285.2	86.5
59	56.5	17.1	119	113.9	34.5	179	171.3	52.0	239	228.7	69.4	299	286.1	86.8
60	57.4	17.4	120	114.8	34.8	180	172.3	52.3	240	229.7	69.7	300	287.1	87.1
Dist.	Dep.	Lat.	Dist.	Dep.	Lat.	Dist.	Dep.	Lat.	Dist.	Dep.	Lat.	Dist.	Dep.	Lat.

FOR 6½ POINTS.

Dist.	Lat.	Dep.	Dist.	Lat.	Dep.	Dist.	Lat.	Dep.	Dist.	Lat.	Dep.	Dist.	Lat.	Dep.
1	00·9	00·3	61	57·4	20·6	121	113·9	40·8	181	170·4	61·0	241	226·9	81·2
2	01·9	00·7	62	58·4	20·9	122	114·9	41·1	182	171·4	61·3	242	227·9	81·5
3	02·8	01·0	63	59·3	21·2	123	115·8	41·4	183	172·3	61·7	243	228·8	81·9
4	03·8	01·4	64	60·3	21·6	124	116·8	41·8	184	173·2	62·0	244	229·7	82·2
5	04·7	01·7	65	61·2	21·9	125	117·7	42·1	185	174·2	62·3	245	230·7	82·5
6	05·7	02·0	66	62·1	22·2	126	118·6	42·5	186	175·1	62·7	246	231·6	82·9
7	06·6	02·4	67	63·1	22·6	127	119·6	42·8	187	176·1	63·0	247	232·6	83·2
8	07·5	02·7	68	64·0	22·9	128	120·5	43·1	188	177·0	63·3	248	233·5	83·6
9	08·5	03·0	69	65·0	23·3	129	121·5	43·5	189	177·9	63·7	249	234·4	83·9
10	09·4	03·4	70	65·9	23·6	130	122·4	43·8	190	178·9	64·0	250	235·4	84·2
11	10·4	03·7	71	66·9	23·9	131	123·3	44·1	191	179·8	64·4	251	236·3	84·6
12	11·3	04·0	72	67·8	24·3	132	124·3	44·5	192	180·8	64·7	252	237·3	84·9
13	12·2	04·4	73	68·7	24·6	133	125·2	44·8	193	181·7	65·0	253	238·2	85·2
14	13·2	04·7	74	69·7	24·9	134	126·2	45·1	194	182·7	65·4	254	239·1	85·6
15	14·1	05·1	75	70·6	25·3	135	127·1	45·5	195	183·6	65·7	255	240·1	85·9
16	15·1	05·4	76	71·6	25·6	136	128·1	45·8	196	184·5	66·0	256	241·0	86·2
17	16·0	05·7	77	72·5	25·9	137	129·0	46·2	197	185·5	66·4	257	242·0	86·6
18	17·0	06·1	78	73·4	26·3	138	129·9	46·5	198	186·4	66·7	258	242·9	86·9
19	17·9	06·4	79	74·4	26·6	139	130·9	46·8	199	187·4	67·0	259	243·9	87·3
20	18·8	06·7	80	75·3	27·0	140	131·8	47·2	200	188·3	67·4	260	244·8	87·6
21	19·8	07·1	81	76·3	27·3	141	132·8	47·5	201	189·3	67·7	261	245·7	87·9
22	20·7	07·4	82	77·2	27·6	142	133·7	47·8	202	190·2	68·1	262	246·7	88·3
23	21·7	07·8	83	78·2	28·0	143	134·6	48·2	203	191·1	68·4	263	247·6	88·6
24	22·6	08·1	84	79·1	28·3	144	135·6	48·5	204	192·1	68·7	264	248·6	88·9
25	23·5	08·4	85	80·0	28·6	145	136·5	48·9	205	193·0	69·1	265	249·5	89·3
26	24·5	08·8	86	81·0	29·0	146	137·5	49·2	206	194·0	69·4	266	250·5	89·6
27	25·4	09·1	87	81·9	29·3	147	138·4	49·5	207	194·9	69·7	267	251·4	90·0
28	26·4	09·4	88	82·9	29·7	148	139·4	49·9	208	195·8	70·1	268	252·3	90·3
29	27·3	09·8	89	83·8	30·0	149	140·3	50·2	209	196·8	70·4	269	253·3	90·6
30	28·3	10·1	90	84·7	30·3	150	141·2	50·5	210	197·7	70·8	270	254·2	91·0
31	29·2	10·4	91	85·7	30·7	151	142·2	50·9	211	198·7	71·1	271	255·2	91·3
32	30·1	10·8	92	86·6	31·0	152	143·1	51·2	212	199·6	71·4	272	256·1	91·6
33	31·1	11·1	93	87·6	31·3	153	144·1	51·5	213	200·6	71·8	273	257·0	92·0
34	32·0	11·5	94	88·5	31·7	154	145·0	51·9	214	201·5	72·1	274	258·0	92·3
35	33·0	11·8	95	89·5	32·0	155	145·9	52·2	215	202·4	72·4	275	258·9	92·6
36	33·9	12·1	96	90·4	32·3	156	146·9	52·6	216	203·4	72·8	276	259·9	93·0
37	34·8	12·5	97	91·3	32·7	157	147·8	52·9	217	204·3	73·1	277	260·8	93·3
38	35·8	12·8	98	92·3	33·0	158	148·8	53·2	218	205·3	73·4	278	261·8	93·7
39	36·7	13·1	99	93·2	33·4	159	149·7	53·6	219	206·2	73·8	279	262·7	94·0
40	37·7	13·5	100	94·2	33·7	160	150·7	53·9	220	207·1	74·1	280	263·6	94·3
41	38·6	13·8	101	95·1	34·0	161	151·6	54·2	221	208·1	74·5	281	264·6	94·7
42	39·5	14·2	102	96·0	34·4	162	152·5	54·6	222	209·0	74·8	282	265·5	95·0
43	40·5	14·5	103	97·0	34·7	163	153·5	54·9	223	210·0	75·1	283	266·5	95·3
44	41·4	14·8	104	97·9	35·0	164	154·4	55·3	224	210·9	75·5	284	267·4	95·7
45	42·4	15·2	105	98·9	35·4	165	155·4	55·6	225	211·9	75·8	285	268·3	96·0
46	43·3	15·5	106	99·8	35·7	166	156·3	55·9	226	212·8	76·1	286	269·3	96·4
47	44·3	15·8	107	100·7	36·1	167	157·2	56·3	227	213·7	76·5	287	270·2	96·7
48	45·2	16·2	108	101·7	36·4	168	158·2	56·6	228	214·7	76·8	288	271·2	97·0
49	46·1	16·5	109	102·6	36·7	169	159·1	56·9	229	215·6	77·2	289	272·1	97·4
50	47·1	16·8	110	103·6	37·1	170	160·1	57·3	230	216·6	77·5	290	273·0	97·7
51	48·0	17·2	111	104·5	37·4	171	161·0	57·6	231	217·5	77·8	291	274·0	98·0
52	49·0	17·5	112	105·5	37·7	172	161·9	58·0	232	218·4	78·2	292	274·9	98·4
53	49·9	17·9	113	106·4	38·1	173	162·9	58·3	233	219·4	78·5	293	275·9	98·7
54	50·8	18·2	114	107·3	38·4	174	163·8	58·6	234	220·3	78·8	294	276·8	99·0
55	51·8	18·5	115	108·3	38·7	175	164·8	59·0	235	221·3	79·2	295	277·8	99·4
56	52·7	18·9	116	109·2	39·1	176	165·7	59·3	236	222·2	79·5	296	278·7	99·7
57	53·7	19·2	117	110·2	39·4	177	166·7	59·6	237	223·1	79·8	297	279·6	100·1
58	54·6	19·5	118	111·1	39·8	178	167·6	60·0	238	224·1	80·2	298	280·6	100·4
59	55·6	19·9	119	112·0	40·1	179	168·5	60·3	239	225·0	80·5	299	281·5	100·7
60	56·5	20·2	120	113·0	40·4	180	169·5	60·6	240	226·0	80·9	300	282·5	101·1
Dist.	Dep.	Lat.	Dist.	Dep.	Lat.	Dist.	Dep.	Lat.	Dist.	Dep.	Lat.	Dist.	Dep.	Lat.

FOR 6¼ POINTS.

Dist.	Lat.	Dep	Dist.	Lat.	Dep	Dist.	Lat.	Dep.	Dist.	Lat.	Dep	Dist.	Lat.	Dep.
1	00·9	00·4	61	56·4	23·3	121	111·8	46·3	181	167·2	69·3	241	222·7	92·2
2	01·9	00·8	62	57·3	23·7	122	112·7	46·7	182	168·2	69·7	242	223·6	92·6
3	02·8	01·2	63	58·2	24·1	123	113·6	47·1	183	169·1	70·0	243	224·5	93·0
4	03·7	01·5	64	59·1	24·5	124	114·6	47·5	184	170·0	70·4	244	225·4	93·4
5	04·6	01·9	65	60·1	24·9	125	115·5	47·8	185	170·9	70·8	245	226·4	93·8
6	05·5	02·3	66	61·0	25·3	126	116·4	48·2	186	171·8	71·2	246	227·3	94·1
7	06·5	02·7	67	61·9	25·6	127	117·3	48·6	187	172·8	71·6	247	228·2	94·5
8	07·4	03·1	68	62·8	26·0	128	118·3	49·0	188	173·7	71·9	248	229·1	94·9
9	08·3	03·4	69	63·8	26·4	129	119·2	49·4	189	174·6	72·3	249	230·1	95·3
10	09·2	03·8	70	64·7	26·8	130	120·1	49·8	190	175·5	72·7	250	231·0	95·6
11	10·2	04·2	71	65·6	27·2	131	121·0	50·1	191	176·5	73·1	251	231·9	96·1
12	11·1	04·6	72	66·5	27·6	132	122·0	50·5	192	177·4	73·5	252	232·8	96·4
13	12·0	05·0	73	67·4	27·9	133	122·9	50·9	193	178·3	73·9	253	233·7	96·8
14	12 9	05·4	74	68·4	28·3	134	123·8	51·3	194	179·2	74·2	254	234·7	97·2
15	13·9	05·7	75	69·3	28·7	135	124·7	51·7	195	180·2	74·6	255	235·6	97·6
16	14·8	06·1	76	70·2	29·1	136	125·7	52·0	196	181·1	75·0	256	236·5	98·0
17	15·7	06·5	77	71·1	29·5	137	126·6	52·4	197	182·0	75·4	257	237·4	98·4
18	16·6	06·9	78	72·1	29·9	138	127·5	52·8	198	182·9	75·8	258	238·4	98·7
19	17·6	07·3	79	73·0	30·2	139	128·4	53·2	199	183·9	76·2	259	239·3	99·1
20	18·5	07·7	80	73·9	30·6	140	129·3	53·6	200	184·8	76·5	260	240·2	99·5
21	19·4	08·0	81	74·8	31·0	141	130·3	54·0	201	185·7	76·9	261	241·1	99·9
22	20·3	08·4	82	75·8	31·4	142	131·2	54·3	202	186·6	77·3	262	242·1	100·3
23	21·3	08·8	83	76·7	31·8	143	132·1	54·7	203	187·6	77·7	263	243·0	100·6
24	22·2	09·2	84	77·6	32·2	144	133·0	55·1	204	188·5	78·1	264	243·9	101·0
25	23·1	09·6	85	78·5	32·5	145	134·0	55·5	205	189·4	78·5	265	244·8	101·4
26	24·0	10·0	86	79·5	32·9	146	134·9	55·9	206	190·3	78·8	266	245·8	101·8
27	24·9	10·3	87	80·4	33·3	147	135·8	56·3	207	191·2	79·2	267	246·7	102·2
28	25·9	10·7	88	81·3	33·7	148	136·7	56·6	208	192·2	79·6	268	247·6	102·6
29	26·8	11·1	89	82·2	34·1	149	137·7	57·0	209	193·1	80·0	269	248·5	102·9
30	27·7	11·5	90	83·2	34·4	150	138·6	57·4	210	194·0	80·4	270	249·5	103·3
31	28·6	11·9	91	84·1	34·8	151	139·5	57·8	211	194·9	80·8	271	250·4	103·7
32	29·6	12·3	92	85·0	35·2	152	140·4	58·2	212	195·9	81·1	272	251·3	104·1
33	30·5	12·6	93	85·9	35·6	153	141·4	58·6	213	196·8	81·5	273	252·2	104·5
34	31·4	13·0	94	86·9	36·0	154	142·3	58·9	214	197·7	81·9	274	253·1	104·9
35	32·3	13·4	95	87·8	36·4	155	143·2	59·3	215	198·6	82·3	275	254·1	105·2
36	33·3	13·8	96	88·7	36·7	156	144·1	59·7	216	199·6	82·7	276	255·0	105·6
37	34·2	14·2	97	89·6	37·1	157	145·1	60·1	217	200·5	83·0	277	255·9	106·0
38	35·1	14·5	98	90·5	37·5	158	146·0	60·5	218	201·4	83·4	278	256·8	106·4
39	36·0	14·9	99	91·5	37·9	159	146·9	60·9	219	202·3	83·8	279	257·8	106·8
40	37·0	15·3	100	92·4	38·3	160	147·8	61·2	220	203·3	84·2	280	258·7	107·2
41	37·9	15·7	101	93·3	38·7	161	148·7	61·6	221	204·2	84·6	281	259·6	107·5
42	38·8	16·1	102	94·2	39·0	162	149·7	62·0	222	205·1	85·0	282	260·5	107·9
43	39·7	16·5	103	95·2	39·4	163	150·6	62·4	223	206·0	85·3	283	261·5	108·3
44	40·7	16·8	104	96·1	39·8	164	151·5	62·8	224	207·0	85·7	284	262·4	108·7
45	41·6	17·2	105	97·0	40·2	165	152·4	63·1	225	207·9	86·1	285	263·3	109·1
46	42·5	17·6	106	97·9	40·6	166	153·4	63·5	226	208·8	86·5	286	264·2	109·5
47	43·4	18·0	107	98·8	41·0	167	154·3	63·9	227	209·7	86·9	287	265·2	109·8
48	44 4	18·4	108	99·8	41·3	168	155·2	64·3	228	210·6	87·3	288	266·1	110·2
49	45·3	18·8	109	100·7	41·7	169	156·1	64·7	229	211·6	87·6	289	267·0	110·6
50	46·2	19·1	110	101·6	42·1	170	157·1	65·1	230	212·5	88·0	290	267·9	111·0
51	47·1	19·5	111	102·6	42·5	171	158·0	65·4	231	213·4	88·4	291	268·9	111·4
52	48·0	19·9	112	103·5	42·9	172	158·9	65·8	232	214·3	88·8	292	269·8	111·7
53	49·0	20·3	113	104·4	43·2	173	159·8	66·2	233	215·3	89·2	293	270·7	112·1
54	49 9	20·7	114	105·3	43·6	174	160·8	66·6	234	216·2	89·6	294	271·6	112·5
55	50·8	21·1	115	106·3	44·0	175	161·7	67·0	235	217·1	89·9	295	272·5	112·9
56	51·7	21·4	116	107·2	44·4	176	162·6	67·4	236	218·0	90·3	296	273·5	113·3
57	52·7	21·8	117	108·1	44·8	177	163·5	67·7	237	219·0	90·7	297	274·4	113·7
58	53·6	22·2	118	109·0	45·2	178	164·5	68·1	238	219·9	91·1	298	275·3	114·0
59	54·5	22·6	119	109·9	45·5	179	165·4	68·5	239	220·8	91·5	299	276·2	114·4
60	55·4	23·0	120	110·9	45·9	180	166·3	68·9	240	221·7	91·8	300	277·2	114·8
Dist.	Dep.	Lat.	Dist.	Dep.	Lat.	Dist.	Dep.	Lat.	Dist.	Dep.	Lat.	Dist.	Dep.	Lat.

FOR 6 POINTS.

Dist.	Lat.	Dep.	Dist.	Lat.	Dep.	Dist.	Lat.	Dep.	Dist.	Lat.	Dep.	Dist.	Lat.	Dep.
1	00·9	00·4	61	55·1	26·1	121	109·4	51·7	181	163·6	77·4	241	217·9	103·0
2	01·8	00·9	62	56·0	26·5	122	110·3	52·2	182	164·5	77·8	242	218·8	103·5
3	02·7	01·3	63	57·0	26·9	123	111·2	52·6	183	165·4	78·3	243	219·7	103·9
4	03·6	01·7	64	57·9	27·4	124	112·1	53·0	184	166·3	78·7	244	220·6	104·3
5	04·5	02·1	65	58·8	27·8	125	113·0	53·5	185	167·2	79·1	245	221·5	104·8
6	05·4	02·6	66	59·7	28·2	126	113·9	53·9	186	168·1	79·5	246	222·4	105·2
7	06·3	03·0	67	60·6	28·7	127	114·8	54·3	187	169·0	80·0	247	223·3	105·6
8	07·2	03·4	68	61·5	29·1	128	115·7	54·7	188	169·9	80·4	248	224·2	106·1
9	08·1	03·8	69	62·4	29·5	129	116·6	55·2	189	170·9	80·8	249	225·1	106·5
10	09·0	04·3	70	63·3	29·9	130	117·5	55·6	190	171·8	81·3	250	226·0	106·9
11	09·9	04·7	71	64·2	30·4	131	118·4	56·0	191	172·7	81·7	251	226·9	107·3
12	10·8	05·1	72	65·1	30·8	132	119·3	56·5	192	173·6	82·1	252	227·8	107·8
13	11·8	05·6	73	66·0	31·2	133	120·2	56·9	193	174·5	82·5	253	228·7	108·2
14	12·7	06·0	74	66·9	31·6	134	121·1	57·3	194	175·4	83·0	254	229·6	108·6
15	13·6	06·4	75	67·8	32·1	135	122·0	57·7	195	176·3	83·4	255	230·5	109·0
16	14·5	06·8	76	68·7	32·5	136	122·9	58·2	196	177·2	83·8	256	231·4	109·5
17	15·4	07·3	77	69·6	32·9	137	123·8	58·6	197	178·1	84·2	257	232·3	109·9
18	16·3	07·7	78	70·5	33·4	138	124·7	59·0	198	179·0	84·7	258	233·2	110·3
19	17·2	08·1	79	71·4	33·8	139	125·7	59·4	199	179·9	85·1	259	234·1	110·8
20	18·1	08·6	80	72·3	34·2	140	126·6	59·9	200	180·8	85·5	260	235·0	111·2
21	19·0	09·0	81	73·2	34·6	141	127·5	60·3	201	181·7	85·9	261	235·9	111·6
22	19·9	09·4	82	74·1	35·1	142	128·4	60·7	202	182·6	86·4	262	236·8	112·0
23	20·8	09·8	83	75·0	35·5	143	129·3	61·2	203	183·5	86·8	263	237·7	112·5
24	21·7	10·3	84	75·9	35·9	144	130·2	61·6	204	184·4	87·2	264	238·7	112·9
25	22·6	10·7	85	76·8	36·3	145	131·1	62·0	205	185·3	87·7	265	239·6	113·3
26	23·5	11·1	86	77·7	36·8	146	132·0	62·4	206	186·2	88·1	266	240·5	113·7
27	24·4	11·5	87	78·7	37·2	147	132·9	62·9	207	187·1	88·5	267	241·4	114·2
28	25·3	12·0	88	79·6	37·6	148	133·8	63·3	208	188·0	88·9	268	242·3	114·6
29	26·2	12·4	89	80·5	38·1	149	134·7	63·7	209	188·9	89·4	269	243·2	115·0
30	27·1	12·8	90	81·4	38·5	150	135·6	64·1	210	189·8	89·8	270	244·1	115·5
31	28·0	13·3	91	82·3	38·9	151	136·5	64·6	211	190·7	90·2	271	245·0	115·9
32	28·9	13·7	92	83·2	39·3	152	137·4	65·0	212	191·6	90·7	272	245·9	116·3
33	29·8	14·1	93	84·1	39·8	153	138·3	65·4	213	192·6	91·1	273	246·8	116·7
34	30·7	14·5	94	85·0	40·2	154	139·2	65·9	214	193·5	91·5	274	247·7	117·2
35	31·6	15·0	95	85·9	40·6	155	140·1	66·3	215	194·4	91·9	275	248·6	117·6
36	32·5	15·4	96	86·8	41·1	156	141·0	66·7	216	195·3	92·4	276	249·5	118·0
37	33·4	15·8	97	87·7	41·5	157	141·9	67·1	217	196·2	92·8	277	250·4	118·5
38	34·4	16·3	98	88·6	41·9	158	142·8	67·6	218	197·1	93·2	278	251·3	118·9
39	35·3	16·7	99	89·5	42·3	159	143·7	68·0	219	198·0	93·7	279	252·2	119·3
40	36·2	17·1	100	90·4	42·8	160	144·6	68·4	220	198·9	94·1	280	253·1	119·7
41	37·1	17·5	101	91·3	43·2	161	145·5	68·8	221	199·8	94·5	281	254·0	120·2
42	38·0	18·0	102	92·2	43·6	162	146·4	69·3	222	200·7	94·9	282	254·9	120·6
43	38·9	18·4	103	93·1	44·1	163	147·3	69·7	223	201·6	95·4	283	255·8	121·0
44	39·8	18·8	104	94·0	44·5	164	148·3	70·1	224	202·5	95·8	284	256·7	121·5
45	40·7	19·2	105	94·9	44·9	165	149·2	70·6	225	203·4	96·2	285	257·6	121·9
46	41·6	19·7	106	95·8	45·3	166	150·1	71·0	226	204·3	96·6	286	258·5	122·3
47	42·5	20·1	107	96·7	45·8	167	151·0	71·4	227	205·2	97·1	287	259·4	122·7
48	43·4	20·5	108	97·6	46·2	168	151·9	71·8	228	206·1	97·5	288	260·3	123·2
49	44·3	21·0	109	98·5	46·6	169	152·8	72·3	229	207·0	97·9	289	261·3	123·6
50	45·2	21·4	110	99·4	47·0	170	153·7	72·7	230	207·9	98·4	290	262·2	124·0
51	46·1	21·8	111	100·3	47·5	171	154·6	73·1	231	208·8	98·8	291	263·1	124·4
52	47·0	22·2	112	101·2	47·9	172	155·5	73·6	232	209·7	99·2	292	264·0	124·9
53	47·9	22·7	113	102·2	48·3	173	156·4	74·0	233	210·6	99·6	293	264·9	125·3
54	48·8	23·1	114	103·1	48·7	174	157·3	74·4	234	211·5	100·1	294	265·8	125·7
55	49·7	23·5	115	104·0	49·2	175	158·2	74·8	235	212·4	100·5	295	266·7	126·2
56	50·6	23·9	116	104·9	49·6	176	159·1	75·3	236	213·3	100·9	296	267·6	126·6
57	51·5	24·4	117	105·8	50·0	177	160·0	75·7	237	214·2	101·4	297	268·5	127·0
58	52·4	24·8	118	106·7	50·5	178	160·9	76·1	238	215·1	101·8	298	269·4	127·4
59	53·3	25·2	119	107·6	50·9	179	161·8	76·5	239	216·1	102·2	299	270·3	127·9
60	54·2	25·7	120	108·5	51·3	180	162·7	77·0	240	217·0	102·6	300	271·2	128·3
Dist.	Dep.	Lat.	Dist.	Dep.	Lat.	Dist.	Dep.	Lat.	Dist.	Dep.	Lat.	Dist.	Dep.	Lat.

FOR 5¾ POINTS.

Dist.	Lat.	Dep.	Dist.	Lat.	Dep.	Dist.	Lat.	Dep.	Dist.	Lat.	Dep.	Dist.	Lat.	Dep.
1	00·9	00·5	61	53·8	28·8	121	106·7	57·0	181	159·6	85·3	241	212·5	113·6
2	01·8	00·9	62	54·7	29·2	122	107·6	57·5	182	160·5	85·8	242	213·4	114·1
3	02·6	01·4	63	55·6	29·7	123	108·5	58·0	183	161·4	86·3	243	214·3	114·6
4	03·5	01·9	64	56·4	30·2	124	109·4	58·4	184	162·3	86·7	244	215·2	115·0
5	04·4	02·4	65	57·3	30·6	125	110·2	58·9	185	163·2	87·2	245	216·1	115·5
6	05·3	02·8	66	58·2	31·1	126	111·1	59·4	186	164·0	87·7	246	217·0	116·0
7	06·2	03·3	67	59·1	31·6	127	112·0	59·9	187	164·9	88·1	247	217·8	116·4
8	07·1	03·8	68	60·0	32·1	128	112·9	60·3	188	165·8	88·6	248	218·7	116·9
9	07·9	04·2	69	60·9	32·5	129	113·8	60·8	189	166·7	89·1	249	219·6	117·4
10	08·8	04·7	70	61·7	33·0	130	114·7	61·3	190	167·6	89·6	250	220·5	117·8
11	09·7	05·2	71	62·6	33·5	131	115·5	61·7	191	168·5	90·0	251	221·4	118·3
12	10·6	05·7	72	63·5	33·9	132	116·4	62·2	192	169·3	90·5	252	222·2	118·8
13	11·5	06·1	73	64·4	34·4	133	117·3	62·7	193	170·2	91·0	253	223·1	119·3
14	12·3	06·6	74	65·3	34·9	134	118·2	63·2	194	171·1	91·4	254	224·0	119·7
15	13·2	07·1	75	66·1	35·4	135	119·1	63·6	195	172·0	91·9	255	224·9	120·2
16	14·1	07·5	76	67·0	35·8	136	119·9	64·1	196	172·9	92·4	256	225·8	120·7
17	15·0	08·0	77	67·9	36·3	137	120·8	64·6	197	173·7	92·9	257	226·7	121·1
18	15·9	08·5	78	68·8	36·8	138	121·7	65·0	198	174·6	93·3	258	227·5	121·6
19	16·8	09·0	79	69·7	37·2	139	122·6	65·5	199	175·5	93·8	259	228·4	122·1
20	17·6	09·4	80	70·6	37·7	140	123·5	66·0	200	176·4	94·3	260	229·3	122·6
21	18·5	09·9	81	71·4	38·2	141	124·4	66·5	201	177·3	94·7	261	230·2	123·0
22	19·4	10·4	82	72·3	38·6	142	125·2	66·9	202	178·2	95·2	262	231·1	123·5
23	20·3	10·8	83	73·2	39·1	143	126·1	67·4	203	179·0	95·7	263	231·9	124·0
24	21·2	11·3	84	74·1	39·6	144	127·0	67·9	204	179·9	96·2	264	232·8	124·4
25	22·1	11·8	85	75·0	40·1	145	127·9	68·3	205	180·8	96·6	265	233·7	124·9
26	22·9	12·3	86	75·9	40·5	146	128·8	68·8	206	181·7	97·1	266	234·6	125·4
27	23·8	12·7	87	76·7	41·0	147	129·6	69·3	207	182·6	97·6	267	235·5	125·9
28	24·7	13·2	88	77·6	41·5	148	130·5	69·8	208	183·4	98·0	268	236·4	126·3
29	25·6	13·7	89	78·5	41·9	149	131·4	70·2	209	184·3	98·5	269	237·2	126·8
30	26·5	14·1	90	79·4	42·4	150	132·3	70·7	210	185·2	99·0	270	238·1	127·3
31	27·3	14·6	91	80·3	42·9	151	133·2	71·2	211	186·1	99·5	271	239·0	127·7
32	28·2	15·1	92	81·1	43·4	152	134·1	71·6	212	187·0	99·9	272	239·9	128·2
33	29·1	15·6	93	82·0	43·8	153	134·9	72·1	213	187·8	100·4	273	240·8	128·7
34	30·0	16·0	94	82·9	44·3	154	135·8	72·6	214	188·7	100·9	274	241·7	129·2
35	30·9	16·5	95	83·8	44·8	155	136·7	73·1	215	189·6	101·3	275	242·5	129·6
36	31·8	17·0	96	84·7	45·2	156	137·6	73·5	216	190·5	101·8	276	243·4	130·1
37	32·6	17·4	97	85·6	45·7	157	138·5	74·0	217	191·4	102·3	277	244·3	130·6
38	33·5	17·9	98	86·4	46·2	158	139·3	74·5	218	192·3	102·8	278	245·2	131·0
39	34·4	18·4	99	87·3	46·7	159	140·2	74·9	219	193·1	103·2	279	246·1	131·5
40	35·3	18·9	100	88·2	47·1	160	141·1	75·4	220	194·0	103·7	280	246·9	132·0
41	36·2	19·3	101	89·1	47·6	161	142·0	75·9	221	194·9	104·2	281	247·8	132·5
42	37·0	19·8	102	90·0	48·1	162	142·9	76·4	222	195·8	104·6	282	248·7	132·9
43	37·9	20·3	103	90·8	48·5	163	143·8	76·8	223	196·7	105·1	283	249·6	133·4
44	38·8	20·7	104	91·7	49·0	164	144·6	77·3	224	197·6	105·6	284	250·5	133·9
45	39·7	21·2	105	92·6	49·5	165	145·5	77·8	225	198·4	106·1	285	251·4	134·3
46	40·6	21·7	106	93·5	50·0	166	146·4	78·2	226	199·3	106·5	286	252·2	134·8
47	41·5	22·2	107	94·4	50·4	167	147·3	78·7	227	200·2	107·0	287	253·1	135·3
48	42·3	22·6	108	95·3	50·9	168	148·2	79·2	228	201·1	107·5	288	254·0	135·8
49	43·2	23·1	109	96·1	51·4	169	149·0	79·7	229	202·0	107·9	289	254·9	136·2
50	44·1	23·6	110	97·0	51·8	170	149·9	80·1	230	202·8	108·4	290	255·8	136·7
51	45·0	24·0	111	97·9	52·3	171	150·8	80·6	231	203·7	108·9	291	256·6	137·2
52	45·9	24·5	112	98·8	52·8	172	151·7	81·1	232	204·6	109·4	292	257·5	137·6
53	46·7	25·0	113	99·7	53·3	173	152·6	81·5	233	205·5	109·8	293	258·4	138·1
54	47·6	25·5	114	100·5	53·7	174	153·5	82·0	234	206·4	110·3	294	259·3	138·6
55	48·5	25·9	115	101·4	54·2	175	154·3	82·5	235	207·3	110·8	295	260·2	139·1
56	49·4	26·4	116	102·3	54·7	176	155·2	83·0	236	208·1	111·2	296	261·1	139·5
57	50·3	26·9	117	103·2	55·1	177	156·1	83·4	237	209·0	111·7	297	261·9	140·0
58	51·2	27·3	118	104·1	55·6	178	157·0	83·9	238	209·9	112·2	298	262·8	140·5
59	52·0	27·8	119	105·0	56·1	179	157·9	84·4	239	210·8	112·7	299	263·7	140·9
60	52·9	28·3	120	105·8	56·6	180	158·8	84·8	240	211·7	113·1	300	264·6	141·4
Dist.	Dep.	Lat.	Dist.	Dep.	Lat.	Dist.	Dep.	Lat.	Dist.	Dep.	Lat.	Dist.	Dep.	Lat.

FOR 5½ POINTS.

Dist.	Lat.	Dep.	Dist.	Lat.	Dep.	Dist.	Lat.	Dep.	Dist.	Lat.	Dep.	Dist.	Lat.	Dep.
1	00·9	00·5	61	52·3	31·4	121	103·8	62·2	181	155·3	93·0	241	206·7	123·9
2	01·7	01·0	62	53·2	31·9	122	104·6	62·7	182	156·1	93·6	242	207·6	124·4
3	02·6	01·5	63	54·0	32·4	123	105·5	63·2	183	157·0	94·1	243	208·4	124·9
4	03·4	02·1	64	54·9	32·9	124	106·4	63·7	184	157·8	94·6	244	209·3	125·4
5	04·3	02·6	65	55·8	33·4	125	107·2	64·3	185	158·7	95·1	245	210.1	125·9
6	05·1	03·1	66	56·6	33·9	126	108·1	64·8	186	159·5	95·6	246	211·0	126·5
7	06·0	03·6	67	57·5	34·4	127	108·9	65·3	187	160·4	96·1	247	211·9	127·0
8	06·9	04·1	68	58·3	35·0	128	109·8	65·8	188	161·2	96·6	248	212·7	127·5
9	07·7	04·6	69	59·2	35·5	129	110·6	66·3	189	162·1	97·2	249	213·6	128·0
10	08·6	05·1	70	60·0	36·0	130	111·5	66·8	190	163·0	97·7	250	214·4	128·5
11	09·4	05·7	71	60·9	36·5	131	112·4	67·3	191	163·8	98·2	251	215·3	129·0
12	10·3	06·2	72	61·8	37·0	132	113·2	67·9	192	164·7	98·7	252	216·1	129·5
13	11·2	06·7	73	62·6	37·5	133	114·1	68·4	193	165·5	99·2	253	217·0	130·1
14	12·0	07·2	74	63·5	38·0	134	114·9	68·9	194	166·4	99·7	254	217·9	130·6
15	12·9	07·7	75	64·3	38·6	135	115·8	69·4	195	167·3	100·2	255	218·7	131·1
16	13·7	08·2	76	65·2	39·1	136	116·6	69·9	196	168·1	100·8	256	219·6	131·6
17	14·6	08·7	77	66·0	39·6	137	117·5	70·4	197	169·0	101·3	257	220·4	132·1
18	15·4	09·3	78	66·9	40·1	138	118·4	70·9	198	169·8	101·8	258	221·3	132·6
19	16·3	09·8	79	67·8	40·6	139	119·2	71·5	199	170·7	102·3	259	222·2	133·1
20	17·2	10·3	80	68·6	41·1	140	120·1	72·0	200	171·5	102·8	260	223·0	133·7
21	18·0	10·8	81	69·5	41·6	141	120·9	72·5	201	172·4	103·3	261	223·9	134·2
22	18·9	11·3	82	70·3	42·1	142	121·8	73·0	202	173·3	103·8	262	224·7	134·7
23	19·7	11·8	83	71·2	42·7	143	122·7	73·5	203	174·1	104·4	263	225·6	135·2
24	20·6	12·3	84	72·0	43·2	144	123·5	74·0	204	175·0	104·9	264	226·4	135·7
25	21·4	12·9	85	72·9	43·7	145	124·4	74·5	205	175·8	105·4	265	227·3	136·2
26	22·3	13·4	86	73·8	44·2	146	125·2	75·1	206	176·7	105·9	266	228·2	136·7
27	23·2	13·9	87	74·6	44·7	147	126·1	75·6	207	177·5	106·4	267	229·0	137·3
28	24·0	14·4	88	75·5	45·2	158	126·9	76·1	208	178·4	106·9	268	229·9	137·8
29	24·9	14·9	89	76·3	45·7	149	127·8	76·6	209	179·3	107·4	269	230·7	138·3
30	25·7	15·4	90	77·2	46·3	150	128·7	77·1	210	180·1	108·0	270	231·6	138·8
31	26·6	15·9	91	78·1	46·8	151	129·5	77·6	211	181·0	108·5	271	232·4	139·3
32	27·4	16·5	92	78·9	47·3	152	130·4	78·1	212	181·8	109·0	272	233·3	139·8
33	28·3	17·0	93	79·8	47·8	153	131·2	78·7	213	182·7	109·5	273	234·2	140·3
34	29·2	17·5	94	80·6	48·3	154	132·1	79·2	214	183·5	110·0	274	235·0	140·9
35	30·0	18·0	95	81·5	48·8	155	132·9	79·7	215	184·4	110·5	275	235·9	141·4
36	30·9	18·5	96	82·3	49·3	156	133·8	80·2	216	185·3	111·0	276	236·7	141·9
37	31·7	19·0	97	83·2	49·9	157	134·7	80·7	217	186·1	111·6	277	237·6	142·4
38	32·6	19·5	98	84·1	50·4	158	135·5	81·2	218	187·0	112·1	278	238·4	142·9
39	33·5	20·1	99	84·9	50·9	159	136·4	81·7	219	187·8	112·6	279	239·3	143·4
40	34·3	20·6	100	85·8	51·4	160	137·2	82·3	220	188·7	113·1	280	240·2	143·9
41	35·2	21·1	101	86·6	51·9	161	138·1	82·8	221	189·6	113·6	281	241·0	144·5
42	36·0	21·6	102	87·5	52·4	162	138·9	83·3	222	190·4	114·1	282	241·9	145·0
43	36·9	22·1	103	88·3	52·9	163	139·8	83·8	223	191·3	114·6	283	242·7	145·5
44	37·7	22·6	104	89·2	53·5	164	140·7	84·3	224	192·1	115·2	284	243·6	146·0
45	38·6	23·1	105	90·1	54·0	165	141·5	84·8	225	193·0	115·7	285	244·4	146·5
46	39·5	23·6	106	90·9	54·5	166	142·4	85·3	226	193·8	116·2	286	245·3	147·0
47	40·3	24·2	107	91·8	55·0	167	143·2	85·8	227	194·7	116·7	287	246·2	147·5
48	41·2	24·7	108	92·6	55·5	168	144·1	86·4	228	195·6	117·2	288	247·0	148·1
49	42·0	25·2	109	93·5	56·0	169	145·0	86·9	229	196·4	117·7	289	247·9	148·6
50	42·9	25·7	110	94·3	56·5	170	145·8	87·4	230	197·3	118·2	290	248·7	149·1
51	43·7	26·2	111	95·2	57·1	171	146·7	87·9	231	198·1	118·8	291	249·6	149·6
52	44·6	26·7	112	96·1	57·6	172	147·5	88·4	232	199·0	119·3	292	250·5	150·1
53	45·5	27·2	113	96·9	58·1	173	148·4	88·9	233	199·8	119·8	293	251·3	150·6
54	46·3	27·8	114	97·8	58·6	174	149·2	89·4	234	200·7	120·3	294	252·2	151·1
55	47·2	28·3	115	98·6	59·1	175	150·1	90·0	235	201·6	120·8	295	253·0	151·7
56	48·0	28·8	116	99·5	59·6	176	151·0	90·5	236	202·4	121·3	296	253·9	152·2
57	48·9	29·3	117	100·4	60·1	177	151·8	91·0	237	203·3	121·8	297	254·7	152·7
58	49·7	29·8	118	101·2	60·7	178	152·7	91·5	238	204·1	122·4	298	255·6	153·2
59	50·6	30·3	119	102·1	61·2	179	153·5	92·0	239	205·0	122·9	299	256·5	153·7
60	51·5	30·8	120	102·9	61·7	180	154·4	92·5	240	205·9	123·4	300	257·3	154·2
Dist.	Dep.	Lat.	Dist.	Dep.	Lat.	Dist.	Dep.	Lat.	Dist.	Dep.	Lat.	Dist.	Dep.	Lat.

FOR 5¼ POINTS.

Dist.	Lat.	Dep.	Dist.	Lat.	Dep.	Dist.	Lat.	Dep.	Dist.	Lat.	Dep.	Dist.	Lat.	Dep.
1	00·8	00·6	61	50·7	33·9	121	100·6	67·2	181	150·5	100·6	241	200·4	133·9
2	01·7	01·1	62	51·5	34·4	122	101·4	67·8	182	151·3	101·1	242	201·2	134·4
3	02·5	01·7	63	52·4	35·0	123	102·3	68·3	183	152·2	101·7	243	202·0	135·0
4	03·3	02·2	64	53·2	35·6	124	103·1	68·9	184	153·0	102·2	244	202·9	135·6
5	04·2	02·8	65	54·0	36·1	125	103·9	69·4	185	153·8	102·8	245	203·7	136·1
6	05·0	03·3	66	54·9	36·7	126	104·8	70·0	186	154·6	103·3	246	204·5	136·7
7	05·8	03·9	67	55·7	37·2	127	105·6	70·6	187	155·5	103·9	247	205·4	137·2
8	06·7	04·4	68	56·5	37·8	128	106·4	71·1	188	156·3	104·4	248	206·2	137·8
9	07·5	05·0	69	57·4	38·3	129	107·3	71·7	189	157·1	105·0	249	207·0	138·3
10	08·3	05·6	70	58·2	38·9	130	108·1	72·2	190	158·0	105·6	250	207·9	138·9
11	09·1	06·1	71	59·0	39·4	131	108·9	72·8	191	158·8	106·1	251	208·7	139·4
12	10·0	06·7	72	59·9	40·0	132	109·7	73·3	192	159·6	106·7	252	209·5	140·0
13	10·8	07·2	73	60·7	40·6	133	110·6	73·9	193	160·5	107·2	253	210·4	140·6
14	11·6	07·8	74	61·5	41·1	134	111·4	74·4	194	161·3	107·8	254	211·2	141·1
15	12·5	08·3	75	62·4	41·7	135	112·2	75·0	195	162·1	108·3	255	212·0	141·7
16	13·3	08·9	76	63·2	42·2	136	113·1	75·6	196	163·0	108·9	256	212·9	142·2
17	14·1	09·4	77	64·0	42·8	137	113·9	76·1	197	163·8	109·4	257	213·7	142·8
18	15·0	10·0	78	64·8	43·3	138	114·7	76·7	198	164·6	110·0	258	214·5	143·3
19	15·8	10·6	79	65·7	43·9	139	115·6	77·2	199	165·5	110·6	259	215·3	143·9
20	16·6	11·1	80	66·5	44·4	140	116·4	77·8	200	166·3	111·1	260	216·2	144·4
21	17·5	11·7	81	67·3	45·0	141	117·2	78·3	201	167·1	111·7	261	217·0	145·0
22	18·3	12·2	82	68·2	45·6	142	118·1	78·9	202	168·0	112·2	262	217·8	145·6
23	19·1	12·8	83	69·0	46·1	143	118·9	79·4	203	168·8	112·8	263	218·7	146·1
24	20·0	13·3	84	69·8	46·7	144	119·7	80·0	204	169·6	113·3	264	219·5	146·7
25	20·8	13·9	85	70·7	47·2	145	120·6	80·6	205	170·4	113·9	265	220·3	147·2
26	21·6	14·4	86	71·5	47·8	146	121·4	81·1	206	171·3	114·4	266	221·2	147·8
27	22·4	15·0	87	72·3	48·3	147	122·2	81·7	207	172·1	115·0	267	222·0	148·3
28	23·3	15·6	88	73·2	48·9	148	123·1	82·2	208	172·9	115·6	268	222·8	148·9
29	24·1	16·1	89	74·0	49·4	149	123·9	82·8	209	173·8	116·1	269	223·7	149·4
30	24·9	16·7	90	74·8	50·0	150	124·7	83·3	210	174·6	116·7	270	224·5	150·0
31	25·8	17·2	91	75·7	50·6	151	125·5	83·9	211	175·4	117·2	271	225·3	150·6
32	26·6	17·8	92	76·5	51·1	152	126·4	84·4	212	176·3	117·8	272	226·2	151·1
33	27·4	18·3	93	77·3	51·7	153	127·2	85·0	213	177·1	118·3	273	227·0	151·7
34	28·3	18·9	94	78·2	52·2	154	128·0	85·6	214	177·9	118·9	274	227·8	152·2
35	29·1	19·4	95	79·0	52·8	155	128·9	86·1	215	178·8	119·4	275	228·6	152·8
36	29·9	20·0	96	79·8	53·3	156	129·7	86·7	216	179·6	120·0	276	229·5	153·3
37	30·8	20·6	97	80·6	53·9	157	130·5	87·2	217	180·4	120·6	277	230·3	153·9
38	31·6	21·1	98	81·5	54·4	158	131·4	87·8	218	181·3	121·1	278	231·1	154·4
39	32·4	21·7	99	82·3	55·0	159	132·2	88·3	219	182·1	121·7	279	232·0	155·0
40	33·3	22·2	100	83·1	55·6	160	133·0	88·9	220	182·9	122·2	280	232·8	155·6
41	34·1	22·8	101	84·0	56·1	161	133·9	89·4	221	183·7	122·8	281	233·6	156·1
42	34·9	23·3	102	84·8	56·7	162	134·7	90·0	222	184·6	123·3	282	234·5	156·7
43	35·8	23·9	103	85·6	57·2	163	135·5	90·6	223	185·4	123·9	283	235·3	157·2
44	36·6	24·4	104	86·5	57·8	164	136·4	91·1	224	186·2	124·4	284	236·1	157·8
45	37·4	25·0	105	87·3	58·3	165	137·2	91·7	225	187·1	125·0	285	237·0	158·3
46	38·2	25·6	106	88·1	58·9	166	138·0	92·2	226	187·9	125·6	286	237·8	158·9
47	39·1	26·1	107	89·0	59·4	167	138·9	92·8	227	188·7	126·1	287	238·6	159·4
48	39·9	26·7	108	89·8	60·0	168	139·7	93·3	228	189·6	126·7	288	239·5	160·0
49	40·7	27·2	109	90·6	60·6	169	140·5	93·9	229	190·4	127·2	289	240·3	160·6
50	41·6	27·8	110	91·5	61·1	170	141·3	94·4	230	191·2	127·8	290	241·1	161·1
51	42·4	28·3	111	92·3	61·7	171	142·2	95·0	231	192·1	128·3	291	242·0	161·7
52	43·2	28·9	112	93·1	62·2	172	143·0	95·6	232	192·9	128·9	292	242·8	162·2
53	44·1	29·4	113	94·0	62·8	173	143·8	96·1	233	193·7	129·4	293	243·6	162·8
54	44·9	30·0	114	94·8	63·3	174	144·7	96·7	234	194·6	130·0	294	244·4	163·3
55	45·7	30·6	115	95·6	63·9	175	145·5	97·2	235	195·4	130·6	295	245·3	163·9
56	46·6	31·1	116	96·4	64·4	176	146·3	97·8	236	196·2	131·1	296	246·1	164·4
57	47·4	31·7	117	97·3	65·0	177	147·2	98·3	237	197·1	131·7	297	246·9	165·0
58	48·2	32·2	118	98·1	65·6	178	148·0	98·9	238	197·9	132·2	298	247·8	165·6
59	49·1	32·8	119	98·9	66·1	179	148·8	99·4	239	198·7	132·8	299	248·6	166·1
60	49·9	33·3	120	99·8	66·7	180	149·7	100·0	240	199·5	133·3	300	249·4	166·7
Dist.	Dep.	Lat.	Dist.	Dep.	Lat.	Dist.	Dep.	Lat.	Dist.	Dep.	Lat.	Dist.	Dep.	Lat.

Dist.	Lat.	Dep.	Dist.	Lat.	Dep.	Dist.	Lat.	Dep.	Dist.	Lat.	Dep.	Dist.	Lat.	Dep.
1	00·8	00·6	61	49·0	36·3	121	97·2	72·1	181	145·4	107·8	241	193·6	143·6
2	01·6	01·2	62	49·8	36·9	122	98·0	72·7	182	146·2	108·4	242	194·4	144·2
3	02·4	01·8	63	50·6	37·5	123	98·8	73·3	183	147·0	109·0	243	195·2	144·8
4	03·2	02·4	64	51·4	38·1	124	99·6	73·9	184	147·8	109·6	244	196·0	145·4
5	04·0	03·0	65	52·2	38·7	125	100·4	74·5	185	148·6	110·2	245	196·8	146·0
6	04·8	03·6	66	53·0	39·3	126	101·2	75·1	186	149·4	110·8	246	197·6	146·5
7	05·6	04·2	67	53·8	39·9	127	102·0	75·7	187	150·2	111·4	247	198·4	147·1
8	06·4	04·8	68	54·6	40·5	128	102·8	76·3	188	151·0	112·0	248	199·2	147·7
9	07·2	05·4	69	55·4	41·1	129	103·6	76·9	189	151·8	112·6	249	200·0	148·3
10	08·0	06·0	70	56·2	41·7	130	104·4	77·4	190	152·6	113·2	250	200·8	148·9
11	08·8	06·6	71	57·0	42·3	131	105·2	78·0	191	153·4	113·8	251	201·6	149·5
12	09·6	07·1	72	57·8	42·9	132	106·0	78·6	192	154·2	114·4	252	202·4	150·1
13	10·4	07·7	73	58·6	43·5	133	106·8	79·2	193	155·0	115·0	253	203·2	150.7
14	11·2	08·3	74	59·4	44·1	134	107·6	79·8	194	155·8	115·6	254	204·0	151·3
15	12·0	08·9	75	60·2	44·7	135	108·4	80·4	195	156·6	116·2	255	204·8	151·9
16	12·9	09·5	76	61·0	45·3	136	109·2	81·0	196	157·4	116·8	256	205·6	152·5
17	13·7	10·1	77	61·8	45·9	137	110·0	81·6	197	158·2	117·4	257	206·4	153·1
18	14·5	10·7	78	62·6	46·5	138	110·8	82·2	198	159·0	118·0	258	207·2	153·7
19	15·3	11·3	79	63·4	47·1	139	111·6	82·8	199	159·8	118·5	259	208·0	154·3
20	16·1	11·9	80	64·3	47·7	140	112·4	83·4	200	160·6	119·1	260	208·8	154·9
21	16·9	12·5	81	65·1	48·3	141	113·2	84·0	201	161·4	119·7	261	209·6	155·5
22	17·7	13·1	82	65·9	48·9	142	114·0	84·6	202	162·2	120·3	262	210·4	156·1
23	18·5	13·7	83	66·7	49·4	143	114·9	85·2	203	163·0	120·9	263	211·2	156·7
24	19·3	14·3	84	67·5	50·0	144	115·7	85·8	204	163·9	121·5	264	212·0	157·3
25	20·1	14·9	85	68·3	50·6	145	116·5	86·4	205	164·7	122·1	265	212·8	157·9
26	20·9	15·5	86	69·1	51·2	146	117·3	87·0	206	165·5	122·7	266	213·6	158·5
27	21·7	16·1	87	69·9	51·8	147	118·1	87·6	207	166·3	123·3	267	214·5	159·1
28	22·5	16·7	88	70·7	52·4	148	118·9	88·2	208	167·1	123·9	268	215·3	159·6
29	23·3	17·3	89	71·5	53·0	149	119·7	88·8	209	167·9	124·5	269	216·1	160·2
30	24·1	17·9	90	72·3	53·6	150	120·5	89·4	210	168·7	125·1	270	216·9	160·8
31	24·9	18·5	91	73·1	54·2	151	121·3	90·0	211	169·5	125·7	271	217·7	161·4
32	25·7	19·1	92	73·9	54·8	152	122·1	90·5	212	170·3	126·3	272	218·5	162·0
33	26·5	19·7	93	74·7	55·4	153	122·9	91·1	213	171·1	126·9	273	219·3	162·6
34	27·3	20·3	94	75·5	56·0	154	123·7	91·7	214	171·9	127·5	274	220·1	163·2
35	28·1	20·9	95	76·3	56·6	155	124·5	92·3	215	172·7	128·1	275	220·9	163·8
36	28·9	21·4	96	77·1	57·2	156	125·3	92·9	216	173·5	128·7	276	221·7	164·4
37	29·7	22·0	97	77·9	57·8	157	126·1	93·5	217	174·3	129·3	277	222·5	165·0
38	30·5	22·6	98	78·7	58·4	158	126·9	94·1	218	175·1	129·9	278	223·3	165·6
39	31·3	23·2	99	79·5	59·0	159	127·7	94·7	219	175·9	130·5	279	224·1	166·2
40	32·1	23·8	100	80·3	59·6	160	128·5	95·3	220	176·7	131·1	280	224·9	166·8
41	32·9	24·4	101	81·1	60·2	161	129·3	95·9	221	177·5	131·7	281	225·7	167·4
42	33·7	25·0	102	81·9	60·8	162	130·1	96·5	222	178·3	132·2	282	226·5	168·0
43	34·5	25·6	103	82·7	61·4	163	130·9	97·1	223	179·1	132·8	283	227·3	168·6
44	35·3	26·2	104	83·5	62·0	164	131·7	97·7	224	179·9	133·4	284	228·1	169·2
45	36·1	26·8	105	84·3	62·6	165	132·5	98·3	225	180·7	134·0	285	228·9	169·8
46	36·9	27·4	106	85·1	63·1	166	133·3	98·9	226	181·5	134·6	286	229·7	170·4
47	37·7	28·0	107	85·9	63·7	167	134·1	99·5	227	182·3	135·2	287	230·5	171·0
48	38·6	28·6	108	86·7	64·3	168	134·9	100·1	228	183·1	135·8	288	231·3	171·6
49	39·4	29·2	109	87·5	64·9	169	135·7	100·7	229	183·9	136·4	289	232·1	172·2
50	40·2	29·8	110	88·4	65·5	170	136·5	101·3	230	184·7	137·0	290	232·9	172·8
51	41·0	30·4	111	89·2	66·1	171	137·3	101·9	231	185·5	137·6	291	233·7	173·3
52	41·8	31·0	112	90·0	66·7	172	138·1	102·5	232	186·3	138·2	292	234·5	173·9
53	42·6	31·6	113	90·8	67·3	173	138·9	103·1	233	187·1	138·8	293	235·3	174·5
54	43·4	32·2	114	91·6	67·9	174	139·8	103·7	234	187·9	139·4	294	236·1	175·1
55	44·2	32·8	115	92·4	68·5	175	140·6	104·2	235	188·8	140·0	295	236·9	175·7
56	45·0	33·4	116	93·2	69·1	176	141·4	104·8	236	189·6	140·6	296	237·7	176·3
57	45·8	34·0	117	94·0	69·7	177	142·2	105·4	237	190·4	141·2	297	238·5	176·9
58	46·6	34·6	118	94·8	70·3	178	143·0	106·0	238	191·2	141·8	298	239·4	177·5
59	47·4	35·1	119	95·6	70·9	179	143·8	106·6	239	192·0	142·4	299	240·2	178·1
60	48·2	35·7	120	96·4	71·5	180	144·6	107·2	240	192·8	143·0	300	241·0	178·7
Dist.	Dep.	Lat.	Dist.	Dep.	Lat.	Dist.	Dep.	Lat.	Dist.	Dep.	Lat.	Dist.	Dep.	Lat.

FOR 4¾ POINTS.

Dist.	Lat.	Dep.	Dist.	Lat.	Dep.	Dist.	Lat.	Dep.	Dist.	Lat.	Dep.	Dist.	Lat.	Dep.
1	00·8	00·6	61	47·1	38·7	121	93·5	76·8	181	139·9	114·8	241	186·3	152·9
2	01·5	01·3	62	47·9	39·3	122	94·3	77·4	182	140·7	115·5	242	187·1	153·5
3	02·3	01·9	63	48·7	40·0	123	95·1	78·0	183	141·5	116·1	243	187·8	154·2
4	03·1	02·5	64	49·5	40·6	124	95·8	78·7	184	142·2	116·7	244	188·6	154·8
5	03·9	03·2	65	50·2	41·2	125	96·6	79·3	185	143·0	117·4	245	189·4	155·4
6	04·6	03·8	66	51·0	41·9	126	97·4	79·9	186	143·8	118·0	246	190·2	156·1
7	05·4	04·4	67	51·8	42·5	127	98·2	80·6	187	144·5	118·6	247	190·9	156·7
8	06·2	05·1	68	52·6	43·1	128	98·9	81·2	188	145·3	119·3	248	191·7	157·3
9	07·0	05·7	69	53·3	43·8	129	99·7	81·8	189	146·1	119·9	249	192·5	158·0
10	07·7	06·3	70	54·1	44·4	130	100·5	82·5	190	146·9	120·5	250	193·2	158·6
11	08·5	07·0	71	54·9	45·0	131	101·3	83·1	191	147·6	121·2	251	194·0	159·2
12	09·3	07·6	72	55·7	45·7	132	102·0	83·7	192	148·4	121·8	252	194·8	159·9
13	10·0	08·2	73	56·4	46·3	133	102·8	84·4	193	149·2	122·4	253	195·6	160·5
14	10·8	08·9	74	57·2	46·9	134	103·6	85·0	194	150·0	123·1	254	196·3	161·1
15	11·6	09·5	75	58·0	47·6	135	104·4	85·6	195	150·7	123·7	255	197·1	161·8
16	12·4	10·1	76	58·7	48·2	136	105·1	86·3	196	151·5	124·3	256	197·9	162·4
17	13·1	10·8	77	59·5	48·8	137	105·9	86·9	197	152·3	125·0	257	198·7	163·0
18	13·9	11·4	78	60·3	49·5	138	106·7	87·5	198	153·1	125·6	258	199·4	163·7
19	14·7	12·0	79	61·1	50·1	139	107·4	88·2	199	153·8	126·2	259	200·2	164·3
20	15·5	12·7	80	61·8	50·7	140	108·2	88·8	200	154·6	126·9	260	201·0	164·9
21	16·2	13·3	81	62·6	51·4	141	109·0	89·4	201	155·4	127·5	261	201·8	165·6
22	17·0	14·0	82	63·4	52·0	142	109·8	90·1	202	156·1	128·1	262	202·5	166·2
23	17·8	14·6	83	64·2	52·7	143	110·5	90·7	203	156·9	128·8	263	203·3	166·8
24	18·6	15·2	84	64·9	53·3	144	111·3	91·3	204	157·7	129·4	264	204·1	167·5
25	19·3	15·9	85	65·7	53·9	145	112·1	92·0	205	158·5	130·0	265	204·8	168·1
26	20·1	16·5	86	66·5	54·6	146	112·9	92·6	206	159·2	130·7	266	205·6	168·7
27	20·9	17·1	87	67·2	55·2	147	113·6	93·3	207	160·0	131·3	267	206·4	169·4
28	21·6	17·8	88	68·0	55·8	148	114·4	93·9	208	160·8	132·0	268	207·2	170·0
29	22·4	18·4	89	68·8	56·5	149	115·2	94·5	209	161·6	132·6	269	207·9	170·6
30	23·2	19·0	90	69·6	57·1	150	115·9	95·2	210	162·3	133·2	270	208·7	171·3
31	24·0	19·7	91	70·3	57·7	151	116·7	95·8	211	163·1	133·9	271	209·5	171·9
32	24·7	20·3	92	71·1	58·4	152	117·5	96·4	212	163·9	134·5	272	210·3	172·6
33	25·5	20·9	93	71·9	59·0	153	118·3	97·1	213	164·6	135·1	273	211·0	173·2
34	26·3	21·6	94	72·7	59·6	154	119·0	97·7	214	165·4	135·8	274	211·8	173·8
35	27·1	22·2	95	73·4	60·3	155	119·8	98·3	215	166·2	136·4	275	212·6	174·5
36	27·8	22·8	96	74·2	60·9	156	120·6	99·0	216	167·0	137·0	276	213·3	175·1
37	28·6	23·5	97	75·0	61·5	157	121·4	99·6	217	167·7	137·7	277	214·1	175·7
38	29·4	24·1	98	75·7	62·2	158	122·1	100·2	218	168·5	138·3	278	214·9	176·4
39	30·1	24·7	99	76·5	62·8	159	122·9	100·9	219	169·3	138·9	279	215·7	177·0
40	30·9	25·4	100	77·3	63·4	160	123·7	101·5	220	170·1	139·6	280	216·4	177·6
41	31·7	26·0	101	78·1	64·1	161	124·4	102·1	221	170·8	140·2	281	217·2	178·3
42	32·5	26·6	102	78·8	64·7	162	125·2	102·8	222	171·6	140·8	282	218·0	178·9
43	33·2	27·3	103	79·6	65·3	163	126·0	103·4	223	172·4	141·5	283	218·8	179·5
44	34·0	27·9	104	80·4	66·0	164	126·8	104·0	224	173·1	142·1	284	219·5	180·2
45	34·8	28·5	105	81·2	66·6	165	127·5	104·7	225	173·9	142·7	285	220·3	180·8
46	35·6	29·2	106	81·9	67·2	166	128·3	105·3	226	174·7	143·4	286	221·1	181·4
47	36·3	29·8	107	82·7	67·9	167	129·1	105·9	227	175·5	144·0	287	221·8	182·1
48	37·1	30·4	108	83·5	68·5	168	129·9	106·6	228	176·2	144·6	288	222·6	182·7
49	37·9	31·1	109	84·3	69·1	169	130·6	107·2	229	177·0	145·3	289	223·4	183·3
50	38·6	31·7	110	85·0	69·8	170	131·4	107·8	230	177·8	145·9	290	224·2	184·0
51	39·4	32·3	111	85·8	70·4	171	132·2	108·5	231	178·6	146·5	291	224·9	184·6
52	40·2	33·0	112	86·6	71·0	172	133·0	109·1	232	179·3	147·2	292	225·7	185·2
53	41·0	33·6	113	87·3	71·7	173	133·7	109·7	233	180·1	147·8	293	226·5	185·9
54	41·7	34·3	114	88·1	72·3	174	134·5	110·4	234	180·9	148·4	294	227·3	186·5
55	42·5	34·9	115	88·9	73·0	175	135·3	111·0	235	181·7	149·1	295	228·0	187·1
56	43·3	35·5	116	89·7	73·6	176	136·0	111·6	236	182·4	149·7	296	228·8	187·8
57	44·1	36·2	117	90·4	74·2	177	136·8	112·3	237	183·2	150·3	297	229·6	188·4
58	44·8	36·8	118	91·2	74·9	178	137·6	112·9	238	184·0	151·0	298	230·4	189·0
59	45·6	37·4	119	92·0	75·5	179	138·4	113·6	239	184·7	151·6	299	231·1	189·7
60	46·4	38·1	120	92·8	76·1	180	139·1	114·2	240	185·5	152·3	300	231·9	190·3
Dist.	Dep.	Lat.	Dist.	Dep.	Lat.	Dist.	Dep.	Lat.	Dist.	Dep.	Lat.	Dist.	Dep.	Lat.

FOR 4½ POINTS.

Dist.	Lat.	Dep.	Dist.	Lat.	Dep.	Dist.	Lat.	Dep.	Dist.	Lat.	Dep.	Dist.	Lat.	Dep.
1	00·7	00·7	61	45·2	41·0	121	89·6	81·3	181	134·1	121·5	241	178·6	161·8
2	01·5	01·3	62	45·9	41·6	122	90·4	81·9	182	134·8	122·2	242	179·3	162·5
3	02·2	02·0	63	46·7	42·3	123	91·1	82·6	183	135·6	122·9	243	180·0	163·2
4	03·0	02·7	64	47·4	43·0	124	91·9	83·3	184	136·3	123·6	244	180·8	163·8
5	03·7	03·4	65	48·2	43·6	125	92·6	83·9	185	137·1	124·2	245	181·5	164·5
6	04·4	04·0	66	48·9	44·3	126	93·4	84·6	186	137·8	124·9	246	182·3	165·2
7	05·2	04·7	67	49·6	45·0	127	94·1	85·3	187	138·6	125·6	247	183·0	165·9
8	05·9	05·4	68	50·4	45·7	128	94·8	86·0	188	139·3	126·2	248	183·8	166·5
9	06·7	06·0	69	51·1	46·3	129	95·6	86·6	189	140·0	126·9	249	184·5	167·2
10	07·4	06·7	70	51·9	47·0	130	96·3	87·3	190	140·8	127·6	250	185·2	167·9
11	08·2	07·4	71	52·6	47·7	131	97·1	88·0	191	141·5	128·3	251	186·0	168·5
12	08·9	08·1	72	53·3	48·3	132	97·8	88·6	192	142·3	128·9	252	186·7	169·2
13	09·6	08·7	73	54·1	49·0	133	98·5	89·3	193	143·0	129·6	253	187·5	169·9
14	10·4	09·4	74	54·8	49·7	134	99·3	90·0	194	143·7	130·3	254	188·2	170·6
15	11·1	10·1	75	55·6	50·4	135	100·0	90·7	195	144·5	130·9	255	188·9	171·2
16	11·9	10·7	76	56·3	51·0	136	100·8	91·3	196	145·2	131·6	256	189·7	171·9
17	12·6	11·4	77	57·0	51·7	137	101·5	92·0	197	146·0	132·3	257	190·4	172·6
18	13·3	12·1	78	57·8	52·4	138	102·2	92·7	198	146·7	133·0	258	191·2	173·2
19	14·1	12·8	79	58·5	53·0	139	103·0	93·3	199	147·4	133·6	259	191·9	173·9
20	14·8	13·4	80	59·3	53·7	140	103·7	94·0	200	148·2	134·3	260	192·6	174·6
21	15·6	14·1	81	60·0	54·4	141	104·5	94·7	201	148·9	135·0	261	193·4	175·3
22	16·3	14·8	82	60·8	55·1	142	105·2	95·4	202	149·7	135·6	262	194·1	175·9
23	17·0	15·4	83	61·5	55·7	143	106·0	96·0	203	150·4	136·3	263	194·9	176·6
24	17·8	16·1	84	62·2	56·4	144	106·7	96·7	204	151·1	137·0	264	195·6	177·3
25	18·5	16·8	85	63·0	57·1	145	107·4	97·4	205	151·9	137·7	265	196·3	178·0
26	19·3	17·5	86	63·7	57·7	146	108·2	98·0	206	152·6	138·3	266	197·1	178·6
27	20·0	18·1	87	64·5	58·4	147	108·9	98·7	207	153·4	139·0	267	197·8	179·3
28	20·7	18·8	88	65·2	59·1	148	109·7	99·4	208	154·1	139·7	268	198·6	180·0
29	21·5	19·5	89	65·9	59·8	149	110·4	100·1	209	154·9	140·3	269	199·3	180·6
30	22·2	20·1	90	66·7	60·4	150	111·1	100·7	210	155·6	141·0	270	200·1	181·3
31	23·0	20·8	91	67·4	61·1	151	111·9	101·4	211	156·3	141·7	271	200·8	182·0
32	23·7	21·5	92	68·2	61·8	152	112·6	102·1	212	157·1	142·4	272	201·5	182·7
33	24·4	22·2	93	68·9	62·4	153	113·4	102·7	213	157·8	143·0	273	202·3	183·3
34	25·2	22·8	94	69·6	63·1	154	114·1	103·4	214	158·6	143·7	274	203·0	184·0
35	25·9	23·5	95	70·4	63·8	155	114·8	104·1	215	159·3	144·4	275	203·8	184·7
36	26·7	24·2	96	71·1	64·5	156	115·6	104·8	216	160·0	145·0	276	204·5	185·3
37	27·4	24·8	97	71·9	65·1	157	116·3	105·4	217	160·8	145·7	277	205·2	186·0
38	28·2	25·5	98	72·6	65·8	158	117·1	106·1	218	161·5	146·4	278	206·0	186·7
39	28·9	26·2	99	73·3	66·5	159	117·8	106·8	219	162·3	147·1	279	206·7	187·4
40	29·6	26·9	100	74·1	67·2	160	118·5	107·4	220	163·0	147·7	280	207·5	188·0
41	30·4	27·5	101	74·8	67·8	161	119·3	108·1	221	163·7	148·4	281	208·2	188·7
42	31·1	28·2	102	75·6	68·5	162	120·0	108·8	222	164·5	149·1	282	208·9	189·4
43	31·9	28·9	103	76·3	69·2	163	120·8	109·5	223	165·2	149·7	283	209·7	190·0
44	32·6	29·5	104	77·1	69·8	164	121·5	110·1	224	166·0	150·4	284	210·4	190·7
45	33·3	30·2	105	77·8	70·5	165	122·3	110·8	225	166·7	151·1	285	211·2	191·4
46	34·1	30·9	106	78·5	71·2	166	123·0	111·5	226	167·4	151·8	286	211·9	192·1
47	34·8	31·6	107	79·3	71·8	167	123·7	112·1	227	168·2	152·4	287	212·6	192·7
48	35·6	32·2	108	80·0	72·5	168	124·5	112·8	228	168·9	153·1	288	213·4	193·4
49	36·3	32·9	109	80·8	73·2	169	125·2	113·5	229	169·7	153·8	289	214·1	194·1
50	37·0	33·6	110	81·5	73·9	170	126·0	114·2	230	170·4	154·5	290	214·9	194·7
51	37·8	34·2	111	82·2	74·5	171	126·7	114·8	231	171·2	155·1	291	215·6	195·4
52	38·5	34·9	112	83·0	75·2	172	127·4	115·5	232	171·9	155·8	292	216·4	196·1
53	39·3	35·6	113	83·7	75·9	173	128·2	116·2	233	172·6	156·5	293	217·1	196·8
54	40·0	36·3	114	84·5	76·5	174	128·9	116·8	234	173·4	157·1	294	217·8	197·4
55	40·7	36·9	115	85·2	77·2	175	129·7	117·5	235	174·1	157·8	295	218·6	198·1
56	41·5	37·6	116	85·9	77·9	176	130·4	118·2	236	174·9	158·5	296	219·3	198·8
57	42·2	38·3	117	86·7	78·6	177	131·1	118·9	237	175·6	159·1	297	220·1	199·4
58	43·0	38·9	118	87·4	79·2	178	131·9	119·5	238	176·3	159·8	298	220·8	200·1
59	43·7	39·6	119	88·2	79·9	179	132·6	120·2	239	177·1	160·5	299	221·5	200·8
60	44·5	40·3	120	88·9	80·6	180	133·4	120·9	240	177·8	161·2	300	222·3	201·5
Dist.	Dep.	Lat.	Dist.	Dep.	Lat.	Dist.	Dep.	Lat.	Dist.	Dep.	Lat.	Dist.	Dep.	Lat.

FOR 4¼ POINTS.

Dist.	Lat.	Dep.	Dist.	Lat.	Dep.	Dist.	Lat.	Dep.	Dist.	Lat.	Dep.	Dist.	Lat.	Dep.
1	00·7	00·7	61	43·1	43·1	121	85·6	85·6	181	128·0	128·0	241	170·4	170·4
2	01·4	01·4	62	43·8	43·8	122	86·3	86·3	182	128·7	128·7	242	171·1	171·1
3	02·1	02·1	63	44·5	44·5	123	87·0	87·0	183	129·4	129·4	243	171·8	171·8
4	02·8	02·8	64	45·3	45·3	124	87·7	87·7	184	130·1	130·1	244	172·5	172·5
5	03·5	03·5	65	46·0	46·0	125	88·4	88·4	185	130·8	130·8	245	173·2	173·2
6	04·2	04·2	66	46·7	46·7	126	89·1	89·1	186	131·5	131·5	246	173·9	173·9
7	04·9	04·9	67	47·4	47·4	127	89·8	89·8	187	132·2	132·2	247	174·7	174·7
8	05·7	05·7	68	48·1	48·1	128	90·5	90·5	188	132·9	132·9	248	175·4	175·4
9	06·4	06·4	69	48·8	48·8	129	91·2	91·2	189	133·6	133·6	249	176·1	176·1
10	07·1	07·1	70	49·5	49·5	130	91·9	91·9	190	134·3	134·3	250	176·8	176·8
11	07·8	07·8	71	50·2	50·2	131	92·6	92·6	191	135·1	135·1	251	177·5	177·5
12	08·5	08·5	72	50·9	50·9	132	93·3	93·3	192	135·8	135·8	252	178·2	178·2
13	09·2	09·2	73	51·6	51·6	133	94·0	94·0	193	136·5	136·5	253	178·9	178·9
14	09·9	09·9	74	52·3	52·3	134	94·8	94·8	194	137·2	137·2	254	179·6	179·6
15	10·6	10·6	75	53·0	53·0	135	95·5	95·5	195	137·9	137·9	255	180·3	180·3
16	11·3	11·3	76	53·7	53·7	136	96·2	96·2	196	138·6	138·6	256	181·0	181·0
17	12·0	12·0	77	54·4	54·4	137	96·9	96·9	197	139·3	139·3	257	181·7	181·7
18	12·7	12·7	78	55·2	55·2	138	97·6	97·6	198	140·0	140·0	258	182·4	182·4
19	13·4	13·4	79	55·9	55·9	139	98·3	98·3	199	140·7	140·7	259	183·1	183·1
20	14·1	14·1	80	56·6	56·6	140	99·0	99·0	200	141·4	141·4	260	183·8	183·8
21	14·8	14·8	81	57·3	57·3	141	99·7	99·7	201	142·1	142·1	261	184·6	184·6
22	15·6	15·6	82	58·0	58·0	142	100·4	100·4	202	142·8	142·8	262	185·3	185·3
23	16·3	16·3	83	58·7	58·7	143	101·1	101·1	203	143·5	143·5	263	186·0	186·0
24	17·0	17·0	84	59·4	59·4	144	101·8	101·8	204	144·2	144·2	264	186·7	186·7
25	17·7	17·7	85	60·1	60·1	145	102·5	102·5	205	145·0	145·0	265	187·4	187·4
26	18·4	18·4	86	60·8	60·8	146	103·2	103·2	206	145·7	145·7	266	188·1	188·1
27	19·1	19·1	87	61·5	61·5	147	103·9	103·9	207	146·4	146·4	267	188·8	188·8
28	19·8	19·8	88	62·2	62·2	148	104·7	104·7	208	147·1	147·1	268	189·5	189·5
29	20·5	20·5	89	62·9	62·9	149	105·4	105·4	209	147·8	147·8	269	190·2	190·2
30	21·2	21·2	90	63·6	63·6	150	106·1	106·1	210	148·5	148·5	270	190·9	190·9
31	21·9	21·9	91	64·3	64·3	151	106·8	106·8	211	149·2	149·2	271	191·6	191·6
32	22·6	22·6	92	65·1	65·1	152	107·5	107·5	212	149·9	149·9	272	192·3	192·3
33	23·3	23·3	93	65·8	65·8	153	108·2	108·2	213	150·6	150·6	273	193·0	193·0
34	24·0	24·0	94	66·5	66·5	154	108·9	108·9	214	151·3	151·3	274	193·7	193·7
35	24·7	24·7	95	67·2	67·2	155	109·6	109·6	215	152·0	152·0	275	194·5	194·5
36	25·5	25·5	96	67·9	67·9	156	110·3	110·3	216	152·7	152·7	276	195·2	195·2
37	26·2	26·2	97	68·6	68·6	157	111·0	111·0	217	153·4	153·4	277	195·9	195·9
38	26·9	26·9	98	69·3	69·3	158	111·7	111·7	218	154·1	154·1	278	196·6	196·6
39	27·6	27·6	99	70·0	70·0	159	112·4	112·4	219	154·9	154·9	279	197·3	197·3
40	28·3	28·3	100	70·7	70·7	160	113·1	113·1	220	155·6	155·6	280	198·0	198·0
41	29·0	29·0	101	71·4	71·4	161	113·8	113·8	221	156·3	156·3	281	198·7	198·7
42	29·7	29·7	102	72·1	72·1	162	114·5	114·5	222	157·0	157·0	282	199·4	199·4
43	30·4	30·4	103	72·8	72·8	163	115·3	115·3	223	157·7	157·7	283	200·1	200·1
44	31·1	31·1	104	73·5	73·5	164	116·0	116·0	224	158·4	158·4	284	200·8	200·8
45	31·8	31·8	105	74·2	74·2	165	116·7	116·7	225	159·1	159·1	285	201·5	201·5
46	32·5	32·5	106	75·0	75·0	166	117·4	117·4	226	159·8	159·8	286	202·2	202·2
47	33·2	33·2	107	75·7	75·7	167	118·1	118·1	227	160·5	160·5	287	202·9	202·9
48	33·9	33·9	108	76·4	76·4	168	118·8	118·8	228	161·2	161·2	288	203·6	203·6
49	34·6	34·6	109	77·1	77·1	169	119·5	119·5	229	161·9	161·9	289	204·3	204·3
50	35·4	35·4	110	77·8	77·8	170	120·2	120·2	230	162·6	162·6	290	205·1	205·1
51	36·1	36·1	111	78·5	78·5	171	120·9	120·9	231	163·3	163·3	291	205·8	205·8
52	36·8	36·8	112	79·2	79·2	172	121·6	121·6	232	164·0	164·0	292	206·5	206·5
53	37·5	37·5	113	79·9	79·9	173	122·3	122·3	233	164·8	164·8	293	207·2	207·2
54	38·2	38·2	114	80·6	80·6	174	123·0	123·0	234	165·5	165·5	294	207·9	207·9
55	38·9	38·9	115	81·3	81·3	175	123·7	123·7	235	166·2	166·2	295	208·6	208·6
56	39·6	39·6	116	82·0	82·0	176	124·4	124·4	236	166·9	166·9	296	209·3	209·3
57	40·3	40·3	117	82·7	82·7	177	125·2	125·2	237	167·6	167·6	297	210·0	210·0
58	41·0	41·0	118	83·4	83·4	178	125·9	125·9	238	168·3	168·3	298	210·7	210·7
59	41·7	41·7	119	84·1	84·1	179	126·6	126·6	239	169·0	169·0	299	211·4	211·4
60	42·4	42·4	120	84·8	84·8	180	127·3	127·3	240	169·7	169·7	300	212·1	212·1
Dist.	Dep.	Lat.	Dist.	Dep.	Lat.	Dist.	Dep.	Lat.	Dist.	Dep.	Lat.	Dist.	Dep.	Lat.

Table 5.

Dist.	Lat.	Dep.	Dist.	Lat.	Dep.	Dist.	Lat.	Dep.	Dist.	Lat.	Dep.	Dist.	Lat.	Dep.
1	01·0	00·0	61	61·0	01·1	121	121·0	02·1	181	181·0	03·2	241	241·0	04·2
2	02·0	00·0	62	62·0	01·1	122	122·0	02·1	182	182·0	03·2	242	242·0	04·2
3	03·0	00·1	63	63·0	01·1	123	123·0	02·1	183	183·0	03·2	243	243·0	04·2
4	04·0	00·1	64	64·0	01·1	124	124·0	02·2	184	184·0	03·2	244	244·0	04·3
5	05·0	00·1	65	65·0	01·1	125	125·0	02·2	185	185·0	03·2	245	245·0	04·3
6	06·0	00·1	66	66·0	01·2	126	126·0	02·2	186	186·0	03·2	246	246·0	04·3
7	07·0	00·1	67	67·0	01·2	127	127·0	02·2	187	187·0	03·3	247	247·0	04·3
8	08·0	00·1	68	68·0	01·2	128	128·0	02·2	188	188·0	03·3	248	248·0	04·3
9	09·0	00·2	69	69·0	01·2	129	129·0	02·2	189	189·0	03·3	249	249·0	04·3
10	10·0	00·2	70	70·0	01·2	130	130·0	02·3	190	190·0	03·3	250	250·0	04·4
11	11·0	00·2	71	71·0	01·2	131	131·0	02·3	191	191·0	03·3	251	251·0	04·4
12	12·0	00·2	72	72·0	01·3	132	132·0	02·3	192	192·0	03·4	252	252·0	04·4
13	13·0	00·2	73	73·0	01·3	133	133·0	02·3	193	193·0	03·4	253	253·0	04·4
14	14·0	00·2	74	74·0	01·3	134	134·0	02·3	194	194·0	03·4	254	254·0	04·4
15	15·0	00·3	75	75·0	01·3	135	135·0	02·4	195	195·0	03·4	255	255·0	04·4
16	16·0	00·3	76	76·0	01·3	136	136·0	02·4	196	196·0	03·4	256	256·0	04·5
17	17·0	00·3	77	77·0	01·3	137	137·0	02·4	197	197·0	03·4	257	257·0	04·5
18	18·0	00·3	78	78·0	01·4	138	138·0	02·4	198	198·0	03·5	258	258·0	04·5
19	19·0	00·3	79	79·0	01·4	139	139·0	02·4	199	199·0	03·5	259	259·0	04·5
20	20·0	00·3	80	80·0	01·4	140	140·0	02·4	200	200·0	03·5	260	260·0	04·5
21	21·0	00·4	81	81·0	01·4	141	141·0	02·5	201	201·0	03·5	261	261·0	04·5
22	22·0	00·4	82	82·0	01·4	142	142·0	02·5	202	202·0	03·5	262	262·0	04·6
23	23·0	00·4	83	83·0	01·4	143	143·0	02·5	203	203·0	03·5	263	263·0	04·6
24	24·0	00·4	84	84·0	01·5	144	144·0	02·5	204	204·0	03·6	264	264·0	04·6
25	25·0	00·4	85	85·0	01·5	145	145·0	02·5	205	205·0	03·6	265	265·0	04·6
26	26·0	00·5	86	86·0	01·5	146	146·0	02·5	206	206·0	03·6	266	266·0	04·6
27	27·0	00·5	87	87·0	01·5	147	147·0	02·6	207	207·0	03·6	267	267·0	04·7
28	28·0	00·5	88	88·0	01·5	148	148·0	02·6	208	208·0	03·6	268	268·0	04·7
29	29·0	00·5	89	89·0	01·6	149	149·0	02·6	209	209·0	03·7	269	269·0	04·7
30	30·0	00·5	90	90·0	01·6	150	150·0	02·6	210	210·0	03·7	270	270·0	04·7
31	31·0	00·5	91	91·0	01·6	151	151·0	02·6	211	211·0	03·7	271	271·0	04·7
32	32·0	00·6	92	92·0	01·6	152	152·0	02·7	212	212·0	03·7	272	272·0	04·7
33	33·0	00·6	93	93·0	01·6	153	153·0	02·7	213	213·0	03·7	273	273·0	04·8
34	34·0	00·6	94	94·0	01·6	154	154·0	02·7	214	214·0	03·7	274	274·0	04·8
35	35·0	00·6	95	95·0	01·7	155	155·0	02·7	215	215·0	03·8	275	275·0	04·8
36	36·0	00·6	96	96·0	01·7	156	156·0	02·7	216	216·0	03·8	276	276·0	04·8
37	37·0	00·6	97	97·0	01·7	157	157·0	02·7	217	217·0	03·8	277	277·0	04·8
38	38·0	00·7	98	98·0	01·7	158	158·0	02·8	218	218·0	03·8	278	278·0	04·9
39	39·0	00·7	99	99·0	01·7	159	159·0	02·8	219	219·0	03·8	279	279·0	04·9
40	40·0	00·7	100	100·0	01·7	160	160·0	02·8	220	220·0	03·8	280	280·0	04·9
41	41·0	00·7	101	101·0	01·8	161	161·0	02·8	221	221·0	03·9	281	281·0	04·9
42	42·0	00·7	102	102·0	01·8	162	162·0	02·8	222	222·0	03·9	282	282·0	04·9
43	43·0	00·8	103	103·0	01·8	163	163·0	02·8	223	223·0	03·9	283	283·0	04·9
44	44·0	00·8	104	104·0	01·8	164	164·0	02·9	224	224·0	03·9	284	284·0	05·0
45	45·0	00·8	105	105·0	01·8	165	165·0	02·9	225	225·0	03·9	285	285·0	05·0
46	46·0	00·8	106	106·0	01·8	166	166·0	02·9	226	226·0	03·9	286	286·0	05·0
47	47·0	00·8	107	107·0	01·9	167	167·0	02·9	227	227·0	04·0	287	287·0	05·0
48	48·0	00·8	108	108·0	01·9	168	168·0	02·9	228	228·0	04·0	288	288·0	05·0
49	49·0	00·9	109	109·0	01·9	169	169·0	02·9	229	229·0	04·0	289	289·0	05·0
50	50·0	00·9	110	110·0	01·9	170	170·0	03·0	230	230·0	04·0	290	290·0	05·1
51	51·0	00·9	111	111·0	01·9	171	171·0	03·0	231	231·0	04·0	291	291·0	05·1
52	52·0	00·9	112	112·0	02·0	172	172·0	03·0	232	232·0	04·0	292	292·0	05·1
53	53·0	00·9	113	113·0	02·0	173	173·0	03·0	233	233·0	04·1	293	293·0	05·1
54	54·0	00·9	114	114·0	02·0	174	174·0	03·0	234	234·0	04·1	294	294·0	05·1
55	55·0	01·0	115	115·0	02·0	175	175·0	03·1	235	235·0	04·1	295	295·0	05·1
56	56·0	01·0	116	116·0	02·0	176	176·0	03·1	236	236·0	04·1	296	296·0	05·2
57	57·0	01·0	117	117·0	02·0	177	177·0	03·1	237	237·0	04·1	297	297·0	05·2
58	58·0	01·0	118	118·0	02·1	178	178·0	03·1	238	238·0	04·2	298	298·0	05·2
59	59·0	01·0	119	119·0	02·1	179	179·0	03·1	239	239·0	04·2	299	299·0	05·2
60	60·0	01·0	120	120·0	02·1	180	180·0	03·1	240	240·0	04·2	300	300·0	05·2
Dist.	Dep.	Lat.	Dist.	Dep.	Lat.	Dist.	Dep.	Lat.	Dist.	Dep.	Lat.	Dist.	Dep.	Lat.

FOR 89 DEGREES.

Dist.	Lat.	Dep.	Dist.	Lat.	Dep.	Dist.	Lat.	Dep.	Dist.	Lat.	Dep.	Dist.	Lat.	Dep.
1	01·0	00·0	61	61·0	02·1	121	120·9	04·2	181	180·9	06·3	241	240·9	08·4
2	02·0	00·1	62	62·0	02·2	122	121·9	04·3	182	181·9	06·4	242	241·9	08·4
3	03·0	00·1	63	63·0	02·2	123	122·9	04·3	183	182·9	06·4	243	242·9	08·5
4	04·0	00·1	64	64·0	02·2	124	123·9	04·3	184	183·9	06·4	244	243·9	08·5
5	05·0	00·2	65	65·0	02·3	125	124·9	04·4	185	184·9	06·5	245	244·9	08·6
6	06·0	00·2	66	66·0	02·3	126	125·9	04·4	186	185·9	06·5	246	245·8	08·6
7	07·0	00·2	67	67·0	02·3	127	126·9	04·4	187	186·9	06·5	247	246·8	08·6
8	08·0	00·3	68	68·0	02·4	128	127·9	04·5	188	187·9	06·6	248	247·8	08·7
9	09·0	00·3	69	69·0	02·4	129	128·9	04·5	189	188·9	06·6	249	248·8	08·7
10	10·0	00·3	70	70·0	02·4	130	129·9	04·5	190	189·9	06·6	250	249·8	08·7
11	11·0	00·4	71	71·0	02·5	131	130·9	04·6	191	190·9	06·7	251	250·8	08·8
12	12·0	00·4	72	72·0	02·5	132	131·9	04·6	192	191·9	06·7	252	251·8	08·8
13	13·0	00·5	73	73·0	02·5	133	132·9	04·6	193	192·9	06·7	253	252·8	08·8
14	14·0	00·5	74	74·0	02·6	134	133·9	04·7	194	193·9	06·8	254	253·8	08·9
15	15·0	00·5	75	75·0	02·6	135	134·9	04·7	195	194·9	06·8	255	254·8	08·9
16	16·0	00·6	76	76·0	02·7	136	135·9	04·7	196	195·9	06·8	256	255·8	08·9
17	17·0	00·6	77	77·0	02·7	137	136·9	04·8	197	196·9	06·9	257	256·8	09·0
18	18·0	00·6	78	78·0	02·7	138	137·9	04·8	198	197·9	06·9	258	257·8	09·0
19	19·0	00·7	79	79·0	02·8	139	138·9	04·9	199	198·9	06·9	259	258·8	09·0
20	20·0	00·7	80	80·0	02·8	140	139·9	04·9	200	199·9	07·0	260	259·8	09·1
21	21·0	00·7	81	81·0	02·8	141	140·9	04·9	201	200·9	07·0	261	260·8	09·1
22	22·0	00·8	82	81·9	02·9	142	141·9	05·0	202	201·9	07·0	262	261·8	09·1
23	23·0	00·8	83	82·9	02·9	143	142·9	05·0	203	202·9	07·1	263	262·8	09·2
24	24·0	00·8	84	83·9	02·9	144	143·9	05·0	204	203·9	07·1	264	263·8	09·2
25	25·0	00·9	85	84·9	03·0	145	144·9	05·1	205	204·9	07·2	265	264·8	09·2
26	26·0	00·9	86	85·9	03·0	146	145·9	05·1	206	205·9	07·2	266	265·8	09·3
27	27·0	00·9	87	86·9	03·0	147	146·9	05·1	207	206·9	07·2	267	266·8	09·3
28	28·0	01·0	88	87·9	03·1	148	147·9	05·2	208	207·9	07·3	268	267·8	09·4
29	29·0	01·0	89	88·9	03·1	149	148·9	05·2	209	208·9	07·3	269	268·8	09·4
30	30·0	01·0	90	89·9	03·1	150	149·9	05·2	210	209·9	07·3	270	269·8	09·4
31	31·0	01·1	91	90·9	03·2	151	150·9	05·3	211	210·9	07·4	271	270·8	09·5
32	32·0	01·1	92	91·9	03·2	152	151·9	05·3	212	211·9	07·4	272	271·8	09·5
33	33·0	01·2	93	92·9	03·2	153	152·9	05·3	213	212·9	07·4	273	272·8	09·5
34	34·0	01·2	94	93·9	03·3	154	153·9	05·4	214	213·9	07·5	274	273·8	09·6
35	35·0	01·2	95	94·9	03·3	155	154·9	05·4	215	214·9	07·5	275	274·8	09·6
36	36·0	01·3	96	95·9	03·4	156	155·9	05·4	216	215·9	07·5	276	275·8	09·6
37	37·0	01·3	97	96·9	03·4	157	156·9	05·5	217	216·9	07·6	277	276·8	09·7
38	38·0	01·3	98	97·9	03·4	158	157·9	05·5	218	217·9	07·6	278	277·8	09·7
39	39·0	01·4	99	98·9	03·5	159	158·9	05·5	219	218·9	07·6	279	278·8	09·7
40	40·0	01·4	100	99·9	03·5	160	159·9	05·6	220	219·9	07·7	280	279·8	09·8
41	41·0	01·4	101	100·9	03·5	161	160·9	05·6	221	220·9	07·7	281	280·8	09·8
42	42·0	01·5	102	101·9	03·6	162	161·9	05·7	222	221·9	07·7	282	281·8	09·8
43	43·0	01·5	103	102·9	03·6	163	162·9	05·7	223	222·9	07·8	283	282·8	09·9
44	44·0	01·5	104	103·9	03·6	164	163·9	05·7	224	223·9	07·8	284	283·8	09·9
45	45·0	01·6	105	104·9	03·7	165	164·9	05·8	225	224·9	07·9	285	284·8	09·9
46	46·0	01·6	106	105·9	03·7	166	165·9	05·8	226	225·9	07·9	286	285·8	10·0
47	47·0	01·6	107	106·9	03·7	167	166·9	05·8	227	226·9	07·9	287	286·8	10·0
48	48·0	01·7	108	107·9	03·8	168	167·9	05·9	228	227·9	08·0	288	287·8	10·1
49	49·0	01·7	109	108·9	03·8	169	168·9	05·9	229	228·9	08·0	289	288·8	10·1
50	50·0	01·7	110	109·9	03·8	170	169·9	05·9	230	229·9	08·0	290	289·8	10·1
51	51·0	01·8	111	110·9	03·9	171	170·9	06·0	231	230·9	08·1	291	290·8	10·2
52	52·0	01·8	112	111·9	03·9	172	171·9	06·0	232	231·9	08·1	292	291·8	10·2
53	53·0	01·8	113	112·9	03·9	173	172·9	06·0	233	232·9	08·1	293	292·8	10·2
54	54·0	01·9	114	113·9	04·0	174	173·9	06·1	234	233·9	08·2	294	293·8	10·3
55	55·0	01·9	115	114·9	04·0	175	174·9	06·1	235	234·9	08·2	295	294·8	10·3
56	56·0	02·0	116	115·9	04·0	176	175·9	06·1	236	235·9	08·2	296	295·8	10·3
57	57·0	02·0	117	116·9	04·1	177	176·9	06·2	237	236·9	08·3	297	296·8	10·4
58	58·0	02·0	118	117·9	04·1	178	177·9	06·2	238	237·9	08·3	298	297·8	10·4
59	59·0	02·1	119	118·9	04·2	179	178·9	06·2	239	238·9	08·3	299	298·8	10·4
60	60·0	02·1	120	119·9	04·2	180	179·9	06·3	240	239·9	08·4	300	299·8	10·5
Dist.	Dep.	Lat.	Dist.	Dep.	Lat.	Dist.	Dep.	Lat.	Dist.	Dep.	Lat.	Dist.	Dep.	Lat.

FOR 88 DEGREES.

Dist.	Lat.	Dep.	Dist.	Lat.	Dep.	Dist.	Lat.	Dep.	Dist.	Lat.	Dep.	Dist.	Lat.	Dep.
1	01·0	00·1	61	60·9	03·2	121	120·8	06·3	181	180·8	09·5	241	240·7	12·6
2	02·0	00·1	62	61·9	03·2	122	121·8	06·4	182	181·8	09·5	242	241·7	12·7
3	03·0	00·2	63	62·9	03·3	123	122·8	06·4	183	182·7	09·6	243	242·7	12·7
4	04·0	00·2	64	63·9	03·3	124	123·8	06·5	184	183·7	09·6	244	243·7	12·8
5	05·0	00·3	65	64·9	03·4	125	124·8	06·5	185	184·7	09·7	245	244·7	12·8
6	06·0	00·3	66	65·9	03·5	126	125·8	06·6	186	185·7	09·7	246	245·7	12·9
7	07·0	00·4	67	66·9	03·5	127	126·8	06·6	187	186·7	09·8	247	246·7	12·9
8	08·0	00·4	68	67·9	03·6	128	127·8	06·7	188	187·7	09·8	248	247·7	13·0
9	09·0	00·5	69	68·9	03·6	129	128·8	06·8	189	188·7	09·9	249	248·7	13·0
10	10·0	00·5	70	69·9	03·7	130	129·8	06·8	190	189·7	09·9	250	249·7	13·1
11	11·0	00·6	71	70·9	03·7	131	130·8	06·9	191	190·7	10·0	251	250·7	13·1
12	12·0	00·6	72	71·9	03·8	132	131·8	06·9	192	191·7	10·0	252	251·7	13·2
13	13·0	00·7	73	72·9	03·8	133	132·8	07·0	193	192·7	10·1	253	252·7	13·2
14	14·0	00·7	74	73·9	03·9	134	133·8	07·0	194	193·7	10·2	254	253·7	13·3
15	15·0	00·8	75	74·9	03·9	135	134·8	07·1	195	194·7	10·2	255	254·7	13·3
16	16·0	00·8	76	75·9	04·0	136	135·8	07·1	196	195·7	10·3	256	255·6	13·4
17	17·0	00·9	77	76·9	04·0	137	136·8	07·2	197	196·7	10·3	257	256·6	13·5
18	18·0	00·9	78	77·9	04·1	138	137·8	07·2	198	197·7	10·4	258	257·6	13·5
19	19·0	01·0	79	78·9	04·1	139	138·8	07·3	199	198·7	10·4	259	258·6	13·6
20	20·0	01·0	80	79·9	04·2	140	139·8	07·3	200	199·7	10·5	260	259·6	13·6
21	21·0	01·1	81	80·9	04·2	141	140·8	07·4	201	200·7	10·5	261	260·6	13·7
22	22·0	01·2	82	81·9	04·3	142	141·8	07·4	202	201·7	10·6	262	261·6	13·7
23	23·0	01·2	83	82·9	04·3	143	142·8	07·5	203	202·7	10·6	263	262·6	13·8
24	24·0	01·3	84	83·9	04·4	144	143·8	07·5	204	203·7	10·7	264	263·6	13·8
25	25·0	01·3	85	84·9	04·4	145	144·8	07·6	205	204·7	10·7	265	264·6	13·9
26	26·0	01·4	86	85·9	04·5	146	145·8	07·6	206	205·7	10·8	266	265·6	13·9
27	27·0	01·4	87	86·9	04·6	147	146·8	07·7	207	206·7	10·8	267	266·6	14·0
28	28·0	01·5	88	87·9	04·6	148	147·8	07·7	208	207·7	10·9	268	267·6	14·0
29	29·0	01·5	89	88·9	04·7	149	148·8	07·8	209	208·7	10·9	269	268·6	14·1
30	30·0	01·6	90	89·9	04·7	150	149·8	07·9	210	209·7	11·0	270	269·6	14·1
31	31·0	01·6	91	90·9	04·8	151	150·8	07·9	211	210·7	11·0	271	270·6	14·2
32	32·0	01·7	92	91·9	04·8	152	151·8	08·0	212	211·7	11·1	272	271·6	14·2
33	33·0	01·7	93	92·9	04·9	153	152·8	08·0	213	212·7	11·1	273	272·6	14·3
34	34·0	01·8	94	93·9	04·9	154	153·8	08·1	214	213·7	11·2	274	273·6	14·3
35	35·0	01·8	95	94·9	05·0	155	154·8	08·1	215	214·7	11·3	275	274·6	14·4
36	36·0	01·9	96	95·9	05·0	156	155·8	08·2	216	215·7	11·3	276	275·6	14·4
37	36·9	01·9	97	96·9	05·1	157	156·8	08·2	217	216·7	11·4	277	276·6	14·5
38	37·9	02·0	98	97·9	05·1	158	157·8	08·3	218	217·7	11·4	278	277·6	14·5
39	38·9	02·0	99	98·9	05·2	159	158·8	08·3	219	218·7	11·5	279	278·6	14·6
40	39·9	02·1	100	99·9	05·2	160	159·8	08·4	220	219·7	11·5	280	279·6	14·7
41	40·9	02·1	101	100·9	05·3	161	160·8	08·4	221	220·7	11·6	281	280·6	14·7
42	41·9	02·2	102	101·9	05·3	162	161·8	08·5	222	221·7	11·6	282	281·6	14·8
43	42·9	02·3	103	102·9	05·4	163	162·8	08·5	223	222·7	11·7	283	282·6	14·8
44	43·9	02·3	104	103·9	05·4	164	163·8	08·6	224	223·7	11·7	284	283·6	14·9
45	44·9	02·4	105	104·9	05·5	165	164·8	08·6	225	224·7	11·8	285	284·6	14·9
46	45·9	02·4	106	105·9	05·5	166	165·8	08·7	226	225·7	11·8	286	285·6	15·0
47	46·9	02·5	107	106·9	05·6	167	166·8	08·7	227	226·7	11·9	287	286·6	15·0
48	47·9	02·5	108	107·9	05·7	168	167·8	08·8	228	227·7	11·9	288	287·6	15·1
49	48·9	02·6	109	108·9	05·7	169	168·8	08·8	229	228·7	12·0	289	288·6	15·1
50	49·9	02·6	110	109·8	05·8	170	169·8	08·9	230	229·7	12·0	290	289·6	15·2
51	50·9	02·7	111	110·8	05·8	171	170·8	08·9	231	230·7	12·1	291	290·6	15·2
52	51·9	02·7	112	111·8	05·9	172	171·8	09·0	232	231·7	12·1	292	291·6	15·3
53	52·9	02·8	113	112·8	05·9	173	172·8	09·1	233	232·7	12·2	293	292·6	15·3
54	53·9	02·8	114	113·8	06·0	174	173·8	09·1	234	233·7	12·2	294	293·6	15·4
55	54·9	02·9	115	114·8	06·0	175	174·8	09·2	235	234·7	12·3	295	294·6	15·4
56	55·9	02·9	116	115·8	06·1	176	175·8	09.2	236	235·7	12·4	296	295·6	15·5
57	56·9	03·0	117	116·8	06·1	177	176·8	09·3	237	236·7	12·4	297	296·6	15·5
58	57·9	03·0	118	117·8	06·2	178	177·8	09·3	238	237·7	12·5	298	297·6	15·6
59	58·9	03·1	119	118·8	06·2	179	178·8	09·4	239	238·7	12·5	299	298·6	15·6
60	59·9	03·1	120	119·8	06·3	180	179·8	09·4	240	239·7	12·6	300	299·6	15·7
Dist.	Dep.	Lat.	Dist.	Dep.	Lat.	Dist.	Dep.	Lat.	Dist.	Dep.	Lat.	Dist.	Dep.	Lat.

Dist.	Lat.	Dep.	Dist.	Lat.	Dep.	Dist.	Lat.	Dep.	Dist.	Lat.	Dep.	Dist.	Lat.	Dep.
1	01·0	00·1	61	60·9	04·3	121	120·7	08·4	181	180·6	12·6	241	240·4	16·8
2	02·0	00·1	62	61·8	04·3	122	121·7	08·5	182	181·6	12·7	242	241·4	16·9
3	03·0	00·2	63	62·8	04·4	123	122·7	08·6	183	182·6	12·8	243	242·4	17·0
4	04·0	00·3	64	63·8	04·5	124	123·7	08·6	184	183·6	12·8	244	243·4	17·0
5	05·0	00·3	65	64·8	04·5	125	124·7	08·7	185	184·5	12·9	245	244·4	17·1
6	06·0	00·4	66	65·8	04·6	126	125·7	08·8	186	185·5	13·0	246	245·4	17·2
7	07·0	00·5	67	66·8	04·7	127	126·7	08·9	187	186·5	13·0	247	246·4	17·2
8	08·0	00·6	68	67·8	04·7	128	127·7	08·9	188	187·5	13·1	248	247·4	17·3
9	09·0	00·6	69	68·8	04·8	129	128·7	09·0	189	188·5	13·2	249	248·4	17·4
10	10·0	00·7	70	69·8	04·9	130	129·7	09·1	190	189·5	13·3	250	249·4	17·4
11	11·0	00·8	71	70·8	05·0	131	130·7	09·1	191	190·5	13·3	251	250·4	17·5
12	12·0	00·8	72	71·8	05·0	132	131·7	09·2	192	191·5	13·4	252	251·4	17·6
13	13·0	00·9	73	72·8	05·1	133	132·7	09·3	193	192·5	13·5	253	252·4	17·6
14	14·0	01·0	74	73·8	05·2	134	133·7	09·3	194	193·5	13·5	254	253·4	17·7
15	15·0	01·0	75	74·8	05·2	135	134·7	09·4	195	194·5	13·6	255	254·4	17·8
16	16·0	01·1	76	75·8	05·3	136	135·7	09·5	196	195·5	13·7	256	255·4	17·9
17	17·0	01·2	77	76·8	05·4	137	136·7	09·6	197	196·5	13·7	257	256·4	17·9
18	18·0	01·3	78	77·8	05·4	138	137·7	09·6	198	197·5	13·8	258	257·4	18·0
19	19·0	01·3	79	78·8	05·5	139	138·7	09·7	199	198·5	13·9	259	258·4	18·1
20	20·0	01·4	80	79·8	05·6	140	139·7	09·8	200	199·5	14·0	260	259·4	18·1
21	20·9	01·5	81	80·8	05·7	141	140·7	09·8	201	200·5	14·0	261	260·4	18·2
22	21·9	01·5	82	81·8	05·7	142	141·7	09·9	202	201·5	14·1	262	261·4	18·3
23	22·9	01·6	83	82·8	05·8	143	142·7	10·0	203	202·5	14·2	263	262·4	18·3
24	23·9	01·7	84	83·8	05·9	144	143·6	10·0	204	203·5	14·2	264	263·4	18·4
25	24·9	01·7	85	84·8	05·9	145	144·6	10·1	205	204·5	14·3	265	264·4	18·5
26	25·9	01·8	86	85·8	06·0	146	145·6	10·2	206	205·5	14·4	266	265·4	18·6
27	26·9	01·9	87	86·8	06·1	147	146·6	10·3	207	206·5	14·4	267	266·3	18·6
28	27·9	02·0	88	87·8	06·1	148	147·6	10·3	208	207·5	14·5	268	267·3	18·7
29	28·9	02·0	89	88·8	06·2	149	148·6	10·4	209	208·5	14·6	269	268·3	18·8
30	29·9	02·1	90	89·8	06·3	150	149·6	10·5	210	209·5	14·6	270	269·3	18·8
31	30·9	02·2	91	90·8	06·3	151	150·6	10·5	211	210·5	14·7	271	270·3	18·9
32	31·9	02·2	92	91·8	06·4	152	151·6	10·6	212	211·5	14·8	272	271·3	19·0
33	32·9	02·3	93	92·8	06·5	153	152·6	10·7	213	212·5	14·9	273	272·3	19·0
34	33·9	02·4	94	93·8	06·6	154	153·6	10·7	214	213·5	14·9	274	273·3	19·1
35	34·9	02·4	95	94·8	06·6	155	154·6	10·8	215	214·5	15·0	275	274·3	19·2
36	35·9	02·5	96	95·8	06·7	156	155·6	10·9	216	215·5	15·1	276	275·3	19·3
37	36·9	02·6	97	96·8	06·8	157	156·6	11·0	217	216·5	15·1	277	276·3	19·3
38	37·9	02·7	98	97·8	06·8	158	157·6	11·0	218	217·5	15·2	278	277·3	19·4
39	38·9	02·7	99	98·8	06·9	159	158·6	11·1	219	218·5	15·3	279	278·3	19·5
40	39·9	02·8	100	99·8	07·0	160	159·6	11·2	220	219·5	15·3	280	279·3	19·5
41	40·9	02·9	101	100·8	07·0	161	160·6	11·2	221	220·5	15·4	281	280·3	19·6
42	41·9	02·9	102	101·8	07·1	162	161·6	11·3	222	221·5	15·5	282	281·3	19·7
43	42·9	03·0	103	102·7	07·2	163	162·6	11·4	223	222·5	15·6	283	282·3	19·7
44	43·9	03·1	104	103·7	07·3	164	163·6	11·4	224	223·5	15·6	284	283·3	19·8
45	44·9	03·1	105	104·7	07·3	165	164·6	11·5	225	224·5	15·7	285	284·3	19·9
46	45·9	03·2	106	105·7	07·4	166	165·6	11·6	226	225·4	15·8	286	285·3	20·0
47	46·9	03·3	107	106·7	07·5	167	166·6	11·6	227	226·4	15·8	287	286·3	20·0
48	47·9	03·3	108	107·7	07·5	168	167·6	11·7	228	227·4	15·9	288	287·3	20·1
49	48·9	03·4	109	108·7	07·6	169	168·6	11·8	229	228·4	16·0	289	288·3	20·2
50	49·9	03·5	110	109·7	07·7	170	169·6	11·9	230	229·4	16·0	290	289·3	20·2
51	50·9	03·6	111	110·7	07·7	171	170·6	11·9	231	230·4	16·1	291	290·3	20·3
52	51·9	03·6	112	111·7	07·8	172	171·6	12·0	232	231·4	16·2	292	291·3	20·4
53	52·9	03·7	113	112·7	07·9	173	172·6	12·1	233	232·4	16·3	293	292·3	20·4
54	53·9	03·8	114	113·7	08·0	174	173·6	12·1	234	233·4	16·3	294	293·3	20·5
55	54·9	03·8	115	114·7	08·0	175	174·6	12·2	235	234·4	16·4	295	294·3	20·6
56	55·9	03·9	116	115·7	08·1	176	175·6	12·3	236	235·4	16·5	296	295·3	20·6
57	56·9	04·0	117	116·7	08·2	177	176·6	12·3	237	236·4	16·5	297	296·3	20·7
58	57·9	04·0	118	117·7	08·2	178	177·6	12·4	238	237·4	16·6	298	297·3	20·8
59	58·9	04·1	119	118·7	08·3	179	178·6	12·5	239	238·4	16·7	299	298·3	20·9
60	59·9	04·2	120	119·7	08·4	180	179·6	12·6	240	239·4	16·7	300	299·3	20·9
Dist.	Dep.	Lat.	Dist.	Dep.	Lat.	Dist.	Dep.	Lat.	Dist.	Dep.	Lat.	Dist.	Dep.	Lat.

Dist.	Lat.	Dep	Dist.	Lat.	Dep	Dist.	Lat.	Dep.	Dist.	Lat.	Dep.	Dist.	Lat.	Dep.
1	01·0	00·1	61	60·8	05·3	121	120·5	10·5	181	180·3	15·8	241	240·1	21·0
2	02·0	00·2	62	61·8	05·4	122	121·5	10·6	182	181·3	15·9	242	241·1	21·1
3	03·0	00·3	63	62·8	05·5	123	122·5	10·7	183	182·3	15·9	243	242·1	21·2
4	04·0	00·3	64	63·8	05·6	124	123·5	10·8	184	183·3	16·0	244	243·1	21·3
5	05·0	00·4	65	64·8	05·7	125	124·5	10·9	185	184·3	16·1	245	244·1	21·4
6	06·0	00·5	66	65·7	05·8	126	125·5	11·0	186	185·3	16·2	246	245·1	21·4
7	07·0	00·6	67	66·7	05·8	127	126·5	11·1	187	186·3	16·3	247	246·1	21·5
8	08·0	00·7	68	67·7	05·9	128	127·5	11·2	188	187·3	16·4	248	247·1	21·6
9	09·0	00·8	69	68·7	06·0	129	128·5	11·2	189	188·3	16·5	249	248·1	21·7
10	10·0	00·9	70	69·7	06·1	130	129·5	11·3	190	189·3	16·6	250	249·0	21·7
11	11·0	01·0	71	70·7	06·2	131	130·5	11·4	191	190·3	16·6	251	250·0	21·9
12	12·0	01·0	72	71·7	03·3	132	131·5	11·5	192	191·3	16·7	252	251·0	22·0
13	13·0	01·1	73	72·7	06·4	133	132·5	11·6	193	192·3	16·8	253	252·0	22·1
14	13·9	01·2	74	73·7	06·4	134	133·5	11·7	194	193·3	16·9	254	253·0	22·1
15	14·9	01·3	75	74·7	06·5	135	134·5	11·8	195	194·3	17·0	255	254·0	22·2
16	15·9	01·4	76	75·7	06·6	136	135·5	11·9	196	195·3	17·1	256	255·0	22·3
17	16·9	01·5	77	76·7	06·7	137	136·5	11·9	197	196·3	17·2	257	256·0	22·4
18	17·9	01·6	78	77·7	06·8	138	137·5	12·0	198	197·2	17·3	258	257·0	22·5
19	18·9	01·7	79	78·7	06·9	139	138·5	12·1	199	198·2	17·3	259	258·0	22·6
20	19·9	01·7	80	79·7	07·0	140	139·5	12·2	200	199·2	17·4	260	259·0	22·7
21	20·9	01·8	81	80·7	07·1	141	140·5	12·3	201	200·2	17·5	261	260·0	22·7
22	21·9	01·9	82	81·7	07·1	142	141·5	12·4	202	201·2	17·6	262	261·0	22·8
23	22·9	02·0	83	82·7	07·2	143	142·5	12·5	203	202·2	17·7	263	262·0	22·9
24	23·9	02·1	84	83·7	07·3	144	143·5	12·6	204	203·2	17·8	264	263·0	23·0
25	24·9	02·2	85	84·7	07·4	145	144·4	12·6	205	204·2	17·9	265	264·0	23·1
26	25·9	02·3	86	85·7	07·5	146	145·4	12·7	206	205·2	18·0	266	265·0	23·2
27	26·9	02·4	87	86·7	07·6	147	146·4	12·8	207	206·2	18·0	267	266·0	23·3
28	27·9	02·4	88	87·7	07·7	148	147·4	12·9	208	207·2	18·1	268	267·0	23·4
29	28·9	02·5	89	88·7	07·8	149	148·4	13·0	209	208·2	18·2	269	268·0	23·4
30	29·9	02·6	90	89·7	07·8	150	149·4	13·1	210	209·2	18·3	270	269·0	23·5
31	30·9	02·7	91	90·7	07·9	151	150·4	13·2	211	210·2	18·4	271	270·0	23·6
32	31·9	02·8	92	91·6	08·0	152	151·4	13·2	212	211·2	18·5	272	271·0	23·7
33	32·9	02·9	93	92·6	08·1	153	152·4	13·3	213	212·2	18·6	273	272·0	23·8
34	33·9	03·0	94	93·6	08·2	154	153·4	13·4	214	213·2	18·7	274	273·0	23·9
35	34·9	03·1	95	94·6	08·3	155	154·4	13·5	215	214·2	18·7	275	274·0	24·0
36	35·9	03·1	96	95·6	08·4	156	155·4	13·6	216	215·2	18·8	276	274·9	24·1
37	36·9	03·2	97	96·6	08·5	157	156·4	13·7	217	216·2	18·9	277	275·9	24·1
38	37·9	03·3	98	97·6	08·5	158	157·4	13·8	218	217·2	19·0	278	276·9	24·2
39	38·9	03·4	99	98·6	08·6	159	158·4	13·9	219	218·2	19·1	279	277·9	24·3
40	39·8	03·5	100	99·6	08·7	160	159·4	13·9	220	219·2	19·2	280	278·9	24·4
41	40·8	03·6	101	100·6	08·8	161	160·4	14·0	221	220·2	19·3	281	279·9	24·5
42	41·8	03·7	102	101·6	08·9	162	161·4	14·1	222	221·2	19·3	282	280·9	24·6
43	42·8	03·7	103	102·6	09·0	163	162·4	14·2	223	222·2	19·4	283	281·9	24·7
44	43·8	03·8	104	103·6	09·1	164	163·4	14·3	224	223·1	19·5	284	282·9	24·8
45	44·8	03·9	105	104·6	09·2	165	164·4	14·4	225	224·1	19·6	285	283·9	24·8
46	45·8	04·0	106	105·6	09·2	166	165·4	14·5	226	225·1	19·7	286	284·9	24·9
47	46·8	04·1	107	106·6	09·3	167	166·4	14·6	227	226·1	19·8	287	285·9	25·0
48	47·8	04·2	108	107·6	09·4	168	167·4	14·6	228	227·1	19·9	288	286·9	25·1
49	48·8	04·3	109	108·6	09·5	169	168·4	14·7	229	228·1	20·0	289	287·9	25·2
50	49·8	04·4	110	109·6	09·6	170	169·4	14·8	230	229·1	20·0	290	288·9	25·3
51	50·8	04·4	111	110·6	09·7	171	170·3	14·9	231	230·1	20·1	291	289·9	25·4
52	51·8	04·5	112	111·6	09·8	172	171·3	15·0	232	231·1	20·2	292	290·9	25·4
53	52·8	04·6	113	112·6	09·8	173	172·3	15·1	233	232·1	20·3	293	291·9	25·5
54	53·8	04·7	114	113·6	09·9	174	173·3	15·2	234	233·1	20·4	294	292·9	25·6
55	54·8	04·8	115	114·6	10·0	175	174·3	15·3	235	234·1	20·5	295	293·9	25·7
56	55·8	04·9	116	115·6	10·1	176	175·3	15·3	236	235·1	20·6	296	294·9	25·8
57	56·8	05·0	117	116·6	10·2	177	176·3	15·4	237	236·1	20·7	297	295·9	25·9
58	57·8	05·1	118	117·6	10·3	178	177·3	15·5	238	237·1	20·7	298	296·9	26·0
59	58·8	05·1	119	118·5	10·4	179	178·3	15·6	239	238·1	20·8	299	297·9	26·1
60	59·8	05·2	120	119·5	10·5	180	179·3	15·7	240	239·1	20·9	300	298·9	26·1
Dist.	Dep.	Lat.	Dist.	Dep.	Lat.	Dist.	Dep.	Lat	Dist.	Dep.	Lat.	Dist.	Dep.	Lat.

FOR 85 DEGREES.

Dist.	Lat.	Dep.	Dist.	Lat.	Dep.	Dist.	Lat.	Dep.	Dist.	Lat.	Dep.	Dist.	Lat.	Dep.
1	01·0	00·1	61	60·7	06·4	121	120·3	12·6	181	180·0	18·9	241	239·7	25·2
2	02·0	00·2	62	61·7	06·5	122	121·3	12·8	182	181·0	19·0	242	240·7	25·3
3	03·0	00·3	63	62·7	06·6	123	122·3	12·9	183	182·0	19·1	243	241·7	25·4
4	04·0	00·4	64	63·6	06·7	124	123·3	13·0	184	183·0	19·2	244	242·7	25·5
5	05·0	00·5	65	64·6	06·8	125	124·3	13·1	185	184·0	19·3	245	243·7	25·6
6	06·0	00·6	66	65·6	06·9	126	125·3	13·2	186	185·0	19·4	246	244·7	25·7
7	07·0	00·7	67	66·6	07·0	127	126·3	13·3	187	186·0	19·5	247	245·6	25·8
8	08·0	00·8	68	67·6	07·1	128	127·3	13·4	188	187·0	19·7	248	246·6	25·9
9	09·0	00·9	69	68·6	07·2	129	128·3	13·5	189	188·0	19·8	249	247·6	26·0
10	09·9	01·0	70	69·6	07·3	130	129·3	13·6	190	189·0	19·9	250	248·6	26·1
11	10·9	01·1	71	70·6	07·4	131	130·3	13·7	191	190·0	20·0	251	249·6	26·2
12	11·9	01·3	72	71·6	07·5	132	131·3	13·8	192	190·9	20·1	252	250·6	26·3
13	12·9	01·4	73	72·6	07·6	133	132·3	13·9	193	191·9	20·2	253	251·6	26·4
14	13·9	01·5	74	73·6	07·7	134	133·3	14·0	194	192·9	20·3	254	252·6	26·6
15	14·9	01·6	75	74·6	07·8	135	134·3	14·1	195	193·9	20·4	255	253·6	26·7
16	15·9	01·7	76	75·6	07·9	136	135·3	14·2	196	194·9	20·5	256	254·6	26·8
17	16·9	01·8	77	76·6	08·0	137	136·2	14·3	197	195·9	20·6	257	255·6	26·9
18	17·9	01·9	78	77·6	08·2	138	137·2	14·4	198	196·9	20·7	258	256·6	27·0
19	18·9	02·0	79	78·6	08·3	139	138·2	14·5	199	197·9	20·8	259	257·6	27·1
20	19·9	02·1	80	79·6	08·4	140	139·2	14·6	200	198·9	20·9	260	258·6	27·2
21	20·9	02·2	81	80·6	08·5	141	140·2	14·7	201	199·9	21·0	261	259·6	27·3
22	21·9	02·3	82	81·6	08·6	142	141·2	14·8	202	200·9	21·1	262	260·6	27·4
23	22·9	02·4	83	82·5	08·7	143	142·2	14·9	203	201·9	21·2	263	261·6	27·5
24	23·9	02·5	84	83·5	08·8	144	143·2	15·1	204	202·9	21·3	264	262·6	27·6
25	24·9	02·6	85	84·5	08·9	145	144·2	15·2	205	203·9	21·4	265	263·5	27·7
26	25·9	02·7	86	85·5	09·0	146	145·2	15·3	206	204·9	21·5	266	264·5	27·8
27	26·9	02·8	87	86·5	09·1	147	146·2	15·4	207	205·9	21·6	267	265·5	27·9
28	27·8	02·9	88	87·5	09·2	148	147·2	15·5	208	206·9	21·7	268	266·5	28·0
29	28·8	03·0	89	88·5	09·3	149	148·2	15·6	209	207·9	21·8	269	267·5	28·1
30	29·8	03·1	90	89·5	09·4	150	149·2	15·7	210	208·8	22·0	270	268·5	28·2
31	30·8	03·2	91	90·5	09·5	151	150·2	15·8	211	209·8	22·1	271	269·5	28·3
32	31·8	03·3	92	91·5	09·6	152	151·2	15·9	212	210·8	22·2	272	270·5	28·4
33	32·8	03·4	93	92·5	09·7	153	152·2	16·0	213	211·8	22·3	273	271·5	28·5
34	33·8	03·6	94	93·5	09·8	154	153·2	16·1	214	212·8	22·4	274	272·5	28·6
35	34·8	03·7	95	94·5	09·9	155	154·2	16·2	215	213·8	22·5	275	273·5	28·7
36	35·8	03·8	96	95·5	10·0	156	155·1	16·3	216	214·8	22·6	276	274·5	28·8
37	36·8	03·9	97	96·5	10·1	157	156·1	16·4	217	215·8	22·7	277	275·5	29·0
38	37·8	04·0	98	97·5	10·2	158	157·1	16·5	218	216·8	22·8	278	276·5	29·1
39	38·8	04·1	99	98·5	10·3	159	158·1	16·6	219	217·8	22·9	279	277·5	29·2
40	39·8	04·2	100	99·5	10·5	160	159·1	16·7	220	218·8	23·0	280	278·5	29·3
41	40·8	04·3	101	100·4	10·6	161	160·1	16·8	221	219·8	23·1	281	279·5	29·4
42	41·8	04·4	102	101·4	10·7	162	161·1	16·9	222	220·8	23·2	282	280·5	29·5
43	42·8	04·5	103	102·4	10·8	163	162·1	17·0	223	221·8	23·3	283	281·4	29.6
44	43·8	04·6	104	103·4	10·9	164	163·1	17·1	224	222·8	23·4	284	282·4	29.7
45	44·8	04·7	105	104·4	11·0	165	164·1	17·2	225	223·8	23·5	285	283·4	29·8
46	45·7	04·8	106	105·4	11·1	166	165·1	17·4	226	224·8	23·6	286	284·4	29·9
47	46·7	04·9	107	106·4	11·2	167	166·1	17·5	227	225·8	23·7	287	285·4	30·0
48	47·7	05·0	108	107·4	11·3	168	167·1	17·6	228	226·8	23·8	288	286·4	30·1
49	48·7	05·1	109	108·4	11·4	169	168·1	17·7	229	227·7	23·9	289	287·4	30·2
50	49·7	05·2	110	109·4	11·5	170	169·1	17·8	230	228·7	24·0	290	288·4	30·3
51	50·7	05·3	111	110·4	11·6	171	170·1	17·9	231	229·7	24·1	291	289·4	30·4
52	51·7	05·4	112	111·4	11·7	172	171·1	18·0	232	230·7	24·3	292	290·4	30·5
53	52·7	05·5	113	112·4	11·8	173	172·1	18·1	233	231·7	24·4	293	291·4	30·6
54	53·7	05·6	114	113·4	11·9	174	173·0	18·2	234	232·7	24·5	294	292·4	30·7
55	54·7	05·7	115	114·4	12·0	175	174·0	18·3	235	233·7	24·6	295	293·4	30·8
56	55·7	05·9	116	115·4	12·1	176	175·0	18·4	236	234·7	24·7	296	294·4	30·9
57	56·7	06·0	117	116·4	12·2	177	176·0	18·5	237	235·7	24·8	297	295·4	31·0
58	57·7	06·1	118	117·4	12·3	178	177·0	18·6	238	236·7	24·9	298	296·4	31·1
59	58·7	06·2	119	118·3	12·4	179	178·0	18·7	239	237·7	25·0	299	297·4	31·3
60	59·7	06·3	120	119·3	12·5	180	179·0	18·8	240	238·7	25·1	300	298·4	31·4
Dist.	Dep.	Lat.	Dist.	Dep.	Lat.	Dist.	Dep.	Lat.	Dist.	Dep.	Lat.	Dist.	Dep.	Lat.

Dist.	Lat.	Dep.	Dist.	Lat.	Dep.	Dist.	Lat.	Dep.	Dist.	Lat.	Dep.	Dist.	Lat.	Dep.
1	01·0	00·1	61	60·5	07·4	121	120·1	14·7	181	179·7	22·1	241	239·2	29·4
2	02·0	00·2	62	61·5	07·6	122	121·1	14·9	182	180·6	22·2	242	240·2	29·5
3	03·0	00·4	63	62·5	07·7	123	122·1	15·0	183	181·6	22·3	243	241·2	29·6
4	04·0	00·5	64	63·5	07·8	124	123·1	15·1	184	182·6	22·4	244	242·2	29·7
5	05·0	00·6	65	64·5	07·9	125	124·1	15·2	185	183·6	22·5	245	243·2	29·9
6	06·0	00·7	66	65·5	08·0	126	125·1	15·4	186	184·6	22·7	246	244·2	30·0
7	06·9	00·9	67	66·5	08·2	127	126·1	15·5	187	185·6	22·8	247	245·2	30·1
8	07·9	01·0	68	67·5	08·3	128	127·0	15·6	188	186·6	22·9	248	246·2	30·2
9	08·9	01·1	69	68·5	08·4	129	128·0	15·7	189	187·6	23·0	249	247·1	30·3
10	09·9	01·2	70	69·5	08·5	130	129·0	15·8	190	188·6	23·2	250	248·1	30·5
11	10·9	01·3	71	70·5	08·7	131	130·0	16·0	191	189·6	23·3	251	249·1	30·6
12	11·9	01·5	72	71·5	08·8	132	131·0	16·1	192	190·6	23·4	252	250·1	30·7
13	12·9	01·6	73	72·5	08·9	133	132·0	16·2	193	191·6	23·5	253	251·1	30·8
14	13·9	01·7	74	73·4	09·0	134	133·0	16·3	194	192·6	23·6	254	252·1	31·0
15	14·9	01·8	75	74·4	09·1	135	134·0	16·5	195	193·5	23·8	255	253·1	31·1
16	15·9	01·9	76	75·4	09·3	136	135·0	16·6	196	194·5	23·9	256	254·1	31·2
17	16·9	02·1	77	76·4	09·4	137	136·0	16·7	197	195·5	24·0	257	255·1	31·3
18	17·9	02·2	78	77·4	09·5	138	137·0	16·8	198	196·5	24·1	258	256·1	31·4
19	18·9	02·3	79	78·4	09·6	139	138·0	16·9	199	197·5	24·3	259	257·1	31·6
20	19·9	02·4	80	79·4	09·7	140	139·0	17·1	200	198·5	24·4	260	258·1	31·7
21	20·8	02·6	81	80·4	09·9	141	139·9	17.2	201	199·5	24·5	261	259·1	31·8
22	21·8	02·7	82	81·4	10·0	142	140·9	17·3	202	200·5	24·6	262	260·0	31·9
23	22·8	02·8	83	82·4	10·1	143	141·9	17·4	203	201·5	24·7	263	261·0	32·1
24	23·8	02·9	84	83·4	10·2	144	142·9	17·5	204	202·5	24·9	264	262·0	32·2
25	24·8	03·0	85	84·4	10·4	145	143·9	17·7	205	203·5	25·0	265	263·0	32·3
26	25·8	03·2	86	85·4	10·5	146	144·9	17·8	206	204·5	25·1	266	264·0	32·4
27	26·8	03·3	87	86·4	10·6	147	145·9	17·9	207	205·5	25·2	267	265·0	32·5
28	27·8	03·4	88	87·3	10·7	148	146·9	18·0	208	206·4	25·3	268	266·0	32·7
29	28·8	03·5	89	88·3	10·8	149	147·9	18·2	209	207·4	25·5	269	267·0	32·8
30	29·8	03·7	90	89·3	11·0	150	148·9	18·3	210	208·4	25·6	270	268·0	32·9
31	30·8	03·8	91	90·3	11·1	151	149·9	18·4	211	209·4	25·7	271	269·0	33·0
32	31·8	03·9	92	91·3	11·2	152	150·9	18·5	212	210·4	25·8	272	270·0	33·1
33	32·8	04·0	93	92·3	11·3	153	151·9	18·6	213	211·4	26·0	273	271·0	33·3
34	33·7	04·1	94	93·3	11·5	154	152·9	18·8	214	212·4	26·1	274	272·0	33·4
35	34·7	04·3	95	94·3	11·6	155	153·8	18·9	215	213·4	26·2	275	273·0	33·5
36	35·7	04·4	96	95·3	11·7	156	154·8	19·0	216	214·4	26·3	276	273·9	33·6
37	36·7	04·5	97	96·3	11·8	157	155·8	19·1	217	215·4	26·4	277	274·9	33·8
38	37·7	04·6	98	97·3	11·9	158	156·8	19·3	218	216·4	26·6	278	275·9	33·9
39	38·7	04·8	99	98·3	12·1	159	157·8	19·4	219	217·4	26·7	279	276·9	34·0
40	39·7	04·9	100	99·3	12·2	160	158·8	19·5	220	218·4	26·8	280	277·9	34·1
41	40·7	05·0	101	100·2	12·3	161	159·8	19·6	221	219·4	26·9	281	278·9	34·2
42	41·7	05·1	102	101·2	12·4	162	160·8	19·7	222	220·3	27·1	282	279·9	34·4
43	42·7	05·2	103	102·2	12·6	163	161·8	19·9	223	221·3	27·2	283	280·9	34·5
44	43·7	05·4	104	103·2	12·7	164	162·8	20·0	224	222·3	27·3	284	281·9	34·6
45	44·7	05·5	105	104·2	12·8	165	163·8	20·1	225	223·3	27·4	285	282·9	34·7
46	45·7	05·6	106	105·2	12·9	166	164·8	20·2	226	224·3	27·5	286	283·9	34·9
47	46·6	05·7	107	106·2	13·0	167	165·8	20·4	227	225·3	27·7	287	284·9	35·0
48	47·6	05·8	108	107·2	13·2	168	166·7	20·5	228	226·3	27·8	288	285·9	35·1
49	48·6	06·0	109	108·2	13·3	169	167·7	20·6	229	227·3	27·9	289	286·8	35·2
50	49·6	06·1	110	109·2	13·4	170	168·7	20·7	230	228·3	28·0	290	287·8	35·3
51	50·6	06·2	111	110·2	13·5	171	169·7	20·8	231	229·3	28·2	291	288·8	35·5
52	51·6	06·3	112	111·2	13·6	172	170·7	21·0	232	230·3	28·3	292	289·8	35·6
53	52·6	06·5	113	112·2	13·8	173	171·7	21·1	233	231·3	28·4	293	290·8	35·7
54	53·6	06·6	114	113·2	13·9	174	172·7	21·2	234	232·3	28·5	294	291·8	35·8
55	54·6	06·7	115	114·1	14·0	175	173·7	21·3	235	233·2	28·6	295	292·8	36·0
56	55·6	06·8	116	115·1	14·1	176	174·7	21·4	236	234·2	28·8	296	293·8	36·1
57	56·6	06·9	117	116·1	14·3	177	175·7	21·6	237	235·2	28·9	297	294·8	36·2
58	57·6	07·1	118	117·1	14·4	178	176·7	21·7	238	236·2	29·0	298	295·8	36·3
59	58·6	07·2	119	118·1	14·5	179	177·7	21·8	239	237·2	29·1	299	296·8	36·4
60	59·6	07·3	120	119·1	14·6	180	178·7	21·9	240	238·2	29·2	300	297·8	36·6
Dist.	Dep.	Lat.	Dist.	Dep.	Lat.	Dist.	Dep.	Lat.	Dist.	Dep.	Lat.	Dist.	Dep.	Lat.

FOR 83 DEGREES.

Dist	Lat.	Dep.	Dist.	Lat.	Dep.	Dist.	Lat.	Dep.	Dist.	Lat.	Dep.	Dist.	Lat.	Dep.
1	01·0	00·1	61	60·4	08·5	121	119·8	16·8	181	179·2	25·2	241	238·7	33·5
2	02·0	00·3	62	61·4	08·6	122	120·8	17·0	182	180·2	25·3	242	239·6	33·7
3	03·0	00·4	63	62·4	08·8	123	121·8	17·1	183	181·2	25·5	243	240·6	33·8
4	04·0	00·6	64	63·4	08·9	124	122·8	17·3	184	182·2	25·6	244	241·6	34·0
5	05·0	00·7	65	64·4	09·0	125	123·8	17·4	185	183·2	25·7	245	242·6	34·1
6	05·9	00·8	66	65·4	09·2	126	124·8	17·5	186	184·2	25·9	246	243·6	34·2
7	06·9	01·0	67	66·3	09·3	127	125·8	17·7	187	185·2	26·0	247	244·6	34·4
8	07·9	01·1	68	67·3	09·5	128	126·8	17·8	188	186·2	26·2	248	245·6	34·5
9	08·9	01·3	69	68·3	09·6	129	127·7	18·0	189	187·2	26·3	249	246·6	34·7
10	09·9	01·4	70	69·3	09·7	130	128·7	18·1	190	188·2	26·4	250	247·6	34·8
11	10·9	01·5	71	70·3	09·9	131	129·7	18·2	191	189·1	26·6	251	248·6	34·9
12	11·9	01·7	72	71·3	10·0	132	130·7	18·4	192	190·1	26·7	252	249·5	35·1
13	12·9	01·8	73	72·3	10·2	133	131·7	18·5	193	191·1	26·9	253	250·5	35·2
14	13·9	01·9	74	73·3	10·3	134	132·7	18·6	194	192·1	27·0	254	251·5	35·3
15	14·8	02·1	75	74·3	10·4	135	133·7	18·8	195	193·1	27·1	255	252·5	35·5
16	15·8	02·2	76	75·3	10·6	136	134·7	18·9	196	194·1	27·3	256	253·5	35·6
17	16·8	02·4	77	76·3	10·7	137	135·7	19·1	197	195·1	27·4	257	254·5	35·8
18	17·8	02·5	78	77·2	10·9	138	136·7	19·2	198	196·1	27·6	258	255·5	35·9
19	18·8	02·6	79	78·2	11·0	139	137·7	19·3	199	197·1	27·7	259	256·5	36·0
20	19·8	02·8	80	79·2	11·1	140	138·6	19·5	200	198·1	27·8	260	257·5	36·2
21	20·8	02·9	81	80·2	11·3	141	139·6	19·6	201	199·0	28·0	261	258·5	36·3
22	21·8	03·1	82	81·2	11·4	142	140·6	19·8	202	200·0	28·1	262	259·5	36·5
23	22·8	03·2	83	82·2	11·6	143	141·6	19·9	203	201·0	28·3	263	260·4	36·6
24	23·8	03·3	84	83·2	11·7	144	142·6	20·0	204	202·0	28·4	264	261·4	36·7
25	24·8	03·5	85	84·2	11·8	145	143·6	20·2	205	203·0	28·5	265	262·4	36·9
26	25·7	03·6	86	85·2	12·0	146	144·6	20·3	206	204·0	28·7	266	263·4	37·0
27	26·7	03·8	87	86·2	12·1	147	145·6	20·5	207	205·0	28·8	267	264·4	37·2
28	27·7	03·9	88	87·1	12·2	148	146·6	20·6	208	206·0	28·9	268	265·4	37·3
29	28·7	04·0	89	88·1	12·4	149	147·5	20·7	209	207·0	29·1	269	266·4	37·4
30	29·7	04·2	90	89·1	12·5	150	148·5	20·9	210	208·0	29·2	270	267·4	37·6
31	30·7	04·3	91	90·1	12·7	151	149·5	21·0	211	208·9	29·4	271	268·4	37·7
32	31·7	04·5	92	91·1	12·8	152	150·5	21·2	212	209·9	29·5	272	269·4	37·9
33	32·7	04·6	93	92·1	12·9	153	151·5	21·3	213	210·9	29·6	273	270·3	38·0
34	33·7	04·7	94	93·1	13·1	154	152·5	21·4	214	211·9	29·8	274	271·3	38·1
35	34·7	04·9	95	94·1	13·2	155	153·5	21·6	215	212·9	29·9	275	272·3	38·3
36	35·6	05·0	96	95·1	13·4	156	154·5	21·7	216	213·9	30·1	276	273·3	38·4
37	36·6	05·1	97	96·1	13·5	157	155·5	21·9	217	214·9	30·2	277	274·3	38·6
38	37·6	05·3	98	97·0	13·6	158	156·5	22·0	218	215·9	30·3	278	275·3	38·7
39	38·6	05·4	99	98·0	13·8	159	157·5	22·1	219	216·9	30·5	279	276·3	38·8
40	39·6	05·6	100	99·0	13·9	160	158·4	22·3	220	217·9	30·6	280	277·3	39·0
41	40·6	05·7	101	100·0	14·1	161	159·4	22·4	221	218·8	30·8	281	278·3	39·1
42	41·6	05·8	102	101·0	14·2	162	160·4	22·5	222	219·8	30·9	282	279·3	39·2
43	42·6	06·0	103	102·0	14·3	163	161·4	22·7	223	220·8	31·0	283	280·2	39·4
44	43·6	06·1	104	103·0	14·5	164	162·4	22·8	224	221·8	31·2	284	281·2	39·5
45	44·6	06·3	105	104·0	14·6	165	163·4	23·0	225	222·8	31·3	285	282·2	39·7
46	45·6	06·4	106	105·0	14·8	166	164·4	23·1	226	223·8	31·5	286	283·2	39·8
47	46·5	06·5	107	106·0	14·9	167	165·4	23·2	227	224·8	31·6	287	284·2	39·9
48	47·5	06·7	108	106·9	15·0	168	166·4	23·4	228	225·8	31·7	288	285·2	40·1
49	48·5	06·8	109	107·9	15·2	169	167·4	23·5	229	226·8	31·9	289	286·2	40·2
50	49·5	07·0	110	108·9	15·3	170	168·3	23·7	230	227·8	32·0	290	287·2	40·4
51	50·5	07·1	111	109·9	15·4	171	169·3	23·8	231	228·8	32·1	291	288·2	40·5
52	51·5	07·2	112	110·9	15·6	172	170·3	23·9	232	229·7	32·3	292	289·2	40·6
53	52·5	07·4	113	111·9	15·7	173	171·3	24·1	233	230·7	32·4	293	290·1	40·8
54	53·5	07·5	114	112·9	15·9	174	172·3	24·2	234	231·7	32·6	294	291·1	40·9
55	54·5	07·7	115	113·9	16·0	175	173·3	24·4	235	232·7	32·7	295	292·1	41·1
56	55·5	07·8	116	114·9	16·1	176	174·3	24·5	236	233·7	32·8	296	293·1	41·2
57	56·4	07·9	117	115·9	16·3	177	175·3	24·6	237	234·7	33·0	297	294·1	41·3
58	57·4	08·1	118	116·9	16·4	178	176·3	24·8	238	235·7	33·1	298	295·1	41·5
59	58·4	08·2	119	117·8	16·6	179	177·3	24·9	239	236·7	33·3	299	296·1	41·6
60	59·4	08·4	120	118·8	16·7	180	178·2	25·1	240	237·7	33·4	300	297·1	41·8
Dist.	Dep.	Lat.	Dist.	Dep.	Lat.	Dist.	Dep.	Lat.	Dist.	Dep.	Lat.	Dist.	Dep.	Lat.

FOR 82 DEGREES.

Dist.	Lat.	Dep.	Dist.	Lat.	Dep.	Dist.	Lat.	Dep.	Dist.	Lat.	Dep.	Dist.	Lat.	Dep.
1	01·0	00·2	61	60·2	09·5	121	119·5	18·9	181	178·8	28·3	241	238·0	37·7
2	02·0	00·3	62	61·2	09·7	122	120·5	19·1	182	179·8	28·5	242	239·0	37·9
3	03·0	00·5	63	62·2	09·9	123	121·5	19·2	183	18[illegible]·7	28·6	243	240·0	38·0
4	04·0	00·6	64	63·2	10·0	124	122·5	19·4	184	181·7	28·8	244	241·0	38·2
5	04·9	00·8	65	64·2	10·2	125	123·5	19·6	185	182·7	28·9	245	242·0	38·3
6	05·9	00·9	66	65·2	10·3	126	124·4	19·7	186	183·7	29·1	246	243·0	38·5
7	06·9	01·1	67	66·2	10·5	127	125·4	19·9	187	184·7	29·3	247	244·0	38·6
8	07·9	01·3	68	67·2	10·6	128	126·4	20·0	188	185·7	29·4	248	244·9	38·8
9	08·9	01·4	69	68·2	10·8	129	127·4	20·2	189	186·7	29·6	249	245·9	39·0
10	09·9	01·6	70	69·1	11·0	130	128·4	20·3	190	187·7	29·7	250	246·9	39·1
11	10·9	01·7	71	70·1	11·1	131	129·4	20·5	191	188·6	29·9	251	247·9	39·3
12	11·9	01·9	72	71·1	11·3	132	130·4	20·6	192	189·6	30·0	252	248·9	39·4
13	12·8	02·0	73	72·1	11·4	133	131·4	20·8	193	190·6	30·2	253	249·9	39·6
14	13·8	02·2	74	73·1	11·6	134	132·4	21·0	194	191·6	3[illegible]·3	254	250·9	39·7
15	14·8	02·3	75	74·1	11·7	135	133·3	21·1	195	192·6	30·5	255	251·9	39·9
16	15·8	02·5	76	75·1	11·9	136	134·3	21·3	196	193·6	30·7	256	252·8	40·0
17	16·8	02·7	77	76·1	12·0	137	135·3	21·4	197	194·6	30·8	257	253·8	40·2
18	17·8	02·8	78	77·0	12·2	138	136·3	21·6	198	195·6	31·0	258	254·8	40·4
19	18·8	03·0	79	78·0	12·4	139	137·3	21·7	199	196·5	31·1	259	2[illegible]·8	40·5
20	19·8	03·1	80	79·0	12·5	140	138·3	21·9	200	197·5	31·3	260	2[illegible]·8	40·7
21	20·7	03·3	81	80·0	12·7	141	139·3	22·1	201	198·5	31·4	261	257·8	40·8
22	21·7	03·4	82	81·0	12·8	142	140·3	22·2	202	199·5	31·6	262	258·8	41·0
23	22·7	03·6	83	82·0	13·0	143	141·2	22·4	203	200·5	31·8	263	259·8	41·1
24	23·7	03·8	84	83·0	13·1	144	142·2	22·5	204	201·5	31·9	264	260·7	41·3
25	24·7	03·9	85	84·0	13·3	145	143·2	22·7	205	202·5	32·1	265	261·7	41·5
26	25·7	04·1	86	84·9	13·5	146	144·2	22·8	206	203·5	32·2	266	262·7	41·6
27	26·7	04·2	87	85·9	13·6	147	145·2	23·0	207	204·5	32·4	267	263·7	41·8
28	27·7	04·4	88	86·9	13·8	148	146·2	23·2	208	205·4	32·5	268	264·7	41·9
29	28·6	04·5	89	87·9	13·9	149	147·2	23·3	209	206·4	32·7	269	265·7	42·1
30	29·6	04·7	90	88·9	14·1	150	148·2	23·5	210	207·4	32·9	270	266·7	42·2
31	30·6	04·8	91	89·9	14·2	151	149·1	23·6	211	208·4	33·0	271	267·7	42·4
32	31·6	05·0	92	90·9	14·4	152	150·1	23·8	212	209·4	33·2	272	268·7	42·6
33	32·6	05·2	93	91·9	14·5	153	151·1	23·9	213	210·4	33·3	273	269·7	42·7
34	33·6	05·3	94	92·8	14·7	154	152·1	24·1	214	211·4	33·5	274	270·6	42·9
35	34·6	05·5	95	93·8	14·9	155	153·1	24·2	215	212·4	33·6	275	271·6	43·0
36	35·6	05·6	96	94·8	15·0	156	154·1	24·4	216	213·3	33·8	276	272·6	43·2
37	36·5	05·8	97	95·8	15·2	157	155·1	24·6	217	214·3	33·9	277	273·6	43·3
38	37·5	05·9	98	96·8	15·3	158	156·1	24·7	218	215·3	34·1	278	274·6	43·5
39	38·5	06·1	99	97·8	15·5	159	157·0	24·9	219	216·3	34·3	279	275·6	43·6
40	39·5	06·3	100	98·8	15·6	160	158·0	25·0	220	217·3	34·4	280	276·6	43·8
41	40·5	06·4	101	99·8	15·8	161	159·0	25·2	221	218·3	34·6	281	277·5	44·0
42	41·5	06·6	102	100·7	16·0	162	160·0	25·3	222	219·3	34·7	282	278·5	44·1
43	42·5	06·7	103	101·7	16·1	163	161·0	25·5	223	220·3	34·9	283	279·5	44·3
44	43·5	06·9	104	102·7	16·3	164	162·0	25·7	224	221·2	35·0	284	280·5	44·4
45	44·4	07·0	105	103·7	16·4	165	163·0	25·8	225	222·2	35·2	285	281·5	44·6
46	45·4	07·2	106	104·7	16·6	166	164·0	26·0	226	223·2	35·4	286	282·5	44·7
47	46·4	07·4	107	105·7	16·7	167	164·9	26·1	227	224·2	35·5	287	283·5	44·9
48	47·4	07·5	108	106·7	16·9	168	165·9	26·3	228	225·2	35·7	288	284·5	45·1
49	48·4	07·7	109	107·7	17·1	169	166·9	26·4	229	226·2	35·8	289	285·4	45·2
50	49·4	07·8	110	108·6	17·2	170	167·9	26·6	230	227·2	36·0	290	286·4	45·4
51	50·4	08·0	111	109·6	17·4	171	168·9	26·8	231	228·2	36·1	291	287·4	45·5
52	51·4	08·1	112	110·6	17·5	172	169·9	26·9	232	229·1	36·3	292	288·4	45·7
53	52·3	08·3	113	111·6	17·7	173	170·9	27·1	233	230·1	36·4	293	289·4	45·8
54	53·3	08·4	114	112·6	17·8	174	171·9	27·2	234	231·1	36·6	294	290·4	46·0
55	54·3	08·6	115	113·6	18·0	175	172·8	27·4	235	232·1	36·8	295	291·4	46·1
56	55·3	08·8	116	114·6	18·1	176	173·8	27·5	236	233·1	36·[illegible]	296	292·4	46·3
57	56·3	08·9	117	115·6	18·3	177	174·8	27·7	237	234·1	37·1	297	29[illegible]·4	46·5
58	57·3	09·1	118	116·5	18·5	178	175·8	27·8	238	235·1	37·2	298	294·3	46·6
59	58·3	09·2	119	117·5	18·6	179	176·8	28·0	239	236·1	37·4	299	295·3	46·8
60	59·3	09·4	120	118·5	18·8	180	177·8	28·2	240	237·0	37·5	300	296·3	46·9
Dist.	Dep.	Lat.	Dist.	Dep.	Lat.	Dist.	Dep.	Lat.	Dist.	Dep.	Lat.	Dist.	Dep.	Lat.

FOR 81 DEGREES

Dist.	Lat.	Dep.	Dist.	Lat.	Dep.	Dist.	Lat.	Dep.	Dist.	Lat.	Dep.	Dist.	Lat.	Dep.
1	01·0	00·2	61	60·1	10·6	121	119·2	21·0	181	178·3	31·4	241	237·3	41·8
2	02·0	00·3	62	61·1	10·8	122	120·1	21·2	182	179·2	31·6	242	238·3	42·0
3	03·0	00·5	63	62·0	10·9	123	121·1	21·4	183	180·2	31·8	243	239·3	42·2
4	03·9	00·7	64	63·0	11·1	124	122·1	21·5	184	181·2	32·0	244	240·3	42·4
5	04·9	00·9	65	64·0	11·3	125	123·1	21·7	185	182·2	32·1	245	241·3	42·5
6	05·9	01·0	66	65·0	11·5	126	124·1	21·9	186	183·2	32·3	246	242·3	42·7
7	06·9	01·2	67	66·0	11·6	127	125·1	22·1	187	184·2	32·5	247	243·2	42·9
8	07·9	01·4	68	67·0	11·8	128	126·1	22·2	188	185·1	32·6	248	244·2	43·1
9	08·9	01·6	69	68·0	12·0	129	127·0	22·4	189	186·1	32·8	249	245·2	43·2
10	09·8	01·7	70	68·9	12·2	130	128·0	22·6	190	187·1	33·0	250	246·2	43·4
11	10·8	01·9	71	69·9	12·3	131	129·0	22·7	191	188·1	33·2	251	247·2	43·6
12	11·8	02·1	72	70·9	12·5	132	130·0	22·9	192	189·1	33·3	252	248·2	43·8
13	12·8	02·3	73	71·9	12·7	133	131·0	23·1	193	190·1	33·5	253	249·2	43·9
14	13·8	02·4	74	72·9	12·8	134	132·0	23·3	194	191·0	33·7	254	250·1	44·1
15	14·8	02·6	75	73·9	13·0	135	132·9	23·4	195	192·0	33·9	255	251·1	44·3
16	15·8	02·8	76	74·8	13·2	136	133·9	23·6	196	193·0	34·0	256	252·1	44·5
17	16·7	03·0	77	75·8	13·4	137	134·9	23·8	197	194·0	34·2	257	253·1	44·6
18	17·7	03·1	78	76·8	13·5	138	135·9	24·0	198	195·0	34·4	258	254·1	44·8
19	18·7	03·3	79	77·8	13·7	139	136·9	24·1	199	196·0	34·6	259	255·1	45·0
20	19·7	03·5	80	78·8	13·9	140	137·9	24·3	200	197·0	34·7	260	256·1	45·1
21	20·7	03·6	81	79·8	14·1	141	138·9	24·5	201	197·9	34·9	261	257·0	45·3
22	21·7	03·8	82	80·8	14·2	142	139·8	24·7	202	198·9	35·1	262	258·0	45·5
23	22·7	04·0	83	81·7	14·4	143	140·8	24·8	203	199·9	35·3	263	259·0	45·7
24	23·6	04·2	84	82·7	14·6	144	141·8	25·0	204	200·9	35·4	264	260·0	45·8
25	24·6	04·3	85	83·7	14·8	145	142·8	25·2	205	201·9	35·6	265	261·0	46·0
26	25·6	04·5	86	84·7	14·9	146	143·8	25·4	206	202·9	35·8	266	262·0	46·2
27	26·6	04·7	87	85·7	15·1	147	144·8	25·5	207	203·9	35·9	267	262·9	46·4
28	27·6	04·9	88	86·7	15·3	148	145·8	25·7	208	204·8	36·1	268	263·9	46·5
29	28·6	05·0	89	87·6	15·5	149	146·7	25·9	209	205·8	36·3	269	264·9	46·7
30	29·5	05·2	90	88·6	15·6	150	147·7	26·0	210	206·8	36·5	270	265·9	46·9
31	30·5	05·4	91	89·6	15·8	151	148·7	26·2	211	207·8	36·6	271	266·9	47·1
32	31·5	05·6	92	90·6	16·0	152	149·7	26·4	212	208·8	36·8	272	267·9	47·2
33	32·5	05·7	93	91·6	16·1	153	150·7	26·6	213	209·8	37·0	273	268·9	47·4
34	33·5	05·9	94	92·6	16·3	154	151·7	26·7	214	210·7	37·2	274	269·8	47·6
35	34·5	06·1	95	93·6	16·5	155	152·6	26·9	215	211·7	37·3	275	270·8	47·8
36	35·5	06·3	96	94·5	16·7	156	153·6	27·1	216	212·7	37·5	276	271·8	47·9
37	36·4	06·4	97	95·5	16·8	157	154·6	27·3	217	213·7	37·7	277	272·8	48·1
38	37·4	06·6	98	96·5	17·0	158	155·6	27·4	218	214·7	37·9	278	273·8	48·3
39	38·4	06·8	99	97·5	17·2	159	156·6	27·6	219	215·7	38·0	279	274·8	48·4
40	39·4	06·9	100	98·5	17·4	160	157·6	27·8	220	216·7	38·2	280	275·7	48·6
41	40·4	07·1	101	99·5	17·5	161	158·6	28·0	221	217·6	38·4	281	276·7	48·8
42	41·4	07·3	102	100·5	17·7	162	159·5	28·1	222	218·6	38·5	282	277·7	49·0
43	42·3	07·5	103	101·4	17·9	163	160·5	28·3	223	219·6	38·7	283	278·7	49·1
44	43·3	07·6	104	102·4	18·1	164	161·5	28·5	224	220·6	38·9	284	279·7	49·3
45	44·3	07·8	105	103·4	18·2	165	162·5	28·7	225	221·6	39·1	285	280·7	49·5
46	45·3	08·0	106	104·4	18·4	166	163·5	28·8	226	222·6	39·2	286	281·7	49·7
47	46·3	08·2	107	105·4	18·6	167	164·5	29·0	227	223·6	39·4	287	282·6	49·8
48	47·3	08·3	108	106·4	18·8	168	165·4	29·2	228	224·5	39·6	288	283·6	50·0
49	48·3	08·5	109	107·3	18·9	169	166·4	29·3	229	225·5	39·8	289	284·6	50·2
50	49·2	08·7	110	108·3	19·1	170	167·4	29·5	230	226·5	39·9	290	285·6	50·4
51	50·2	08·9	111	109·3	19·3	171	168·4	29·7	231	227·5	40·1	291	286·6	50·5
52	51·2	09·0	112	110·3	19·4	172	169·4	29·9	232	228·5	40·3	292	287·6	50·7
53	52·2	09·2	113	111·3	19·6	173	170·4	30·0	233	229·5	40·5	293	288·5	50·9
54	53·2	09·4	114	112·3	19·8	174	171·4	30·2	234	230·4	40·6	294	289·5	51·1
55	54·2	09·6	115	113·3	20·0	175	172·3	30·4	235	231·4	40·8	295	290·5	51·2
56	55·1	09·7	116	114·2	20·1	176	173·3	30·6	236	232·4	41·0	296	291·5	51·4
57	56·1	09·9	117	115·2	20·3	177	174·3	30·7	237	233·4	41·2	297	292·5	51·6
58	57·1	10·1	118	116·2	20·5	178	175·3	30·9	238	234·4	41·3	298	293·5	51·7
59	58·1	10·2	119	117·2	20·7	179	176·3	31·1	239	235·4	41·5	299	294·5	51·9
60	59·1	10·4	120	118·2	20·8	180	177·3	31·3	240	236·4	41·7	300	295·4	52·1
Dist.	Dep.	Lat.	Dist.	Dep.	Lat.	Dist.	Dep.	Lat.	Dist.	Dep.	Lat.	Dist.	Dep.	Lat.

Dist.	Lat.	Dep.	Dist.	Lat.	Dep.	Dist.	Lat.	Dep.	Dist.	Lat.	Dep.	Dist.	Lat.	Dep.
1	01·0	00·2	61	59·9	11·6	121	118·8	23·1	181	177·7	34·5	241	236·6	46·0
2	02·0	00·4	62	60·9	11·8	122	119·8	23·3	182	178·7	34·7	242	237·6	46·2
3	02·9	00·6	63	61·8	12·0	123	120·7	23·5	183	179·6	34·9	243	238·5	46·4
4	03·9	00·8	64	62·8	12·2	124	121·7	23·7	184	180·6	35·1	244	239·5	46·6
5	04·9	01·0	65	63·8	12·4	125	122·7	23·9	185	181·6	35·3	245	240·5	46·7
6	05·9	01·1	66	64·8	12·6	126	123·7	24·0	186	182·6	35·5	246	241·5	46·9
7	06·9	01·3	67	65·8	12·8	127	124·7	24·2	187	183·6	35·7	247	242·5	47·1
8	07·9	01·5	68	66·8	13·0	128	125·6	24·4	188	184·5	35·9	248	243·4	47·3
9	08·8	01·7	69	67·7	13·2	129	126·6	24·6	189	185·5	36·1	249	244·4	47·5
10	09·8	01·9	70	68·7	13·4	130	127·6	24·8	190	186·5	36·3	250	245·4	47·7
11	10·8	02·1	71	69·7	13·5	131	128·6	25·0	191	187·5	36·4	251	246·4	47·9
12	11·8	02·3	72	70·7	13·7	132	129·6	25·2	192	188·5	36·6	252	247·4	48·1
13	12·8	02·5	73	71·7	13·9	133	130·6	25·4	193	189·5	36·8	253	248·4	48·3
14	13·7	02·7	74	72·6	14·1	134	131·5	25·6	194	190·4	37·0	254	249·3	48·5
15	14·7	02·9	75	73·6	14·3	135	132·5	25·8	195	191·4	37·2	255	250·3	48·7
16	15·7	03·1	76	74·6	14·5	136	133·5	26·0	196	192·4	37·4	256	251·3	48·8
17	16·7	03·2	77	75·6	14·7	137	134·5	26·1	197	193·4	37·6	257	252·3	49·0
18	17·7	03·4	78	76·6	14·9	138	135·5	26·3	198	194·4	37·8	258	253·3	49·2
19	18·7	03·6	79	77·5	15·1	139	136·4	26·5	199	195·3	38·0	259	254·2	49·4
20	19·6	03·8	80	78·5	15·3	140	137·4	26·7	200	196·3	38·2	260	255·2	49·6
21	20·6	04·0	81	79·5	15·5	141	138·4	26·9	201	197·3	38·4	261	256·2	49·8
22	21·6	04·2	82	80·5	15·6	142	139·4	27·1	202	198·3	38·5	262	257·2	50·0
23	22·6	04·4	83	81·5	15·8	143	140·4	27·3	203	199·3	38·7	263	558·2	50·2
24	23·6	04·6	84	82·5	16·0	144	141·4	27·5	204	200·3	38·9	264	259·1	50·4
25	24·5	04·8	85	83·4	16·2	145	142·3	27·7	205	201·2	39·1	265	260·1	50·6
26	25·5	05·0	86	84·4	16·4	146	143·3	27·9	206	202·2	39·3	266	261·1	50·8
27	26·5	05·2	87	85·4	16·6	147	144·3	28·0	207	203·2	39·5	267	262·1	50·9
28	27·5	05·3	88	86·4	16·8	148	145·3	28·2	208	204·2	39·7	268	263·1	51·1
29	28·5	05·5	89	87·4	17·0	149	146·3	28·4	209	205·2	39·9	269	264·1	51·3
30	29·4	05·7	90	88·3	17·2	150	147·2	28·6	210	206·1	40·1	270	265·0	51·5
31	30·4	05·9	91	89·3	17·4	151	148·2	28·8	211	207·1	40·3	271	266·0	51·7
32	31·4	06·1	92	90·3	17·6	152	149·2	29·0	212	208·1	40·5	272	267·0	51·9
33	32·4	06·3	93	91·3	17·7	153	150·2	29·2	213	209·1	40·6	273	268·0	52·1
34	33·4	06·5	94	92·3	17·9	154	151·2	29·4	214	210·1	40·8	274	269·0	52·3
35	34·4	06·7	95	93·3	18·1	155	152·2	29·6	215	211·0	41·0	275	269·9	52·5
36	35·3	06·9	96	94·2	18·3	156	153·1	29·8	216	212·0	41·2	276	270·9	52·7
37	36·3	07·1	97	95·2	18·5	157	154·1	30·0	217	213·0	41·4	277	271·9	52·9
38	37·3	07·3	98	96·2	18·7	158	155·1	30·1	218	214·0	41·6	278	272·9	53·0
39	38·3	07·4	99	97·2	18·9	159	156·1	30·3	219	215·0	41·8	279	273·9	53·2
40	39·3	07·6	100	98·2	19·1	160	157·1	30·5	220	216·0	42·0	280	274·9	53·4
41	40·2	07·8	101	99·1	19·3	161	158·0	30·7	221	216·9	42·2	281	275·8	53·6
42	41·2	08·0	102	100·1	19·5	162	159·0	30·9	222	217·9	42·4	282	276·8	53·8
43	42·2	08·2	103	101·1	19·7	163	160·0	31·1	223	218·9	42·6	283	277·8	54·0
44	43·2	08·4	104	102·1	19·8	164	161·0	31·3	224	219·9	42·7	284	278·8	54·2
45	44·2	08·6	105	103·1	20·0	165	162·0	31·5	225	220·9	42·9	285	279·8	54·4
46	45·2	08·8	106	104·1	20·2	166	163·0	31·7	226	221·8	43·1	286	280·7	54·6
47	46·1	09·0	107	105·0	20·4	167	163·9	31·9	227	222·8	43·3	287	281·7	54·8
48	47·1	09·2	108	106·0	20·6	168	164·9	32·1	228	223·8	43·5	288	282·7	55·0
49	48·1	09·3	109	107·0	20·8	169	165·9	32·2	229	224·8	43·7	289	283·7	55·1
50	49·1	09·5	110	108·0	21·0	170	166·9	32·4	230	225·8	43·9	290	284·7	55·3
51	50·1	09·7	111	109·0	21·2	171	167·9	32·6	231	226·8	44·1	291	285·7	55·5
52	51·0	09·9	112	109·9	21·4	172	168·8	32·8	232	227·7	44·3	292	286·6	55·7
53	52·0	10·1	113	110·9	21·6	173	169·8	33·0	233	228·7	44·5	293	287·6	55·9
54	53·0	10·3	114	111·9	21·8	174	170·8	33·2	234	229·7	44·6	294	288·6	56·1
55	54·0	10·5	115	112·9	21·9	175	171·8	33·4	235	230·7	44·8	295	289·6	56·3
56	55·0	10·7	116	113·9	22·1	176	172·8	33·6	236	231·7	45·0	206	290·6	56·5
57	56·0	10·9	117	114·9	22·3	177	173·7	33·8	237	232·6	45·2	297	291·5	56·7
58	56·9	11·1	118	115·8	22·5	178	174·7	34·0	238	233·6	45·4	298	292·5	56·9
59	57·9	11·3	119	116·8	22·7	179	175·7	34·2	239	234·6	45·6	299	293·5	57·1
60	58·9	11·4	120	117·8	22·9	180	176·7	34·3	240	235·6	45·8	300	294·5	57·2
Dist.	Dep.	Lat.	Dist.	Dep.	Lat.	Dist.	Dep.	Lat.	Dist.	Dep.	Lat.	Dist.	Dep.	Lat.

FOR 79 DEGREES.

Dist.	Lat.	Dep.	Dist.	Lat.	Dep.	Dist.	Lat.	Dep.	Dist.	Lat.	Dep.	Dist.	Lat.	Dep.
1	01·0	00·2	61	59·7	12·7	121	118·4	25·2	181	177·0	37·6	241	235·7	50·1
2	02·0	00·4	62	60·6	12·9	122	119·3	25·4	182	178·0	37·8	242	236·7	50·3
3	02·9	00·6	63	61·6	13·1	123	120·3	25·6	183	179·0	38·0	243	237·7	50·5
4	03·9	00·8	64	62·6	13·3	124	121·3	25·8	184	180·0	38·3	244	238·7	50·7
5	04·9	01·0	65	63·6	13·5	125	122·3	26·0	185	181·0	38·5	245	239·6	50·9
6	05·9	01·2	66	64·6	13·7	126	123·2	26·2	186	181·9	38·7	246	240·6	51·1
7	06·8	01·5	67	65·5	13·9	127	124·2	26·4	187	182·9	38·9	247	241·6	51·4
8	07·8	01·7	68	66·5	14·1	128	125·2	26·6	188	183·9	39·1	248	242·6	51·6
9	08·8	01·9	69	67·5	14·3	129	126·2	26·8	189	184·9	39·3	249	243·6	51·8
10	09·8	02·1	70	68·5	14·6	130	127·2	27·0	190	185·8	39·5	250	244·5	52·0
11	10·8	02·3	71	69·4	14·8	131	128·1	27·2	191	186·8	39·7	251	245·5	52·2
12	11·7	02·5	72	70·4	15·0	132	129·1	27·4	192	187·8	39·9	252	246·5	52·4
13	12·7	02·7	73	71·4	15·2	133	130·1	27·7	193	188·8	40·1	253	247·5	52·6
14	13·7	02·9	74	72·4	15·4	134	131·1	27·9	194	189·8	40·3	254	248·4	52·8
15	14·7	03·1	75	73·4	15·6	135	132·0	28·1	195	190·7	40·5	255	249·4	53·0
16	15·7	03·3	76	74·3	15·8	136	133·0	28·3	196	191·7	40·8	256	250·4	53·2
17	16·6	03·5	77	75·3	16·0	137	134·0	28·5	197	192·7	41·0	257	251·4	53·4
18	17·6	03·7	78	76·3	16·2	138	135·0	28·7	198	193·7	41·2	258	252·4	53·6
19	18·6	04·0	79	77·3	16·4	139	136·0	28·9	199	194·7	41·4	259	253·3	53·8
20	19·6	04·2	80	78·3	16·6	140	136·9	29·1	200	195·6	41·6	260	254·3	54·1
21	20·5	04·4	81	79·2	16·8	141	137·9	29·3	201	196·6	41·8	261	255·3	54·3
22	21·5	04·6	82	80·2	17·0	142	138·9	29·5	202	197·6	42·0	262	256·3	54·5
23	22·5	04·8	83	81·2	17·3	143	139·9	29·7	203	198·6	42·2	263	257·3	54·7
24	23·5	05·0	84	82·2	17·5	144	140·9	29·9	204	199·5	42·4	264	258·2	54·9
25	24·5	05·2	85	83·1	17·7	145	141·8	30·1	205	200·5	42·6	265	259·2	55·1
26	25·4	05·4	86	84·1	17·9	146	142·8	30·4	206	201·5	42·8	266	260·2	55·3
27	26·4	05·6	87	85·1	18·1	147	143·8	30·6	207	202·5	43·0	267	261·2	55·5
28	27·4	05·8	88	86·1	18·3	148	144·8	30·8	208	203·5	43·2	268	262·1	55·7
29	28·4	06·0	89	87·1	18·5	149	145·7	31·0	209	204·4	43·5	269	263·1	55·9
30	29·3	06·2	90	88·0	18·7	150	146·7	31·2	210	205·4	43·7	270	264·1	56·1
31	30·3	06·4	91	89·0	18·9	151	147·7	31·4	211	206·4	43·9	271	265·1	56·3
32	31·3	06·7	92	90·0	19·1	152	148·7	31·6	212	207·4	44·1	272	266·1	56·6
33	32·3	06·9	93	91·0	19·3	153	149·7	31·8	213	208·3	44·3	273	267·0	56·8
34	33·3	07·1	94	91·9	19·5	154	150·6	32·0	214	209·3	44·5	274	268·0	57·0
35	34·2	07·3	95	92·9	19·8	155	151·6	32·2	215	210·3	44·7	275	269·0	57·2
36	35·2	07·5	96	93·9	20·0	156	152·6	32·4	216	211·3	44·9	276	270·0	57·4
37	36·2	07·7	97	94·9	20·2	157	153·6	32·6	217	212·3	45·1	277	270·9	57·6
38	37·2	07·9	98	95·9	20·4	158	154·5	32·9	218	213·2	45·3	278	271·9	57·8
39	38·1	08·1	99	96·8	20·6	159	155·5	33·1	219	214·2	45·5	279	272·9	58·0
40	39·1	08·3	100	97·8	20·8	160	156·5	33·3	220	215·2	45·7	280	273·9	58·2
41	40·1	08·5	101	98·8	21·0	161	157·5	33·5	221	216·2	45·9	281	274·9	58·4
42	41·1	08·7	102	99·8	21·2	162	158·5	33·7	222	217·1	46·2	282	275·8	58·6
43	42·1	08·9	103	100·7	21·4	163	159·4	33·9	223	218·1	46·4	283	276·8	58·8
44	43·0	09·1	104	101·7	21·6	164	160·4	34·1	224	219·1	46·6	284	277·8	59·0
45	44·0	09·4	105	102·7	21·8	165	161·4	34·3	225	220·1	46·8	285	278·8	59·3
46	45·0	09·6	106	103·7	22·0	166	162·4	34·5	226	221·1	47·0	286	279·8	59·5
47	46·0	09·8	107	104·7	22·2	167	163·4	34·7	227	222·0	47·2	287	280·7	59·7
48	47·0	10·0	108	105·6	22·5	168	164·3	34·9	228	223·0	47·4	288	281·7	59·9
49	47·9	10·2	109	106·6	22·7	169	165·3	35·1	229	224·0	47·6	289	282·7	60·1
50	48·9	10·4	110	107·6	22·9	170	166·3	35·3	230	225·0	47·8	290	283·7	60·3
51	49·9	10·6	111	108·6	23·1	171	167·3	35·6	231	226·0	48·0	291	284·6	60·5
52	50·9	10·8	112	109·6	23·3	172	168·2	35·8	232	226·9	48·2	292	285·6	60·7
53	51·8	11·0	113	110·5	23·5	173	169·2	36·0	233	227·9	48·4	293	286·6	60·9
54	52·8	11·2	114	111·5	23·7	174	170·2	36·2	234	228·9	48·7	294	287·6	61·1
55	53·8	11·4	115	112·5	23·9	175	171·2	36·4	235	229·9	48·9	295	288·6	61·3
56	54·8	11·6	116	113·5	24·1	176	172·2	36·6	236	230·8	49·1	296	289·5	61·5
57	55·8	11·9	117	114·4	24·3	177	173·1	36·8	237	231·8	49·3	297	290·5	61·7
58	56·7	12·1	118	115·4	24·5	178	174·1	37·0	238	232·8	49·5	298	291·5	62·0
59	57·7	12·3	119	116·4	24·7	179	175·1	37·2	239	233·8	49·7	299	292·5	62·2
60	58·7	12·5	120	117·4	24·9	180	176·1	37·4	240	234·8	49·9	300	293·4	62·4
Dist.	Dep.	Lat.	Dist.	Dep.	Lat.	Dist.	Dep.	Lat.	Dist.	Dep.	Lat.	Dist.	Dep.	Lat.

Dist.	Lat.	Dep.	Dist.	Lat.	Dep.	Dist.	Lat.	Dep.	Dist.	Lat.	Dep.	Dist.	Lat.	Dep.
1	01·0	00·2	61	59·4	13·7	121	117·9	27·2	181	176·4	40·7	241	234·8	54·2
2	01·9	00·4	62	60·4	13·9	122	118·9	27·4	182	177·3	40·9	242	235·8	54·4
3	02·9	00·7	63	61·4	14·2	123	119·8	27·7	183	178·3	41·2	243	236·8	54·7
4	03·9	00·9	64	62·4	14·4	124	120·8	27·9	184	179·3	41·4	244	237·7	54·9
5	04·9	01·1	65	63·3	14·6	125	121·8	28·1	185	180·3	41·6	245	238·7	55·1
6	05·8	01·3	66	64·3	14·8	126	122·8	28·3	186	181·2	41·8	246	239·7	55·3
7	06·8	01·6	67	65·3	15·1	127	123·7	28·6	187	182·2	42·1	247	240·7	55·6
8	07·8	01·8	68	66·3	15·3	128	124·7	28·8	188	183·2	42·3	248	241·6	55·8
9	08·8	02·0	69	67·2	15·5	129	125·7	29·0	189	184·2	42·5	249	242·6	56·0
10	09·7	02·2	70	68·2	15·7	130	126·7	29·2	190	185·1	42·7	250	243·6	56·2
11	10·7	02·5	71	69·2	16·0	131	127·6	29·5	191	186·1	43·0	251	244·6	56·5
12	11·7	02·7	72	70·2	16·2	132	128·6	29·7	192	187·1	43·2	252	245·5	56·7
13	12·7	02·9	73	71·1	16·4	133	129·6	29·9	193	188·1	43·4	253	246·5	56·9
14	13·6	03·1	74	72·1	16·6	134	130·6	30·1	194	189·0	43·6	254	247·5	57·1
15	14·6	03·4	75	73·1	16·9	135	131·5	30·4	195	190·0	43·9	255	248·5	57·4
16	15·6	03·6	76	74·1	17·1	136	132·5	30·6	196	191·0	44·1	256	249·4	57·6
17	16·6	03·8	77	75·0	17·3	137	133·5	30·8	197	192·0	44·3	257	250·4	57·8
18	17·5	04·0	78	76·0	17·5	138	134·5	31·0	198	192·9	44·5	258	251·4	58·0
19	18·5	04·3	79	77·0	17·8	139	135·4	31·3	199	193·9	44·8	259	252·4	58·3
20	19·5	04·5	80	77·9	18·0	140	136·4	31·5	200	194·9	45·0	260	253·3	58·5
21	20·5	04·7	81	78·9	18·2	141	137·4	31·7	201	195·8	45·2	261	254·3	58·7
22	21·4	04·9	82	79·9	18·4	142	138·4	31·9	202	196·8	45·4	262	255·3	58·9
23	22·4	05·2	83	80·9	18·7	143	139·3	32·2	203	197·8	45·7	263	256·3	59·2
24	23·4	05·4	84	81·8	18·9	144	140·3	32·4	204	198·8	45·9	264	257·2	59·4
25	24·4	05·6	85	82·8	19·1	145	141·3	32·6	205	199·7	46·1	265	258·2	59·6
26	25·3	05·8	86	83·8	19·3	146	142·3	32·8	206	200·7	46·3	266	259·2	59·8
27	26·3	06·1	87	84·8	19·6	147	143·2	33·1	207	201·7	46·6	267	260·2	60·1
28	27·3	06·3	88	85·7	19·8	148	144·2	33·3	208	202·7	46·8	268	261·1	60·3
29	28·3	06·5	89	86·7	20·0	149	145·2	33·5	209	203·6	47·0	269	262·1	60·5
30	29·2	06·7	90	87·7	20·2	150	146·2	33·7	210	204·6	47·2	270	263·1	60·7
31	30·2	07·0	91	88·7	20·5	151	147·1	34·0	211	205·6	47·5	271	264·1	61·0
32	31·2	07·2	92	89·6	20·7	152	148·1	34·2	212	206·6	47·7	272	265·0	61·2
33	32·2	07·4	93	90·6	20·9	153	149·1	34·4	213	207·5	47·9	273	266·0	61·4
34	33·1	07·6	94	91·6	21·1	154	150·1	34·6	214	208·5	48·1	274	267·0	61·6
35	34·1	07·9	95	92·6	21·4	155	151·0	34·9	215	209·5	48·4	275	268·0	61·9
36	35·1	08·1	96	93·5	21·6	156	152·0	35·1	216	210·5	48·6	276	268·9	62·1
37	36·1	08·3	97	94·5	21·8	157	153·0	35·3	217	211·4	48·8	277	269·9	62·3
38	37·0	08·5	98	95·5	22·0	158	154·0	35·5	218	212·4	49·0	278	270·9	62·5
39	38·0	08·8	99	96·5	22·3	159	154·9	35·8	219	213·4	49·3	279	271·8	62·8
40	39·0	09·0	100	97·4	22·5	160	155·9	36·0	220	214·4	49·5	280	272·8	63·0
41	39·9	09·2	101	98·4	22·7	161	156·9	36·2	221	215·3	49·7	281	273·8	63·2
42	40·9	09·4	102	99·4	22·9	162	157·8	36·4	222	216·3	49·9	282	274·8	63·4
43	41·9	09·7	103	100·4	23·2	163	158·8	36·7	223	217·3	50·2	283	275·7	63·7
44	42·9	09·9	104	101·3	23·4	164	159·8	36·9	224	218·3	50·4	284	276·7	63·9
45	43·8	10·1	105	102·3	23·6	165	160·8	37·1	225	219·2	50·6	285	277·7	64·1
46	44·8	10·3	106	103·3	23·8	166	161·7	37·3	226	220·2	50·8	286	278·7	64·3
47	45·8	10·6	107	104·3	24·1	167	162·7	37·6	227	221·2	51·1	287	279·6	64·6
48	46·8	10·8	108	105·2	24·3	168	163·7	37·8	228	222·2	51·3	288	280·6	64·8
49	47·7	11·0	109	106·2	24·5	169	164·7	38·0	229	223·1	51·5	289	281·6	65·0
50	48·7	11·2	110	107·2	24·7	170	165·6	38·2	230	224·1	51·7	290	282·6	65·2
51	49·7	11·5	111	108·2	25·0	171	166·6	38·5	231	225·1	52·0	291	283·5	65·5
52	50·7	11·7	112	109·1	25·2	172	167·6	38·7	232	226·1	52·2	292	284·5	65·7
53	51·6	11·9	113	110·1	25·4	173	168·6	38·9	233	227·0	52·4	293	285·5	65·9
54	52·6	12·1	114	111·1	25·6	174	169·5	39·1	234	228·0	52·6	294	286·5	66·1
55	53·6	12·4	115	112·1	25·9	175	170·5	39·4	235	229·0	52·9	295	287·4	66·4
56	54·6	12·6	116	113·0	26·1	176	171·5	39·6	236	230·0	53·1	296	288·4	66·6
57	55·5	12·8	117	114·0	26·3	177	172·5	39·8	237	230·9	53·3	297	289·4	66·8
58	56·5	13·0	118	115·0	26·5	178	173·4	40·0	238	231·9	53·5	298	290·4	67·0
59	57·5	13·3	119	116·0	26·8	179	174·4	40·3	239	232·9	53·8	299	291·3	67·3
60	58·5	13·5	120	116·9	27·0	180	175·4	40·5	240	233·8	54·0	300	292·3	67·5
Dist.	Dep.	Lat.	Dist.	Dep.	Lat.	Dist.	Dep.	Lat.	Dist.	Dep.	Lat.	Dist.	Dep.	Lat.

Dist.	Lat.	Dep.	Dist.	Lat.	Dep.	Dist.	Lat.	Dep.	Dist.	Lat.	Dep.	Dist.	Lat.	Dep.
1	01·0	00·2	61	59·2	14·8	121	117·4	29·3	181	175·6	43·8	241	233·8	58·3
2	01·9	00·5	62	60·2	15·0	122	118·4	29·5	182	176·6	44·0	242	234·8	58·5
3	02·9	00·7	63	61·1	15·2	123	119·3	29·8	183	177·6	44·3	243	235·8	58·8
4	03·9	01·0	64	62·1	15·5	124	120·3	30·0	184	178·5	44·5	244	236·8	59·0
5	04·9	01·2	65	63·1	15·7	125	121·3	30·2	185	179·5	44·8	245	237·7	59·3
6	05·8	01·5	66	64·0	16·0	126	122·3	30·5	186	180·5	45·0	246	238·7	59·5
7	06·8	01·7	67	65·0	16·2	127	123·2	30·7	187	181·4	45·2	247	239·7	59·8
8	07·8	01·9	68	66·0	16·5	128	124·2	31·0	188	182·4	45·5	248	240·6	60·0
9	08·7	02·2	69	67·0	16·7	129	125·2	31·2	189	183·4	45·7	249	241·6	60·2
10	09·7	02·4	70	67·9	16·9	130	126·1	31·4	190	184·4	46·0	250	242·6	60·5
11	10·7	02·7	71	68·9	17·2	131	127·1	31·7	191	185·3	46·2	251	243·5	60·7
12	11·6	02·9	72	69·9	17·4	132	128·1	31·9	192	186·3	46·4	252	244·5	61·0
13	12·6	03·1	73	70·8	17·7	133	129·0	32·2	193	187·3	46·7	253	245·5	61·2
14	13·6	03·4	74	71·8	17·9	134	130·0	32·4	194	188·2	46·9	254	246·5	61·4
15	14·6	03·6	75	72·8	18·1	135	131·0	32·7	195	189·2	47·2	255	247·4	61·7
16	15·5	03·9	76	73·7	18·4	136	132·0	32·9	196	190·2	47·4	256	248·4	61·9
17	16·5	04·1	77	74·7	18·6	137	132·9	33·1	197	191·1	47·7	257	249·4	62·2
18	17·5	04·4	78	75·7	18·9	138	133·9	33·4	198	192·1	47·9	258	250·3	62·4
19	18·4	04·6	79	76·7	19·1	139	134·9	33·6	199	193·1	48·1	259	251·3	62·7
20	19·4	04·8	80	77·6	19·4	140	135·8	33·9	200	194·1	48·4	260	252·3	62·9
21	20·4	05·1	81	78·6	19·6	141	136·8	34·1	201	195·0	48·6	261	253·2	63·1
22	21·3	05·3	82	79·6	19·8	142	137·8	34·4	202	196·0	48·9	262	254·2	63·4
23	22·3	05·6	83	80·5	20·1	143	138·8	34·6	203	197·0	49·1	263	255·2	63·6
24	23·3	05·8	84	81·5	20·3	144	139·7	34·8	204	197·9	49·4	264	256·2	63·9
25	24·3	06·0	85	82·5	20·6	145	140·7	35·1	205	198·9	49·6	265	257·1	64·1
26	25·2	06·3	86	83·4	20·8	146	141·7	35·3	206	199·9	49·8	266	258·1	64·4
27	26·2	06·5	87	84·4	21·0	147	142·6	35·6	207	200·9	50·1	267	259·1	64·6
28	27·2	06·8	88	85·4	21·3	148	143·6	35·8	208	201·8	50·3	268	260·0	64·8
29	28·1	07·0	89	86·4	21·5	149	144·6	36·0	209	202·8	50·6	269	261·0	65·1
30	29·1	07·3	90	87·3	21·8	150	145·5	36·3	210	203·8	50·8	270	262·0	65·3
31	30·1	07·5	91	88·3	22·0	151	146·5	36·5	211	204·7	51·0	271	263·0	65·6
32	31·0	07·7	92	89·3	22·3	152	147·5	36·8	212	205·7	51·3	272	263·9	65·8
33	32·0	08·0	93	90·2	22·5	153	148·5	37·0	213	206·7	51·5	273	264·9	66·0
34	33·0	08·2	94	91·2	22·7	154	149·4	37·3	214	207·6	51·8	274	265·9	66·3
35	34·0	08·5	95	92·2	23·0	155	150·4	37·5	215	208·6	52·0	275	266·8	66·5
36	34·9	08·7	96	93·1	23·2	156	151·4	37·7	216	209·6	52·3	276	267·8	66·8
37	35·9	09·0	97	94·1	23·5	157	152·3	38·0	217	210·6	52·5	277	268·8	67·0
38	36·9	09·2	98	95·1	23·7	158	153·3	38·2	218	211·5	52·7	278	269·7	67·3
39	37·8	09·4	99	96·1	24·0	159	154·3	38·5	219	212·5	53·0	279	270·7	67·5
40	38·8	09·7	100	97·0	24·2	160	155·2	38·7	220	213·5	53·2	280	271·7	67·7
41	39·8	09·9	101	98·0	24·4	161	156·2	38·9	221	214·4	53·5	281	272·7	68·0
42	40·8	10·2	102	99·0	24·7	162	157·2	39·2	222	215·4	53·7	282	273·6	68·2
43	41·7	10·4	103	99·9	24·9	163	158·2	39·4	223	216·4	53·9	283	274·6	68·5
44	42·7	10·6	104	100·9	25·2	164	159·1	39·7	224	217·3	54·2	284	275·6	68·7
45	43·7	10·9	105	101·9	25·4	165	160·1	39·9	225	218·3	54·4	285	276·5	68·9
46	44·6	11·1	106	102·9	25·6	166	161·1	40·2	226	219·3	54·7	286	277·5	69·2
47	45·6	11·4	107	103·8	25·9	167	162·0	40·4	227	220·3	54·9	287	278·5	69·4
48	46·6	11·6	108	104·8	26·1	168	163·0	40·6	228	221·2	55·2	288	279·4	69·7
49	47·5	11·9	109	105·8	26·4	169	164·0	40·9	229	222·2	55·4	289	280·4	69·9
50	48·5	12·1	110	106·7	26·6	170	165·0	41·1	230	223·2	55·6	290	281·4	70·2
51	49·5	12·3	111	107·7	26·9	171	165·9	41·4	231	224·1	55·9	291	282·4	70·4
52	50·5	12·6	112	108·7	27·1	172	166·9	41·6	232	225·1	56·1	292	283·3	70·6
53	51·4	12·8	113	109·6	27·3	173	167·9	41·9	233	226·1	56·4	293	284·3	70·9
54	52·4	13·1	114	110·6	27·6	174	168·8	42·1	234	227·0	56·6	294	285·3	71·1
55	53·4	13·3	115	111·6	27·8	175	169·8	42·3	235	228·0	56·9	295	286·2	71·4
56	54·3	13·5	116	112·6	28·1	176	170·8	42·6	236	229·0	57·1	296	287·2	71·6
57	55·3	13·8	117	113·5	28·3	177	171·7	42·8	237	230·0	57·3	297	288·2	71·9
58	56·3	14·0	118	114·5	28·6	178	172·7	43·1	238	230·9	57·6	298	289·1	72·1
59	57·2	14·3	119	115·5	28·8	179	173·7	43·3	239	231·9	57·8	299	290·1	72·3
60	58·2	14·5	120	116·4	29·0	180	174·7	43·5	240	232·9	58·1	300	291·1	72·6
Dist.	Dep.	Lat.	Dist.	Dep.	Lat.	Dist.	Dep.	Lat.	Dist.	Dep.	Lat.	Dist.	Dep.	Lat.

FOR 76 DEGREES.

Dist.	Lat.	Dep.	Dist.	Lat.	Dep.	Dist.	Lat.	Dep	Dist.	Lat.	Dep.	Dist.	Lat.	Dep.
1	01·0	00·3	61	58·9	15·8	121	116·9	31·3	181	174·8	46·8	241	232·8	62·4
2	01·9	00·5	62	59·9	16·0	122	117·8	31·6	182	175·8	47·1	242	233·8	62·6
3	02·9	00·8	63	60·9	16·3	123	118·8	31·8	183	176·8	47·4	243	234·7	62·9
4	03·9	01·0	64	61·8	16·6	124	119·8	32·1	184	177·7	47·6	244	235·7	63·2
5	04·8	01·3	65	62·8	16·8	125	120·7	32·4	185	178·7	47·9	245	236·7	63·4
6	05·8	01·6	66	63·8	17·1	126	121·7	32·6	186	179·7	48·1	246	237·6	63·7
7	06·8	01·8	67	64·7	17·3	127	122·7	32·9	187	180·6	48·4	247	238·6	63·9
8	07·7	02·1	68	65·7	17·6	128	123·6	33·1	188	181·6	48·7	248	239·5	64·2
9	08·7	02·3	69	66·6	17·9	129	124·6	33·4	189	182·6	48·9	249	240·5	64·4
10	09·7	02·6	70	67·6	18·1	130	125·6	33·6	190	183·5	49·2	250	241·5	64·7
11	10·6	02·8	71	68·6	18·4	131	126·5	33·9	191	184·5	49·4	251	242·4	65·0
12	11·6	03·1	72	69·5	18·6	132	127·5	34·2	192	185·5	49·7	252	243·4	65·2
13	12·6	03·4	73	70·5	18·9	133	128·5	34·4	193	186·4	50·0	253	244·4	65·5
14	13·5	03·6	74	71·5	19·2	134	129·4	34·7	194	187·4	50·2	254	245·3	65·7
15	14·5	03·9	75	72·4	19·4	135	130·4	34·9	195	188·4	50·5	255	246·3	66·0
16	15·5	04·1	76	73·4	19·7	136	131·4	35·2	196	189·3	50·7	256	247·3	66·3
17	16·4	04·4	77	74·4	19·9	137	132·3	35·5	197	190·3	51·0	257	248·2	66·5
18	17·4	04·7	78	75·3	20·2	138	133·3	35·7	198	191·3	51·2	258	249·2	66·8
19	18·4	04·9	79	76·3	20·4	139	134·3	36·0	199	192·2	51·5	259	250·2	67·0
20	19·3	05·2	80	77·3	20·7	140	135·2	36·2	200	193·2	51·8	260	251·1	67·3
21	20·3	05·4	81	78·2	21·0	141	136·2	36·5	201	194·2	52·0	261	252·1	67·6
22	21·3	05·7	82	79·2	21·2	142	137·2	36·8	202	195·1	52·3	262	253·1	67·8
23	22·2	06·0	83	80·2	21·5	143	138·1	37·0	203	196·1	52·5	263	254·0	68·1
24	23·2	06·2	84	81·1	21·7	144	139·1	37·3	204	197·0	52·8	264	255·0	68·3
25	24·1	06·5	85	82·1	22·0	145	140·1	37·5	205	198·0	53·1	265	256·0	68·6
26	25·1	06·7	86	83·1	22·3	146	141·0	37·8	206	199·0	53·3	266	256·9	68·8
27	26·1	07·0	87	84·0	22·5	147	142·0	38·0	207	199·9	53·6	267	257·9	69·1
28	27·0	07·2	88	85·0	22·8	148	143·0	38·3	208	200·9	53·8	268	258·9	69·4
29	28·0	07·5	89	86·0	23·0	149	143·9	38·6	209	201·9	54·1	269	259·8	69·6
30	29·0	07·8	90	86·9	23·3	150	144·9	38·8	210	202·8	54·4	270	260·8	69·9
31	29·9	08·0	91	87·9	23·6	151	145·9	39·1	211	203·8	54·6	271	261·8	70·1
32	30·9	08·3	92	88·9	23·8	152	146·8	39·3	212	204·8	54·9	272	262·7	70·4
33	31·9	08·5	93	89·8	24·1	153	147·8	39·6	213	205·7	55·1	273	263·7	70·7
34	32·8	08·8	94	90·8	24·3	154	148·8	39·9	214	206·7	55·4	274	264·7	70·9
35	33·8	09·1	95	91·8	24·6	155	149·7	40·1	215	207·7	55·6	275	265·6	71·2
36	34·8	09·3	96	92·7	24·8	156	150·7	40·4	216	208·6	55·9	276	266·6	71·4
37	35·7	09·6	97	93·7	25·1	157	151·7	40·6	217	209·6	56 2	277	267·6	71·7
38	36·7	09·8	98	94·7	25·4	158	152·6	40·9	218	210·6	56·4	278	268·5	72·0
39	37·7	10·1	99	95·6	25·6	159	153·6	41·2	219	211·5	56·7	279	269·5	72·2
40	38·6	10·4	100	96·6	25·9	160	154·5	41·4	220	212·5	56·9	280	270·5	72·5
41	39·6	10·6	101	97·6	26·1	161	155·5	41·7	221	213·5	57·2	281	271·4	72·7
42	40·6	10·9	102	98·5	26·4	162	156·5	41·9	222	214·4	57·5	282	272·4	73·0
43	41·5	11·1	103	99·5	26·7	163	157·4	42·2	223	215·4	57·7	283	273·4	73·2
44	42·5	11·4	104	100·5	26·9	164	158·4	42·4	224	216·4	58·0	284	274·3	73·5
45	43·5	11·6	105	101·4	27·2	165	159·4	42·7	225	217·3	58·2	285	275·3	73·8
46	44·4	11·9	106	102·4	27·4	166	160·3	43·0	226	218·3	58·5	286	276·3	74·0
47	45·4	12·2	107	103·4	27·7	167	161·3	43·2	227	219·3	58·8	287	277·2	74·3
48	46·4	12·4	108	104·3	28·0	168	162·3	43·5	228	220·2	59·0	288	278·2	74·5
49	47·3	12·7	109	105·3	28·2	169	163·2	43·7	229	221·2	59·3	289	279·2	74·8
50	48·3	12·9	110	106·3	28·5	170	164·2	44·0	230	222·2	59·5	290	280·1	75·1
51	49·3	13·2	111	107·2	28·7	171	165·2	44·3	231	223·1	59·8	291	281·1	75·3
52	50·2	13·5	112	108·2	29·0	172	166·1	44·5	232	224·1	60 0	292	282·1	75·6
53	51·2	13·7	113	109·1	29·2	173	167·1	44·8	233	225·1	60·3	293	283·0	75·8
54	52·2	14·0	114	110·1	29·5	174	168·1	45·0	234	226·0	60 6	294	284·0	76·1
55	53·1	14·2	115	111·1	29·8	175	169·0	45·3	235	227·0	60·8	295	284·9	76·4
56	54·1	14·5	116	112·0	30·0	176	170·0	45·6	236	228·0	61·1	296	285·9	76·6
57	55·1	14·8	117	113·0	30·3	177	171·0	45·8	237	228·9	61·3	297	286·9	76·9
58	56·0	15·0	118	114·0	30·5	178	171·9	46·1	238	229·9	61·6	298	287·8	77·1
59	57·0	15·3	119	114·9	30·8	179	172·9	46·3	239	230·9	61·9	299	288·8	77·4
60	58·0	15·5	120	115·9	31·1	180	173·9	46·6	240	231·8	62·1	300	289·8	77·6
Dist.	Dep.	Lat.	Dist.	Dep.	Lat.	Dist.	Dep.	Lat.	Dist.	Dep.	Lat.	Dist.	Dep.	Lat.

FOR 75 DEGREES.

Dist.	Lat.	Dep.	Dist.	Lat.	Dep.	Dist.	Lat.	Dep.	Dist.	Lat.	Dep.	Dist.	Lat.	Dep.
1	01·0	00·3	61	58·6	16·8	121	116·3	33·4	181	174·0	49·9	241	231·7	66·4
2	01·9	00·6	62	59·6	17·1	122	117·3	33·6	182	174·9	50·2	242	232·6	66·7
3	02·9	00·8	63	60·6	17·4	123	118·2	33·9	183	175·9	50·4	243	233·6	67·0
4	03·8	01·1	64	61·5	17·6	124	119·2	34·2	184	176·9	50·7	244	234·5	67·3
5	04·8	01·4	65	62·5	17·9	125	120·2	34·5	185	177·8	51·0	245	235·5	67·5
6	05·8	01·7	66	63·4	18·2	126	121·1	34·7	186	178·8	51·3	246	236·5	67·8
7	06·7	01·9	67	64·4	18·5	127	122·1	35·0	187	179·8	51·5	247	237·4	68·1
8	07·7	02·2	68	65·4	18·7	128	123·0	35·3	188	180·7	51·8	248	238·4	68·4
9	08·7	02·5	69	66·3	19·0	129	124·0	35·6	189	181·7	52·1	249	239·4	68·6
10	09·6	02·8	70	67·3	19·3	130	125·0	35·8	190	182·6	52·4	250	240·3	68·9
11	10·6	03·0	71	68·2	19·6	131	125·9	36·1	191	183·6	52·6	251	241·3	69·2
12	11·5	03·3	72	69·2	19·8	132	126·9	36·4	192	184·6	52·9	252	242·2	69·5
13	12·5	03·6	73	70·2	20·1	133	127·8	36·7	193	185·5	53·2	253	243·2	69·7
14	13·5	03·9	74	71·1	20·4	134	128·8	36·9	194	186·5	53·5	254	244·2	70·0
15	14·4	04·1	75	72·1	20·7	135	129·8	37·2	195	187·4	53·7	255	245·1	70·3
16	15·4	04·4	76	73·1	20·9	136	130·7	37·5	196	188·4	54·0	256	246·1	70·6
17	16·3	04·7	77	74·0	21·2	137	131·7	37·8	197	189·4	54·3	257	247·0	70·8
18	17·3	05·0	78	75·0	21·5	138	132·7	38·0	198	190·3	54·6	258	248·0	71·1
19	18·3	05·2	79	75·9	21·8	139	133·6	38·3	199	191·3	54·9	259	249·0	71·4
20	19·2	05·5	80	76·9	22·1	140	134·6	38·6	200	192·3	55·1	260	249·9	71·7
21	20·2	05·8	81	77·9	22·3	141	135·5	38·9	201	193·2	55·4	261	250·9	71·9
22	21·1	06·1	82	78·8	22·6	142	136·5	39·1	202	194·2	55·7	262	251·9	72·2
23	22·1	06·3	83	79·8	22·9	143	137·5	39·4	203	195·1	56·0	263	252·8	72·5
24	23·1	06·6	84	80·7	23·2	144	138·4	39·7	204	196·1	56·2	264	253·8	72·8
25	24·0	06·9	85	81·7	23·4	145	139·4	40·0	205	197·1	56·5	265	254·7	73·0
26	25·0	07·2	86	82·7	23·7	146	140·3	40·2	206	198·0	56·8	266	255·7	73·3
27	26·0	07·4	87	83·6	24·0	147	141·3	40·5	207	199·0	57·1	267	256·7	73·6
28	26·9	07·7	88	84·6	24·3	148	142·3	40·8	208	199·9	57·3	268	257·6	73·9
29	27·9	08·0	89	85·6	24·5	149	143·2	41·1	209	200·9	57·6	269	258·6	74·1
30	28·8	08·3	90	86·5	24·8	150	144·2	41·3	210	201·9	57·9	270	259·5	74·4
31	29·8	08·5	91	87·5	25·1	151	145·2	41·6	211	202·8	58·2	271	260·5	74·7
32	30·8	08·8	92	88·4	25·4	152	146·1	41·9	212	203·8	58·4	272	261·5	75·0
33	31·7	09·1	93	89·4	25·6	153	147·1	42·2	213	204·7	58·7	273	262·4	75·2
34	32·7	09·4	94	90·4	25·9	154	148·0	42·4	214	205·7	59·0	274	263·4	75·5
35	33·6	09·6	95	91·3	26·2	155	149·0	42·7	215	206·7	59·3	275	264·3	75·8
36	34·6	09·9	96	92·3	26·5	156	150·0	43·0	216	207·6	59·5	276	265·3	76·1
37	35·6	10·2	97	93·2	26·7	157	150·9	43·3	217	208·6	59·8	277	266·3	76·4
38	36·5	10·5	98	94·2	27·0	158	151·9	43·6	218	209·6	60·1	278	267·2	76·6
39	37·5	10·7	99	95·2	27·3	159	152·8	43·8	219	210·5	60·4	279	268·2	76·9
40	38·5	11·0	100	96·1	27·6	160	153·8	44·1	220	211·5	60·6	280	269·2	77·2
41	39·4	11·3	101	97·1	27·8	161	154·8	44·4	221	212·4	60·9	281	270·1	77·5
42	40·4	11·6	102	98·0	28·1	162	155·7	44·7	222	213·4	61·2	282	271·1	77·7
43	41·3	11·9	103	99·0	28·4	163	156·7	44·9	223	214·4	61·5	283	272·0	78·0
44	42·3	2·1	104	100·0	28·7	164	157·6	45·2	224	215·3	61·7	284	273·0	78·3
45	43·3	12·4	105	100·9	28·9	165	158·6	45·5	225	216·3	62·0	285	274·0	78·6
46	44·2	12·7	106	101·9	29·2	166	159·6	45·8	226	217·2	62·3	286	274·9	78·8
47	45·2	13·0	107	102·9	29·5	167	160·5	46·0	227	218·2	62·6	287	275·9	79·1
48	46·1	13·2	108	103·8	29·8	168	161·5	46·3	228	219·2	62·8	288	276·8	79·4
49	47·1	13·5	109	104·8	30·0	169	162·5	46·6	229	220·1	63·1	289	277·8	79·7
50	48·1	13·8	110	105·7	30·3	170	163·4	46·9	230	221·1	63·4	290	278·8	79·9
51	49·0	14·1	111	106·7	30·6	171	164·4	47·1	231	222·1	63·7	291	279·7	80·2
52	50·0	14·3	112	107·7	30·9	172	165·3	47·4	232	223·0	63·9	292	280·7	80·5
53	50·9	14·6	113	108·6	31·1	173	166·3	47·7	233	224·0	64·2	293	281·6	80·8
54	51·9	14·9	114	109·6	31·4	174	167·3	48·0	234	224·9	64·5	294	282·6	81·0
55	52·9	15·2	115	110·5	31·7	175	168·2	48·2	235	225·9	64·8	295	283·6	81·3
56	53·8	15·4	116	111·5	32·0	176	169·2	48·5	236	226·9	65·1	296	284·5	81·6
57	54·8	15·7	117	112·5	32·2	177	170·1	48·8	237	227·8	65·3	297	285·5	81·9
58	55·8	16·0	118	113·4	32·5	178	171·1	49·1	238	228·8	65·6	298	286·5	82·1
59	56·7	16·3	119	114·4	32·8	179	172·1	49·3	239	229·7	65·9	299	287·4	82·4
60	57·7	16·5	120	115·4	33·1	180	173·0	49·6	240	230·7	66·2	300	288·4	82·7
Dist.	Dep.	Lat.	Dist.	Dep.	Lat.	Dist.	Dep.	Lat.	Dist.	Dep.	Lat.	Dist.	Dep.	Lat.

Dist.	Lat.	Dep.	Dist.	Lat.	Dep.	Dist.	Lat.	Dep.	Dist.	Lat.	Dep.	Dist.	Lat.	Dep.
1	01·0	00·3	61	58·3	17·8	121	115·7	35·4	181	173·1	52·9	241	230·5	70·5
2	01·9	00·6	62	59·3	18·1	122	116·7	35·7	182	174·0	53·2	242	231·4	70·8
3	02·9	00·9	63	60·2	18·4	123	117·6	36·0	183	175·0	53·5	243	232·4	71·0
4	03·8	01·2	64	61·2	18·7	124	118·6	36·3	184	176·0	53·8	244	233·3	71·3
5	04·8	01·5	65	62·2	19·0	125	119·5	36·5	185	176·9	54·1	245	234·3	71·6
6	05·7	01·8	66	63·1	19·3	126	120·5	36·8	186	177·9	54·4	246	235·3	71·9
7	06·7	02·0	67	64·1	19·6	127	121·5	37·1	187	178·8	54·7	247	236·2	72·2
8	07·7	02·3	68	65·0	19·9	128	122·4	37·4	188	179·8	55·0	248	237·2	72·5
9	08·6	02·6	69	66·0	20·2	129	123·4	37·7	189	180·7	55·3	249	238·1	72·8
10	09·6	02·9	70	66·9	20·5	130	124·3	38·0	190	181·7	55·6	250	239·1	73·1
11	10·5	03·2	71	67·9	20·8	131	125·3	38·3	191	182·7	55·8	251	240·0	73·4
12	11·5	03·5	72	68·9	21·1	132	126·2	38·6	192	183·6	56·1	252	241·0	73·7
13	12·4	03·8	73	69·8	21·3	133	127·2	38·9	193	184·6	56·4	253	241·9	74·0
14	13·4	04·1	74	70·8	21·6	134	128·1	39·2	194	185·5	56·7	254	242·9	74·3
15	14·3	04·4	75	71·7	21·9	135	129·1	39·5	195	186·5	57·0	255	243·9	74·6
16	15·3	04·7	76	72·7	22·2	136	130·1	39·8	196	187·4	57·3	256	244·8	74·8
17	16·3	05·0	77	73·6	22·5	137	131·0	40·1	197	188·4	57·6	257	245·8	75·1
18	17·2	05·3	78	74·6	22·8	138	132·0	40·3	198	189·3	57·9	258	246·7	75·4
19	18·2	05·6	79	75·5	23·1	139	132·9	40·6	199	190·3	58·2	259	247·7	75·7
20	19·1	05·8	80	76·5	23·3	140	133·9	40·9	200	191·3	58·5	260	248·6	76·0
21	20·1	06·1	81	77·5	23·7	141	134·8	41·2	201	192·2	58·8	261	249·6	76·3
22	21·0	06·4	82	78·4	24·0	142	135·8	41·5	202	193·2	59·1	262	250·6	76·6
23	22·0	06·7	83	79·4	24·3	143	136·8	41·8	203	194·1	59·4	263	251·5	76·9
24	23·0	07·0	84	80·3	24·6	144	137·7	42·1	204	195·1	59·6	264	252·5	77·2
25	23·9	07·3	85	81·3	24·9	145	138·7	42·4	205	196·0	59·9	265	253·4	77·5
26	24·9	07·6	86	82·2	25·1	146	139·6	42·7	206	197·0	60·2	266	254·4	77·8
27	25·8	07·9	87	83·2	25·4	147	140·6	43·0	207	198·0	60·5	267	255·3	78·1
28	26·8	08·2	88	84·2	25·7	148	141·5	43·3	208	198·9	60·8	268	256·3	78·4
29	27·7	08·5	89	85·1	26·0	149	142·5	43·6	209	199·9	61·1	269	257·2	78·6
30	28·7	08·8	90	86·1	26·3	150	143·4	43·9	210	200·8	61·4	270	258·2	78·9
31	29·6	09·1	91	87·0	26·6	151	144·4	44·1	211	201·8	61·7	271	259·2	79·2
32	30·6	09·4	92	88·0	26·9	152	145·4	44·4	212	202·7	62·0	272	260·1	79·5
33	31·6	09·6	93	88·9	27·2	153	146·3	44·7	213	203·7	62·3	273	261·1	79·8
34	32·5	09·9	94	89·9	27·5	154	147·3	45·0	214	204·6	62·6	274	262·0	80·1
35	33·5	10·2	95	90·8	27·8	155	148·2	45·3	215	205·6	62·9	275	263·0	80·4
36	34·4	10·5	96	91·8	28·1	156	149·2	45·6	216	206·6	63·2	276	263·9	80·7
37	35·4	10·8	97	92·8	28·4	157	150·1	45·9	217	207·5	63·4	277	264·9	81·0
38	36·3	11·1	98	93·7	28·7	158	151·1	46·2	218	208·5	63·7	278	265·9	81·3
39	37·3	11·4	99	94·7	28·9	159	152·1	46·5	219	209·4	64·0	279	266·8	81·6
40	38·3	11·7	100	95·6	29·2	160	153·0	46·8	220	210·4	64·3	280	267·8	81·9
41	39·2	12·0	101	96·6	29·5	161	154·0	47·1	221	211·3	64·6	281	268·7	82·2
42	40·2	12·3	102	97·5	29·8	162	154·9	47·4	222	212·3	64·9	282	269·7	82·4
43	41·1	12·6	103	98·5	30·1	163	155·9	47·7	223	213·3	65·2	283	270·6	82·7
44	42·1	12·9	104	99·5	30·4	164	156·8	47·9	224	214·2	65·5	284	271·6	83·0
45	43·0	13·2	105	100·4	30·7	165	157·8	48·2	225	215·2	65·8	285	272·5	83·3
46	44·0	13·4	106	101·4	31·0	166	158·7	48·5	226	216·1	66·1	286	273·5	83·6
47	44·9	13·7	107	102·3	31·3	167	159·7	48·8	227	217·1	66·4	287	274·5	83·9
48	45·9	14·0	108	103·3	31·6	168	160·7	49·1	228	218·0	66·7	288	275·4	84·2
49	46·9	14·3	109	104·2	31·9	169	161·6	49·4	229	219·0	67·0	289	276·4	84·5
50	47·8	14·6	110	105·2	32·2	170	162·6	49·7	230	220·0	67·2	290	277·3	84·8
51	46·8	14·9	111	106·1	32·5	171	163·5	50·0	231	220·9	67·5	291	278·3	85·1
52	49·7	15·2	112	107·1	32·7	172	164·5	50·3	232	221·9	67·8	292	279·2	85·4
53	50·7	15·5	113	108·1	33·0	173	165·4	50·6	233	222·8	68·1	293	280·2	85·7
54	51·6	15·8	114	109·0	33·3	174	166·4	50·9	234	223·8	68·4	294	281·2	86·0
55	52·6	16·1	115	110·0	33·6	175	167·4	51·2	235	224·7	68·7	295	282·1	86·2
56	53·6	16·4	116	110·9	33·9	176	168·3	51·5	236	225·7	69·0	296	283·1	86·5
57	54·5	16·7	117	111·9	34·2	177	169·3	51·7	237	226·6	69·3	297	284·0	86·8
58	55·5	17·0	118	112·8	34·5	178	170·2	52·0	238	227·6	69·6	298	285·0	87·1
59	56·4	17·2	119	113·8	34·8	179	171·2	52·3	239	228·6	69·9	299	285·9	87·4
60	57·4	17·5	120	114·8	35·1	180	172·1	52·6	240	229·5	70·2	300	286·9	87·7
Dist.	Dep.	Lat.	Dist.	Dep.	Lat.	Dist.	Dep.	Lat.	Dist.	Dep.	Lat.	Dist.	Dep.	Lat.

FOR 73 DEGREES

Dist	Lat.	Dep.	Dist.	Lat.	Dep.	Dist.	Lat.	Dep.	Dist.	Lat.	Dep.	Dist.	Lat.	Dep.
1	01·0	00·3	61	58·0	18·9	121	115·1	37·4	181	172·1	55·9	241	229·2	74·5
2	01·9	00·6	62	59·0	19·2	122	116·0	37·7	182	173·1	56·2	242	230·2	74·8
3	02·9	00·9	63	59·9	19·5	123	117·0	38·0	183	174·0	56·6	243	231·1	75·1
4	03·8	01·2	64	60·9	19·8	124	117·9	38·3	184	175·0	56·9	244	232·1	75·4
5	04·8	01·5	65	61·8	20·1	125	118·9	38·6	185	175·9	57·2	245	233·0	75·7
6	05·7	01·9	66	62·8	20·4	126	119·8	38·9	186	176·9	57·5	246	234·0	76·0
7	06·7	02·2	67	63·7	20·7	127	120·8	39·2	187	177·8	57·8	247	234·9	76·3
8	07·6	02·5	68	64·7	21·0	128	121·7	39·6	188	178·8	58·1	248	235·9	76·6
9	08·6	02·8	69	65·6	21·3	129	122·7	39·9	189	179·7	58·4	249	236·8	76·9
10	09·5	03·1	70	66·6	21·6	130	123·6	40·2	190	180·7	58·7	250	237·8	77·3
11	10·5	03·4	71	67·5	21·9	131	124·6	40·5	191	181·7	59·0	251	238·7	77·6
12	11·4	03·7	72	68·5	22·2	132	125·5	40·8	192	182·6	59·3	252	239·7	77·9
13	12·4	04·0	73	69·4	22·6	133	126·5	41·1	193	183·6	59·6	253	240·6	78·2
14	13·3	04·3	74	70·4	22·9	134	127·4	41·4	194	184·5	59·9	254	241·6	78·5
15	14·3	04·6	75	71·3	23·2	135	128·4	41·7	195	185·5	60·3	255	242·5	78·8
16	15·2	04·9	76	72·3	23·5	136	129·3	42·0	196	186·4	60·6	256	243·5	79·1
17	16·2	05·3	77	73·2	23·8	137	130·3	42·3	197	187·4	60·9	257	244·4	79·4
18	17·1	05·6	78	74·2	24·1	138	131·2	42·6	198	188·3	61·2	258	245·4	79·7
19	18·1	05·9	79	75·1	24·4	139	132·2	43·0	199	189·3	61·5	259	246·3	80·0
20	19·0	06·2	80	76·1	24·7	140	133·1	43·3	200	190·2	61·8	260	247·3	80·3
21	20·0	06·5	81	77·0	25·0	141	134·1	43·6	201	191·2	62·1	261	248·2	80·7
22	20·9	06·8	82	78·0	25·3	142	135·1	43·9	202	192·1	62·4	262	249·2	81·0
23	21·9	07·1	83	78·9	25·6	143	136·0	44·2	203	193·1	62·7	263	250·1	81·3
24	22·8	07·4	84	79·9	26·0	144	137·0	44·5	204	194·0	63·0	264	251·1	81·6
25	23·8	07·7	85	80·8	26·3	145	137·9	44·8	205	195·0	63·3	265	252·0	81·9
26	24·7	08·0	86	81·8	26·6	146	138·9	45·1	206	195·9	63·7	266	253·0	82·2
27	25·7	08·3	87	82·7	26·9	147	139·8	45·4	207	196·9	64·0	267	253·9	82·5
28	26·6	08·7	88	83·7	27·2	148	140·8	45·7	208	197·8	64·3	268	254·9	82·8
29	27·6	09·0	89	84·6	27·5	149	141·7	46·0	209	198·8	64·6	269	255·8	83·1
30	28·5	09·3	90	85·6	27·8	150	142·7	46·4	210	199·7	64·9	270	256·8	83·4
31	29·5	09·6	91	86·5	28·1	151	143·6	46·7	211	200·7	65·2	271	257·7	83·7
32	30·4	09·9	92	87·5	28·4	152	144·6	47·0	212	201·6	65·5	272	258·7	84·1
33	31·4	10·2	93	88·4	28·7	153	145·5	47·3	213	202·6	65·8	273	259·6	84·4
34	32·3	10·5	94	89·4	29·0	154	146·5	47·6	214	203·5	66·1	274	260·6	84·7
35	33·3	10·8	95	90·4	29·4	155	147·4	47·9	215	204·5	66·4	275	261·5	85·0
36	34·2	11·1	96	91·3	29·7	156	148·4	48·2	216	205·4	66·7	276	262·5	85·3
37	35·2	11·4	97	92·3	30·0	157	149·3	48·5	217	206·4	67·1	277	263·4	85·6
38	36·1	11·7	98	93·2	30·3	158	150·3	48·8	218	207·3	67·4	278	264·4	85·9
39	37·1	12·1	99	94·2	30·6	159	151·2	49·1	219	208·3	67·7	279	265·3	86·2
40	38·0	12·4	100	95·1	30·9	160	152·2	49·4	220	209·2	68·0	280	266·3	86·5
41	39·0	12·7	101	96·1	31·2	161	153·1	49·8	221	210·2	68·3	281	267·2	86·8
42	39·9	13·0	102	97·0	31·5	162	154·1	50·1	222	211·1	68·6	282	268·2	87·1
43	40·9	13·3	103	98·0	31·8	163	155·0	50·4	223	212·1	68·9	283	269·1	87·5
44	41·8	13·6	104	98·9	32·1	164	156·0	50·7	224	213·0	69·2	284	270·1	87·8
45	42·8	13·9	105	99·9	32·4	165	156·9	51·0	225	214·0	69·5	285	271·1	88·1
46	43·7	14·2	106	100·8	32·8	166	157·9	51·3	226	214·9	69·8	286	272·0	88·4
47	44·7	14·5	107	101·8	33·1	167	158·8	51·6	227	215·9	70·1	287	273·0	88·7
48	45·7	14·8	108	102·7	33·4	168	159·8	51·9	228	216·8	70·5	288	273·9	89·0
49	46·6	15·1	109	103·7	33·7	169	160·7	52·2	229	217·8	70·8	289	274·9	89·3
50	47·6	15·5	110	104·6	34·0	170	161·7	52·5	230	218·7	71·1	290	275·8	89·6
51	48·5	15·8	111	105·6	34·3	171	162·6	52·8	231	219·7	71·4	291	276·8	89·9
52	49·5	16·1	112	106·5	34·6	172	163·6	53·2	232	220·6	71·7	292	277·7	90·2
53	50·4	16·4	113	107·5	34·9	173	164·5	53·5	233	221·6	72·0	293	278·7	90·5
54	51·4	16·7	114	108·4	35·2	174	165·5	53·8	234	222·5	72·3	294	279·6	90·9
55	52·3	17·0	115	109·4	35·5	175	166·4	54·1	235	223·5	72·6	295	280·6	91·2
56	53·3	17·3	116	110·3	35·8	176	167·4	54·4	236	224·4	72·9	296	281·5	91·5
57	54·2	17·6	117	111·3	36·2	177	168·3	54·7	237	225·4	73·2	297	282·5	91·8
58	55·2	17·9	118	112·2	36·5	178	169·3	55·0	238	226·4	73·5	298	283·4	92·1
59	56·1	18·2	119	113·2	36·8	179	170·2	55·3	239	227·3	73·9	299	284·4	92·4
60	57·1	18·5	120	114·1	37·1	180	171·2	55·6	240	228·3	74·2	300	285·3	92·7
Dist.	Dep.	Lat.	Dist.	Dep.	Lat.	Dist.	Dep.	Lat.	Dist.	Dep.	Lat.	Dist.	Dep.	Lat.

FOR 72 DEGREES.

Dist.	Lat.	Dep.	Dist.	Lat.	Dep.	Dist.	Lat.	Dep.	Dist.	Lat.	Dep.	Dist.	Lat.	Dep.
1	00·9	00·3	61	57·7	19·9	121	114·4	39·4	181	171·1	58·9	241	227·9	78·5
2	01·9	00·7	62	58·6	20·2	122	115·4	39·7	182	172·1	59·3	242	228·8	78·8
3	02·8	01·0	63	59·6	20·5	123	116·3	40·0	183	173·0	59·6	243	229·8	79·1
4	03·8	01·3	64	60·5	20·8	124	117·2	40·4	184	174·0	59·9	244	230·7	79·4
5	04·7	01·6	65	61·5	21·2	125	118·2	40·7	185	174·9	60·2	245	231·7	79·8
6	05·7	02·0	66	62·4	21·5	126	119·1	41·0	186	175·9	60·6	246	232·6	80·1
7	06·6	02·3	67	63·3	21·8	127	120·1	41·3	187	176·8	60·9	247	233·5	80·4
8	07·6	02·6	68	64·3	22·1	128	121·0	41·7	188	177·8	61·2	248	234·5	80·7
9	08·5	02·9	69	65·2	22·5	129	122·0	42·0	189	178·7	61·5	249	235·4	81·1
10	09·5	03·3	70	66·2	22·8	130	122·9	42·3	190	179·6	61·9	250	236·4	81·4
11	10·4	03·6	71	67·1	23·1	131	123·9	42·6	191	180·6	62·2	251	237·3	81·7
12	11·3	03·9	72	68·1	23·4	132	124·8	43·0	192	181·5	62·5	252	238·3	82·0
13	12·3	04·2	73	69·0	23·8	133	125·8	43·3	193	182·5	62·8	253	239·2	82·4
14	13·2	04·6	74	70·0	24·1	134	126·7	43·6	194	183·4	63·2	254	240·2	82·7
15	14·2	04·9	75	70·9	24·4	135	127·6	44·0	195	184·4	63·5	255	241·1	83·0
16	15·1	05·2	76	71·9	24·7	136	128·6	44·3	196	185·3	63·8	256	242·1	83·3
17	16·1	05·5	77	72·8	25·1	137	129·5	44·6	197	186·3	64·1	257	243·0	83·7
18	17·0	05·9	78	73·8	25·4	138	130·5	44·9	198	187·2	64·5	258	243·9	84·0
19	18·0	06·2	79	74·7	25·7	139	131·4	45·3	199	188·2	64·8	259	244·9	84·3
20	18·9	06·5	80	75·6	26·0	140	132·4	45·6	200	189·1	65·1	260	245·8	84·6
21	19·9	06·8	81	76·6	26·4	141	133·3	45·9	201	190·0	65·4	261	246·8	85·0
22	20·8	07·2	82	77·5	26·7	142	134·3	46·2	202	191·0	65·8	262	247·7	85·3
23	21·7	07·5	83	78·5	27·0	143	135·2	46·6	203	191·9	66·1	263	248·7	85·6
24	22·7	07·8	84	79·4	27·3	144	136·2	46·9	204	192·9	66·4	264	249·6	86·0
25	23·6	08·1	85	80·4	27·7	145	137·1	47·2	205	193·8	66·7	265	250·6	86·3
26	24·6	08·5	86	81·3	28·0	146	138·0	47·5	206	194·8	67·1	266	251·5	86·6
27	25·5	08·8	87	82·3	28·3	147	139·0	47·9	207	195·7	67·4	267	252·5	86·9
28	26·5	09·1	88	83·2	28·7	148	139·9	48·2	208	196·7	67·7	268	253·4	87·3
29	27·4	09·4	89	84·2	29·0	149	140·9	48·5	209	197·6	68·0	269	254·3	87·6
30	28·4	09·8	90	85·1	29·3	150	141·8	48·8	210	198·6	68·4	270	255·3	87·9
31	29·3	10·1	91	86·0	29·6	151	142·8	49·2	211	199·5	68·7	271	256·2	88·2
32	30·3	10·4	92	87·0	30·0	152	143·7	49·5	212	200·4	69·0	272	257·2	88·6
33	31·2	10·7	93	87·9	30·3	153	144·7	49·8	213	201·4	69·3	273	258·1	88·9
34	32·1	11·1	94	88·9	30·6	154	145·6	50·1	214	202·3	69·7	274	259·1	89·2
35	33·1	11·4	95	89·8	30·9	155	146·6	50·5	215	203·3	70·0	275	260·0	89·5
36	34·0	11·7	96	90·8	31·3	156	147·5	50·8	216	204·2	70·3	276	261·0	89·9
37	35·0	12·0	97	91·7	31·6	157	148·4	51·1	217	205·2	70·6	277	261·9	90·2
38	35·9	12·4	98	92·7	31·9	158	149·4	51·4	218	206·1	71·0	278	262·9	90·5
39	36·9	12·7	99	93·6	32·2	159	150·3	51·8	219	207·1	71·3	279	263·8	90·8
40	37·8	13·0	100	94·6	32·6	160	151·3	52·1	220	208·0	71·6	280	264·7	91·2
41	38·8	13·3	101	95·5	32·9	161	152·2	52·4	221	209·0	72·0	281	265·7	91·5
42	39·7	13·7	102	96·4	33·2	162	153·2	52·7	222	209·9	72·3	282	266·6	91·8
43	40·7	14·0	103	97·4	33·5	163	154·1	53·1	223	210·9	72·6	283	267·6	92·1
44	41·6	14·3	104	98·3	33·9	164	155·1	53·4	224	211·8	72·9	284	268·5	92·5
45	42·5	14·7	105	99·3	34·2	165	156·0	53·7	225	212·7	73·3	285	269·5	92·8
46	43·5	15·0	106	100·2	34·5	166	157·0	54·0	226	213·7	73·6	286	270·4	93·1
47	44·4	15·3	107	101·2	34·8	167	157·9	54·4	227	214·6	73·9	287	271·4	93·4
48	45·4	15·6	108	102·1	35·2	168	158·8	54·7	228	215·6	74·2	288	272·3	93·8
49	46·3	16·0	109	103·1	35·5	169	159·8	55·0	229	216·5	74·6	289	273·3	94·1
50	47·3	16·3	110	104·0	35·8	170	160·7	55·3	230	217·5	74·9	290	274·2	94·4
51	48·2	16·6	111	105·0	36·1	171	161·7	55·7	231	218.4	75·2	291	275·1	94·7
52	49·2	16·9	112	105·9	36·5	172	162·6	56·0	232	219·4	75·5	292	276·1	95·1
53	50·1	17·3	113	106·8	36·8	173	163·6	56·3	233	220·3	75·9	293	277·0	95·4
54	51·1	17·6	114	107·8	37·1	174	164·5	56·6	234	221·3	76·2	294	278·0	95·7
55	52·0	17·9	115	108·7	37·4	175	165·5	57·0	235	222·2	76·5	295	278·9	96·0
56	52·9	18·2	116	109·7	37·8	176	166·4	57·3	236	223·1	76·8	296	279·9	96·4
57	53·9	18·6	117	110·6	38·1	177	167·4	57·6	237	224·1	77·2	297	280·8	96·7
58	54·8	18·9	118	111·6	38·4	178	168·3	58·0	238	225·0	77·5	298	281·8	97·0
59	55·8	19·2	119	112·5	38·7	179	169·2	58·3	239	226·0	77·8	299	282·7	97·3
60	56·7	19·5	120	113·5	39·1	180	170·2	58·6	240	226·9	78·1	300	283·7	97·7
Dist.	Dep.	Lat.	Dist.	Dep.	Lat.	Dist.	Dep.	Lat.	Dist.	Dep.	Lat.	Dist.	Dep.	Lat.

FOR 71 DEGREES.

Dist.	Lat.	Dep.	Dist.	Lat.	Dep.	Dist.	Lat.	Dep.	Dist.	Lat.	Dep.	Dist.	Lat.	Dep.
1	0·9	0·3	61	57·3	20·9	121	113·7	41·4	181	170·1	61·9	241	226·5	82·4
2	1·9	0·7	62	58·3	21·2	122	114·6	41·7	182	171·0	62·2	242	227·4	82·8
3	2·8	1·0	63	59·2	21·5	123	115·6	42·1	183	172·0	62·6	243	228·3	83·1
4	3·8	1·4	64	60·1	21·9	124	116·5	42·4	184	172·9	62·9	244	229·3	83·5
5	4·7	1·7	65	61·1	22·2	125	117·5	42·8	185	173·8	63·3	245	230·2	83·8
6	5·6	2·1	66	62·0	22·6	126	118·4	43·1	186	174·8	63·6	246	231·2	84·2
7	6·6	2·4	67	63·0	22·9	127	119·3	43·4	187	175·7	64·0	247	232·1	84·5
8	7·5	2·7	68	63·9	23·3	128	120·3	43·8	188	176·7	64·3	248	233·0	84·8
9	8·5	3·1	69	64·8	23·6	129	121·2	44·1	189	177·6	64·6	249	234·0	85·2
10	9·4	3·4	70	65·8	23·9	130	122·2	44·5	190	178·5	65·0	250	234·9	85·5
11	10·3	3·8	71	66·7	24·3	131	123·1	44·8	191	179·5	65·3	251	235·9	85·8
12	11·3	4·1	72	67·7	24·6	132	124·0	45·1	192	180·4	65·7	252	236·8	86·2
13	12·2	4·4	73	68·6	25·0	133	125·0	45·5	193	181·4	66·0	253	237·7	86·5
14	13·2	4·8	74	69·5	25·3	134	125·9	45·8	194	182·3	66·4	254	238·7	86·9
15	14·1	5·1	75	70·5	25·7	135	126·9	46·2	195	183·2	66·7	255	239·6	87·2
16	15·0	5·5	76	71·4	26·0	136	127·8	46·5	196	184·2	67·0	256	240·6	87·6
17	16·0	5·8	77	72·4	26·3	137	128·7	46·9	197	185·1	67·4	257	241·5	87·9
18	16·9	6·2	78	73·3	26·7	138	129·7	47·2	198	186·1	67·7	258	242·4	88·2
19	17·9	6·5	79	74·2	27·0	139	130·6	47·5	199	187·0	68·1	259	243·4	88·6
20	18·8	6·8	80	75·2	27·4	140	131·6	47·9	200	187·9	68·4	260	244·3	88·9
21	19·7	7·2	81	76·1	27·7	141	132·5	48·2	201	188·9	68·7	261	245·3	89·3
22	20·7	7·5	82	77·1	28·0	142	133·4	48·6	202	189·8	69·1	262	246·2	89·6
23	21·6	7·9	83	78·0	28·4	143	134·4	48·9	203	190·8	69·4	263	247·1	90·0
24	22·6	8·2	84	78·9	28·7	144	135·3	49·3	204	191·7	69·8	264	248·1	90·3
25	23·5	8·6	85	79·9	29·1	145	136·3	49·6	205	192·6	70·1	265	249·0	90·6
26	24·4	8·9	86	80·8	29·4	146	137·2	49·9	206	193·6	70·5	266	250·0	91·0
27	25·4	9·2	87	81·8	29·8	147	138·1	50·3	207	194·5	70·8	267	250.9	91·3
28	26·3	9·6	88	82·7	30·1	148	139·1	50·6	208	195·5	71·1	268	251·8	91·7
29	27·3	9·9	89	83·6	30·4	149	140·0	51·0	209	196·4	71·5	269	252·8	92·0
30	28·2	10·3	90	84·6	30·8	150	141·0	51·3	210	197·3	71·8	270	253·7	92·3
31	29·1	10·6	91	85·5	31·1	151	141·9	51·6	211	198·3	72·2	271	254·7	92·7
32	30·1	10·9	92	86·5	31·5	152	142·8	52·0	212	199·2	72·5	272	255·6	93·0
33	31·0	11·3	93	87·4	31·8	153	143·8	52·3	213	200·2	72·9	273	256·5	93·4
34	31·9	11·6	94	88·3	32·1	154	144·7	52·7	214	201·1	73·2	274	257·5	93·7
35	32·9	12·0	95	89·3	32·5	155	145·7	53·0	215	202·0	73·5	275	258·4	94·1
36	33·8	12·3	96	90·2	32·8	156	146·6	53·4	216	203·0	73·9	276	259·4	94·4
37	34·8	12·7	97	91·2	33·2	157	147·5	53·7	217	203·9	74·2	277	260·3	94·7
38	35·7	13·0	98	92·1	33·5	158	148·5	54·0	218	204·9	74·6	278	261·2	95·1
39	36·6	13·3	99	93·0	33·9	159	149·4	54·4	219	205·8	74·9	279	262·2	95·4
40	37·6	13·7	100	94·0	34·2	160	150·4	54·7	220	206·7	75·2	280	263·1	95·8
41	38·5	14·0	101	94·9	34·5	161	151·3	55·1	221	207·7	75·6	281	264·1	96·1
42	39·5	14·4	102	95·8	34·9	162	152·2	55·4	222	208·6	75·9	282	265·0	96·4
43	40·4	14·7	103	96·8	35·2	163	153·2	55·7	223	209·6	76·3	283	265·9	96·8
44	41·3	15·0	104	97·7	35·6	164	154·1	56·1	224	210·5	76·6	284	266·9	97·1
45	42·3	15·4	105	98·7	35·9	165	155·0	56·4	225	211·4	77·0	285	267·8	97·5
46	43·2	15·7	106	99·6	36·3	166	156·0	56·8	226	212·4	77·3	286	268·8	97·8
47	44·2	16·1	107	100·5	36·6	167	156·9	57·1	227	213·3	77·6	287	269·7	98·2
48	45·1	16·4	108	101·5	36·9	168	157·9	57·5	228	214·2	78·0	288	270·6	98·5
49	46·0	16·8	109	102·4	37·3	169	158·8	57·8	229	215·2	78·3	289	271·6	98·8
50	47·0	17·1	110	103·4	37·6	170	159·7	58·1	230	216·1	78·7	290	272·5	99·2
51	47·9	17·4	111	104·3	38·0	171	160·7	58·5	231	217·1	79·0	291	273·5	99·5
52	48·9	17·8	112	105·2	38·3	172	161·6	58·8	232	218·0	79·3	292	274·4	99·9
53	49·8	18·1	113	106·2	38·6	173	162·6	59·2	233	218·9	79·7	293	275·3	100·2
54	50.7	18·5	114	107·1	39·0	174	163·5	59·5	234	219·9	80·0	294	276·3	100·6
55	51·7	18·8	115	108·1	39·3	175	164·4	59·9	235	220·8	80·4	295	277·2	100·9
56	52·6	19·2	116	109·0	39·7	176	165·4	60·2	236	221·8	80·7	296	278·1	101·2
57	53·6	19·5	117	109·9	40·0	177	166·3	60·5	237	222·7	81·1	297	279·1	101·6
58	54·5	19·8	118	110·9	40·4	178	167·3	60·9	238	223·6	81·4	298	280·0	101·9
59	55·4	20·2	119	111·8	40·7	179	168·2	61·2	239	224·6	81·7	299	281·0	102·3
60	56·4	20·5	120	112·8	41·0	180	169·1	61·6	240	225·5	82·1	300	281·9	102·6
Dist.	Dep.	Lat.	Dist.	Dep.	Lat.	Dist.	Dep.	Lat.	Dist.	Dep.	Lat.	Dist.	Dep.	Lat.

FOR 70 DEGREES.

Dist.	Lat.	Dep.	Dist.	Lat.	Dep.	Dist.	Lat.	Dep.	Dist.	Lat.	Dep.	Dist.	Lat.	Dep.
1	00·9	00·4	61	56·9	21·9	121	113·0	43·4	181	169·0	64·9	241	225·0	86·4
2	01·9	00·7	62	57·9	22·2	122	113·9	43·7	182	169·9	65·2	242	225·9	86·7
3	02·8	01·1	63	58·8	22·6	123	114·8	44·1	183	170·8	65·6	243	226·9	87·1
4	03·7	01·4	64	59·7	22·9	124	115·8	44·4	184	171·8	65·9	244	227·8	87·4
5	04·7	01·8	65	60·7	23·3	125	116·7	44·8	185	172·7	66·3	245	228·7	87·8
6	05·6	02·2	66	61·6	23·7	126	117·6	45·2	186	173·6	66·7	246	229·7	88·2
7	06·5	02·5	67	62·5	24·0	127	118·6	45·5	187	174·6	67·0	247	230·6	88·5
8	07·5	02·9	68	63·5	24·4	128	119·5	45·9	188	175·5	67·4	248	231·5	88·9
9	08·4	03·2	69	64·4	24·7	129	120·4	46·2	189	176·4	67·7	249	232·5	89·2
10	09·3	03·6	70	65·4	25·1	130	121·4	46·6	190	177·4	68·1	250	233·4	89·6
11	10·3	03·9	71	66·3	25·4	131	122·3	46·9	191	178·3	68·4	251	234·3	90·0
12	11·2	04·3	72	67·2	25·8	132	123·2	47·3	192	179·2	68·8	252	235·3	90·3
13	12·1	04·7	73	68·2	26·2	133	124·2	47·7	193	180·2	69·2	253	236·2	90·7
14	13·1	05·0	74	69·1	26·5	134	125·1	48·0	194	181·1	69·5	254	237·1	91·0
15	14·0	05·4	75	70·0	26·9	135	126·0	48·4	195	182·0	69·9	255	238·1	91·4
16	14·9	05·7	76	71·0	27·2	136	127·0	48·7	196	183·0	70·2	256	239·0	91·7
17	15·9	06·1	77	71·9	27·6	137	127·9	49·1	197	183·9	70·6	257	239·9	92·1
18	16·8	06·5	78	72·8	28·0	138	128·8	49·5	198	184·8	71·0	258	240·9	92·5
19	17·7	06·8	79	73·8	28·3	139	129·8	49·8	199	185·8	71·3	259	241·8	92·8
20	18·7	07·2	80	74·7	28·7	140	130·7	50·2	200	186·7	71·7	260	242·7	93·2
21	19·6	07·5	81	75·6	29·0	141	131·6	50·5	201	187·6	72·0	261	243·7	93·5
22	20·5	07·9	82	76·6	29·4	142	132·6	50·9	202	188·6	72·4	262	244·6	93·9
23	21·5	08·2	83	77·5	29·7	143	133·5	51·2	203	189·5	72·7	263	245·5	94·3
24	22·4	08·6	84	78·4	30·1	144	134·4	51·6	204	190·4	73·1	264	246·5	94·6
25	23·3	09·0	85	79·4	30·5	145	135·4	52·0	205	191·4	73·5	265	247·4	95·0
26	24·3	09·3	86	80·3	30·8	146	136·3	52·3	206	192·3	73·8	266	248·3	95·3
27	25·2	09·7	87	81·2	31·2	147	137·2	52·7	207	193·3	74·2	267	249·3	95·7
28	26·1	10·0	88	82·2	31·5	148	138·2	53·0	208	194·2	74·5	268	250·2	96·0
29	27·1	10·4	89	83·1	31·9	149	139·1	53·4	209	195·1	74·9	269	251·1	96·4
30	28·0	10·8	90	84·0	32·3	150	140·0	53·8	210	196·1	75·3	270	252·1	96·8
31	28·9	11·1	91	85·0	32·6	151	141·0	54·1	211	197·0	75·6	271	253·0	97·1
32	29·9	11·5	92	85·9	33·0	152	141·9	54·5	212	197·9	76·0	272	253·9	97·5
33	30·8	11·8	93	86·8	33·3	153	142·8	54·8	213	198·9	76·3	273	254·9	97·8
34	31·7	12·2	94	87·8	33·7	154	143·8	55·2	214	199·8	76·7	274	255·8	98·2
35	32·7	12·5	95	88·7	34·0	155	144·7	55·5	215	200·7	77·0	275	256·7	98·6
36	33·6	12·9	96	89·6	34·4	156	145·6	55·9	216	201·7	77·4	276	257·7	98·9
37	34·5	13·3	97	90·6	34·8	157	146·6	56·3	217	202·6	77·8	277	258·6	99·3
38	35·5	13·6	98	91·5	35·1	158	147·5	56·6	218	203·5	78·1	278	259·5	99·6
39	36·4	14·0	99	92·4	35·5	159	148·4	57·0	219	204·5	78·5	279	260·5	100·0
40	37·3	14·3	100	93·4	35·8	160	149·4	57·3	220	205·4	78·8	280	261·4	100·3
41	38·3	14·7	101	94·3	36·2	161	150·3	57·7	221	206·3	79·2	281	262·3	100·7
42	39·2	15·1	102	95·2	36·6	162	151·2	58·1	222	207·3	79·6	282	263·3	101·1
43	40·1	15·4	103	96·2	36·9	163	152·2	58·4	223	208·2	79·9	283	264·2	101·4
44	41·1	15·8	104	97·1	37·3	164	153·1	58·8	224	209·1	80·3	284	265·1	101·8
45	42·0	16·1	105	98·0	37·6	165	154·0	59·1	225	210·1	80·6	285	266·1	102·1
46	42·9	16·5	106	99·0	38·0	166	155·0	59·5	226	211·0	81·0	286	267·0	102·5
47	43·9	16·8	107	99·9	38·3	167	155·9	59·8	227	211·9	81·3	287	267·9	102·9
48	44·8	17·2	108	100·8	38·7	168	156·8	60·2	228	212·9	81·7	288	268·9	103·2
49	45·7	17·6	109	101·8	39·1	169	157·8	60·6	229	213·8	82·1	289	269·8	103·6
50	46·7	17·9	110	102·7	39·4	170	158·7	60·9	230	214·7	82·4	290	270·7	103·9
51	47·6	18·3	111	103·6	39·8	171	159·6	61·3	231	215·7	82·8	291	271·7	104·3
52	48·5	18·6	112	104·6	40·1	172	160·6	61·6	232	216·6	83·1	292	272·6	104·6
53	49·5	19·0	113	105·5	40·5	173	161·5	62·0	233	217·5	83·5	293	273·5	105·0
54	50·4	19·4	114	106·4	40·9	174	162·4	62·4	134	218·5	83·9	294	274·5	105·4
55	51·3	19·7	115	107·4	41·2	175	163·4	62·7	235	219·4	84·2	295	275·4	105·7
56	52·3	20·1	116	108·3	41·6	176	164·3	63·1	236	220·3	84·6	296	276·3	106·1
57	53·2	20·4	117	109·2	41·9	177	165·2	63·4	237	221·3	84·9	297	277·3	106·4
58	54·1	20·8	118	110·2	42·3	178	166·2	63·8	238	222·2	85·3	298	278·2	106·8
59	55·1	21·1	119	111·1	42·6	179	167·1	64·1	239	223·1	85·6	299	279·1	107·2
60	56·0	21·5	120	112·0	43·0	180	168·0	64·5	240	224·1	86·0	300	280·1	107·5
Dist.	Dep.	Lat.	Dist.	Dep.	Lat.	Dist.	Dep.	Lat.	Dist.	Dep.	Lat.	Dist.	Dep.	Lat.

FOR 69 DEGREES.

Dist.	Lat.	Dep.	Dist.	Lat.	Dep.	Dist.	Lat.	Dep.	Dist.	Lat.	Dep.	Dist.	Lat.	Dep.
1	00·9	00·4	61	56·6	22·9	121	112·2	45·3	181	167·8	67·8	241	223·5	90·3
2	01·9	00·7	62	57·5	23·2	122	113·1	45·7	182	168·7	68·2	242	224·4	90·7
3	02·8	01·1	63	58·4	23·6	123	114·0	46·1	183	169·7	68·6	243	225·3	91·0
4	03·7	01·5	64	59·3	24·0	124	115·0	46·5	184	170·6	68·9	244	226·2	91·4
5	04·6	01·9	65	60·3	24·3	125	115·9	46·8	185	171·5	69·3	245	227·2	91·8
6	05·6	02·2	66	61·2	24·7	126	116·8	47·2	186	172·5	69·7	246	228·1	92·2
7	06·5	02·6	67	62·1	25·1	127	117·8	47·6	187	173·4	70·1	247	229·0	92·5
8	07·4	03·0	68	63·0	25·5	128	118·7	47·9	188	174·3	70·4	248	229·9	92·9
9	08·3	03·4	69	64·0	25·8	129	119·6	48·3	189	175·2	70·8	249	230·9	93·3
10	09·3	03·7	70	64·9	26·2	130	120·5	48·7	190	176·2	71·2	250	231·8	93·7
11	10·2	04·1	71	65·8	26·6	131	121·5	49·1	191	177·1	71·5	251	232·7	94·0
12	11·1	04·5	72	66·8	27·0	132	122·4	49·4	192	178·0	71·9	252	233·7	94·4
13	12·1	04·9	73	67·7	27·3	133	123·3	49·8	193	178·9	72·3	253	234·6	94·8
14	13·0	05·2	74	68·6	27·7	134	124·2	50·2	194	179·9	72·7	254	235·5	95·2
15	13·9	05·6	75	69·5	28·1	135	125·2	50·6	195	180·8	73·0	255	236·4	95·5
16	14·8	06·0	76	70·5	28·5	136	126·1	50·9	196	181·7	73·4	256	237·4	95·9
17	15·8	06·4	77	71·4	28·8	137	127·0	51·3	197	182·7	73·8	257	238·3	96·3
18	16·7	06·7	78	72·3	29·2	138	128·0	51·7	198	183·6	74·2	258	239·2	96·6
19	17·6	07·1	79	73·2	29·6	139	128·9	52·1	199	184·5	74·5	259	240·1	97·0
20	18·5	07·5	80	74·2	30·0	140	129·8	52·4	200	185·4	74·9	260	241·1	97·4
21	19·5	07·9	81	75·1	30·3	141	130·7	52·8	201	186·4	75·3	261	242·0	97·8
22	20·4	08·2	82	76·0	30·7	142	131·7	53·2	202	187·3	75·7	262	242·9	98·1
23	21·3	08·6	83	77·0	31·1	143	132·6	53·6	203	188·2	76·0	263	243·8	98·5
24	22·3	09·0	84	77·9	31·5	144	133·5	53·9	204	189·1	76·4	264	244·8	98·9
25	23·2	09·4	85	78·8	31·8	145	134·4	54·3	205	190·1	76·8	265	245·7	99·3
26	24·1	09·7	86	79·7	32·2	146	135·4	54·7	206	191·0	77·2	266	246·6	99·6
27	25·0	10·1	87	80·7	32·6	147	136·3	55·1	207	191·9	77·5	267	247·6	100·0
28	26·0	10·5	88	81·6	33·0	148	137·2	55·4	208	192·9	77·9	268	248·5	100·4
29	26·9	10·9	89	82·5	33·3	149	138·2	55·8	209	193·8	78·3	269	249·4	100·8
30	27·8	11·2	90	83·4	33·7	150	139·1	56·2	210	194·7	78·7	270	250·3	101·1
31	28·7	11·6	91	84·4	34·1	151	140·0	56·6	211	195·6	79·0	271	251·3	101·5
32	29·7	12·0	92	85·3	34·5	152	140·9	56·9	212	196·6	79·4	272	252·2	101·9
33	30·6	12·4	93	86·2	34·8	153	141·9	57·3	213	197·5	79·8	273	253·1	102·3
34	31·5	12·7	94	87·2	35·2	154	142·8	57·7	214	198·4	80·2	274	254·0	102·6
35	32·5	13·1	95	88·1	35·6	155	143·7	58·1	215	199·3	80·5	275	255·0	103·0
36	33·4	13·5	96	89·0	36·0	156	144·6	58·4	216	200·3	80·9	276	255·9	103·4
37	34·3	13·9	97	89·9	36·3	157	145·6	58·8	217	201·2	81·3	277	256·8	103·8
38	35·2	14·2	98	90·9	36·7	158	146·5	59·2	218	202·1	81·7	278	257·8	104·1
39	36·2	14·6	99	91·8	37·1	159	147·4	59·6	219	203·1	82·0	279	258·7	104·5
40	37·1	15·0	100	92·7	37·5	160	148·3	59·9	220	204·0	82·4	280	259·6	104·9
41	38·0	15·4	101	93·6	37·8	161	149·3	60·3	221	204·9	82·8	281	260·5	105·3
42	38·9	15·7	102	94·6	38·2	162	150·2	60·7	222	205·8	83·2	282	261·5	105·6
43	39·9	16·1	103	95·5	38·6	163	151·1	61·1	223	206·8	83·5	283	262·4	106·0
44	40·8	16·5	104	96·4	39·0	164	152·1	61·4	224	207·7	83·9	284	263·3	106·4
45	41·7	16·9	105	97·4	39·3	165	153·0	61·8	225	208·6	84·3	285	264·2	106·8
46	42·7	17·2	106	98·3	39·7	166	153·9	62·2	226	209·5	84·7	286	265·2	107·1
47	43·6	17·6	107	99·2	40·1	167	154·8	62·6	227	210·5	85·0	287	266·1	107·5
48	44·5	18·0	108	100·1	40·5	168	155·8	62·9	228	211·4	85·4	288	267·0	107·9
49	45·4	18·4	109	101·1	40·8	169	156·7	63·3	229	212·3	85·8	289	268·0	108·3
50	46·4	18·7	110	102·0	41·2	170	157·6	63·7	230	213·3	86·2	290	268·9	108·6
51	47·3	19·1	111	102·9	41·6	171	158·5	64·1	231	214·2	86·5	291	269·8	109·0
52	48·2	19·5	112	103·8	42·0	172	159·5	64·4	232	215·1	86·9	292	270·7	109·4
53	49·1	19·9	113	104·8	42·3	173	160·4	64·8	233	216·0	87·3	293	271·7	109·8
54	50·1	20·2	114	105·7	42·7	174	161·3	65·2	234	217·0	87·7	294	272·6	110·1
55	51·0	20·6	115	106·6	43·1	175	162·3	65·6	235	217·9	88·0	295	273·5	110·5
56	51·9	21·0	116	107·6	43·5	176	163·2	65·9	236	218·8	88·4	296	274·4	110·9
57	52·8	21·4	117	108·5	43·8	177	164·1	66·3	237	219·7	88·8	297	275·4	111·3
58	53·8	21·7	118	109·4	44·2	178	165·0	66·7	238	220·7	89·2	298	276·3	111·6
59	54·7	22·1	119	110·3	44·6	179	166·0	67·1	239	221·6	89·5	299	277·2	112·0
60	55·6	22·5	120	111·3	45·0	180	166·9	67·4	240	222·5	89·9	300	278·2	112·4
Dist.	Dep.	Lat.	Dist.	Dep.	Lat.	Dist.	Dep.	Lat.	Dist.	Dep.	Lat.	Dist.	Dep.	Lat.

FOR 68 DEGREES.

Dist.	Lat.	Dep.	Dist.	Lat.	Dep.	Dist.	Lat.	Dep.	Dist.	Lat.	Dep.	Dist.	Lat.	Dep.
1	00·9	00·4	61	56·2	23·8	121	111·4	47·3	181	166·6	70·7	241	221·8	94·2
2	01·8	00·8	62	57·1	24·2	122	112·3	47·7	182	167·5	71·1	242	222·8	94·5
3	02·8	01·2	63	58·0	24·6	123	113·2	48·1	183	168·5	71·5	243	223·7	94·9
4	03·7	01·6	64	58·9	25·0	124	114·1	48·5	184	169·4	71·9	244	224·6	95·3
5	04·6	02·0	65	59·8	25·4	125	115·1	48·8	185	170·3	72·3	245	225·5	95·7
6	05·5	02·3	66	60·8	25·8	126	116·0	49·2	186	171·2	72·7	246	226·4	96·1
7	06·4	02·7	67	61·7	26·2	127	116·9	49·6	187	172·1	73·1	247	227·4	96·5
8	07·4	03·1	68	62·6	26·6	128	117·8	50·0	188	173·1	73·5	248	228·3	96·9
9	08·3	03·5	69	63·5	27·0	129	118·7	50·4	189	174·0	73·8	249	229·2	97·3
10	09·2	03·9	70	64·4	27·4	130	119·7	50·8	190	174·9	74·2	250	230·1	97·7
11	10·1	04·3	71	65·4	27·7	131	120·6	51·2	191	175·8	74·6	251	231·0	98·1
12	11·0	04·7	72	66·3	28·1	132	121·5	51·6	192	176·7	75·0	252	232·0	98·5
13	12·0	05·1	73	67·2	28·5	133	122·4	52·0	193	177·7	75·4	253	232·9	98·9
14	12·9	05·5	74	68·1	28·9	134	123·3	52·4	194	178·6	75·8	254	233·8	99·2
15	13·8	05·9	75	69·0	29·3	135	124·3	52·7	195	179·5	76·2	255	234·7	99·6
16	14·7	06·3	76	70·0	29·7	136	125·2	53·1	196	180·4	76·6	256	235·6	100·0
17	15·6	06·6	77	70·9	30·1	137	126·1	53·5	197	181·3	77·0	257	236·6	100·4
18	16·6	07·0	78	71·8	30·5	138	127·0	53·9	198	182·3	77·4	258	237·5	100·8
19	17·5	07·4	79	72·7	30·9	139	128·0	54·3	199	183·2	77·8	259	238·4	101·2
20	18·4	07·8	80	73·6	31·3	140	128·9	54·7	200	184·1	78·1	260	239·3	101·6
21	19·3	08·2	81	74·6	31·6	141	129·8	55·1	201	185·0	78·5	261	240·3	102·0
22	20·3	08·6	82	75·5	32·0	142	130·7	55·5	202	185·9	78·9	262	241·2	102·4
23	21·2	09·0	83	76·4	32·4	143	131·6	55·9	203	186·9	79·3	263	242·1	102·8
24	22·1	09·4	84	77·3	32·8	144	132·6	56·3	204	187·8	79·7	264	243·0	103·2
25	23·0	09·8	85	78·2	33·2	145	133·5	56·7	205	188·7	80·1	265	243·9	103·5
26	23·9	10·2	86	79·2	33·6	146	134·4	57·0	206	189·6	80·5	266	244·9	103·9
27	24·9	10·5	87	80·1	34·0	147	135·3	57·4	207	190·5	80·9	267	245·8	104·3
28	25·8	10·9	88	81·0	34·4	148	136·2	57·8	208	191·5	81·3	268	246·7	104·7
29	26·7	11·3	89	81·9	34·8	149	137·2	58·2	209	192·4	81·7	269	247·6	105·1
30	27·6	11·7	90	82·8	35·2	150	138·1	58·6	210	193·3	82·1	270	248·5	105·5
31	28·5	12·1	91	83·8	35·6	151	139·0	59·0	211	194·2	82·4	271	249·5	105·9
32	29·5	12·5	92	84·7	35·9	152	139·9	59·4	212	195·1	82·8	272	250·4	106·3
33	30·4	12·9	93	85·6	36·3	153	140·8	59·8	213	196·1	83·2	273	251·3	106·7
34	31·3	13·3	94	86·5	36·7	154	141·8	60·2	214	197·0	83·6	274	252·2	107·1
35	32·2	13·7	95	87·4	37·1	155	142·7	60·6	215	197·9	84·0	275	253·1	107·5
36	33·1	14·1	96	88·4	37·5	156	143·6	61·0	216	198·8	84·4	276	254·1	107·8
37	34·1	14·5	97	89·3	37·9	157	144·5	61·3	217	199·7	84·8	277	255·0	108·2
38	35·0	14·8	98	90·2	38·3	158	145·4	61·7	218	200·7	85·2	278	255·9	108·6
39	35·9	15·2	99	91·1	38·7	159	146·4	62·1	219	201·6	85·6	279	256·8	109·0
40	36·8	15·6	100	92·1	39·1	160	147·3	62·5	220	202·5	86·0	280	257·7	109·4
41	37·7	16·0	101	93·0	39·5	161	148·2	62·9	221	203·4	86·4	281	258·7	109·8
42	38·7	16·4	102	93·9	39·9	162	149·1	63·3	222	204·4	86·7	282	259·6	110·2
43	39·6	16·8	103	94·8	40·2	163	150·0	63·7	223	205·3	87·1	283	260·5	110·6
44	40·5	17·2	104	95·7	40·6	164	151·0	64·1	224	206·2	87·5	284	261·4	111·0
45	41·4	17·6	105	96·7	41·0	165	151·9	64·5	225	207·1	87·9	285	262·3	111·4
46	42·3	18·0	106	97·6	41·4	166	152·8	64·9	226	208·0	88·3	286	263·3	111·7
47	43·3	18·4	107	98·5	41·8	167	153·7	65·3	227	209·0	88·7	287	264·2	112·1
48	44·2	18·8	108	99·4	42·2	168	154·6	65·6	228	209·9	89·1	288	265·1	112·5
49	45·1	19·1	109	100·3	42·6	169	155·6	66·0	229	210·8	89·5	289	266·0	112·9
50	46·0	19·5	110	101·3	43·0	170	156·5	66·4	230	211·7	89·9	290	266·9	113·3
51	46·9	19·9	111	102·2	43·4	171	157·4	66·8	231	212·6	90·3	291	267·9	113·7
52	47·9	20·3	112	103·1	43·8	172	158·3	67·2	232	213·6	90·6	292	268·8	114·1
53	48·8	20·7	113	104·0	44·2	173	159·2	67·6	233	214·5	91·0	293	269·7	114·5
54	49·7	21·1	114	104·9	44·5	174	160·2	68·0	234	215·4	91·4	294	270·6	114·9
55	50·6	21·5	115	105·9	44·9	175	161·1	68·4	235	216·3	91·8	295	271·5	115·3
56	51·5	21·9	116	106·8	45·3	176	162·0	68·8	236	217·2	92·2	296	272·5	115·7
57	52·5	22·3	117	107·7	45·7	177	162·9	69·2	237	218·2	92·6	297	273·4	116·0
58	53·4	22·7	118	108·6	46·1	178	163·8	69·6	238	219·1	93·0	298	274·3	116·4
59	54·3	23·1	119	109·5	46·5	179	164·8	69·9	239	220·0	93·4	299	275·2	116·8
60	55·2	23·4	120	110·5	46·9	180	165·7	70·3	240	220·9	93·8	300	276·2	117·2
Dist.	Dep.	Lat.	Dist.	Dep.	Lat.	Dist.	Dep.	Lat.	Dist.	Dep.	Lat.	Dist.	Dep.	Lat.

FOR 67 DEGREES.

Dist	Lat.	Dep.	Dist.	Lat.	Dep.	Dist.	Lat.	Dep.	Dist.	Lat.	Dep.	Dist.	Lat.	Dep.
1	00·9	00·4	61	55·7	24·8	121	110·5	49·2	181	165·4	73·6	241	220·2	98·0
2	01·8	00·8	62	56·6	25·2	122	111·5	49·6	182	166·3	74·0	242	221·1	98·4
3	02·7	01·2	63	57·6	25·6	123	112·4	50·0	183	167·2	74·4	243	222·0	98·8
4	03·7	01·6	64	58·5	26·0	124	113·3	50·4	184	168·1	74·8	244	222·9	99·2
5	04·6	02·0	65	59·4	26·4	125	114·2	50·8	185	169·0	75·2	245	223·8	99·7
6	05·5	02·4	66	60·3	26·8	126	115·1	51·2	186	169·9	75·7	246	224·7	100·1
7	06·4	02·8	67	61·2	27·3	127	116·0	51·7	187	170·8	76·1	247	225·6	100·5
8	07·3	03·3	68	62·1	27·7	128	116·9	52·1	188	171·7	76·5	248	226·6	100·9
9	08·2	03·7	69	63·0	28·1	129	117·8	52·5	189	172·7	76·9	249	227·5	101·3
10	09·1	04·1	70	63·9	28·5	130	118·8	52·9	190	173·6	77·3	250	228·4	101·7
11	10·0	04·5	71	64·9	28·9	131	119·7	53·3	191	174·5	77·7	251	229·3	102·1
12	11·0	04·9	72	65·8	29·3	132	120·6	53·7	192	175·4	78·1	252	230·2	102·5
13	11·9	05·3	73	66·7	29·7	133	121·5	54·1	193	176·3	78·5	253	231·1	102·9
14	12·8	05·7	74	67·6	30·1	134	122·4	54·5	194	177·2	78·9	254	232·0	103·3
15	13·7	06·1	75	68·5	30·5	135	123·3	54·9	195	178·1	79·3	255	233·0	103·7
16	14·6	06·5	76	69·4	30·9	136	124·2	55·3	196	179·1	79·7	256	233·9	104·1
17	15·5	06·9	77	70·3	31·3	137	125·2	55·7	197	180·0	80·1	257	234·8	104·5
18	16·4	07·3	78	71·3	31·7	138	126·1	56·1	198	180·9	80·5	258	235·7	104·9
19	17·4	07·7	79	72·2	32·1	139	127·0	56·5	199	181·8	80·9	259	236·6	105·3
20	18·3	08·1	80	73·1	32·5	140	127·9	56·9	200	182·7	81·3	260	237·5	105·8
21	19·2	08·5	81	74·0	32·9	141	128·8	57·3	201	183·6	81·8	261	238·4	106·2
22	20·1	08·9	82	74·9	33·4	142	129·7	57·8	202	184·5	82·2	262	239·3	106·6
23	21·0	09·4	83	75·8	33·8	143	130·6	58·2	203	185·4	82·6	263	240·3	107·0
24	21·9	09·8	84	76·7	34·2	144	131·6	58·6	204	186·4	83·0	264	241·2	107·4
25	22·8	10·2	85	77·7	34·6	145	132·5	59·0	205	187·3	83·4	265	242·1	107·8
26	23·8	10·6	86	78·6	35·0	146	133·4	59·4	206	188·2	83·8	266	243·0	108·2
27	24·7	11·0	87	79·5	35·4	147	134·3	59·8	207	189·1	84·2	267	243·9	108·6
28	25·6	11·4	88	80·4	35·8	148	135·2	60·2	208	190·0	84·6	268	244·8	109·0
29	26·5	11·8	89	81·3	36·2	149	136·1	60·6	209	190·9	85·0	269	245·7	109·4
30	27·4	12·2	90	82·2	36·6	150	137·0	61·0	210	191·8	85·4	270	246·7	109·8
31	28·3	12·6	91	83·1	37·0	151	137·9	61·4	211	192·8	85·8	271	247·6	110·2
32	29·2	13·0	92	84·0	37·4	152	138·9	61·8	212	193·7	86·2	272	248·5	110·6
33	30·1	13·4	93	85·0	37·8	153	139·8	62·2	213	194·6	86·6	273	249·4	111·0
34	31·1	13·8	94	85·9	38·2	154	140·7	62·6	214	195·5	87·0	274	250·3	111·4
35	32·0	14·2	95	86·8	38·6	155	141·6	63·0	215	196·4	87·4	275	251·2	111·9
36	32·9	14·6	96	87·7	39·0	156	142·5	63·5	216	197·3	87·9	276	252·1	112·3
37	33·8	15·0	97	88·6	39·5	157	143·4	63·9	217	198·2	88·3	277	253·1	112·7
38	34·7	15·5	98	89·5	39·9	158	144·3	64·3	218	199·2	88·7	278	254·0	113·1
39	35·6	15·9	99	90·4	40·3	159	145·3	64·7	219	200·1	89·1	279	254·9	113·5
40	36·5	16·3	100	91·4	40·7	160	146·2	65·1	220	201·0	89·5	280	255·8	113·9
41	37·5	16·7	101	92·3	41·1	161	147·1	65·5	221	201·9	89·9	281	256·7	114·3
42	38·4	17·1	102	93·2	41·5	162	148·0	65·9	222	202·8	90·3	282	257·6	114·7
43	39·3	17·5	103	94·1	41·9	163	148·9	66·3	223	203·7	90·7	283	258·5	115·1
44	40·2	17·9	104	95·0	42·3	164	149·8	66·7	224	204·6	91·1	284	259·4	115·5
45	41·1	18·3	105	95·9	42·7	165	150·7	67·1	225	205·5	91·5	285	260·4	115·9
46	42·0	18·7	106	96·8	43·1	166	151·6	67·5	226	206·5	91·9	286	261·3	116·3
47	42·9	19·1	107	97·7	43·5	167	152·6	67·9	227	207·4	92·3	287	262·2	116·7
48	43·9	19·5	108	98·7	43·9	168	153·5	68·3	228	208·3	92·7	288	263·1	117·1
49	44·8	19·9	109	99·6	44·3	169	154·4	68·7	229	209·2	93·1	289	264·0	117·5
50	45·7	20·3	110	100·5	44·7	170	155·3	69·1	230	210·1	93·5	290	264·9	118·0
51	46·6	20·7	111	101·4	45·1	171	156·2	69·6	231	211·0	94·0	291	265·8	118·4
52	47·5	21·2	112	102·3	45·6	172	157·1	70·0	232	211·9	94·4	292	266·8	118·8
53	48·4	21·6	113	103·2	46·0	173	158·0	70·4	233	212·9	94·8	293	267·7	119·2
54	49·3	22·0	114	104·1	46·4	174	159·0	70·8	234	213·8	95·2	294	268·6	119·6
55	50·2	22·4	115	105·1	46·8	175	159·9	71·2	235	214·7	95·6	295	269·5	120·0
56	51·2	22·8	116	106·0	47·2	176	160·8	71·6	236	215·6	96·0	296	270·4	120·4
57	52·1	23·2	117	106·9	47·6	177	161·7	72·0	237	216·5	96·4	297	271·3	120·8
58	53·0	23·6	118	107·8	48·0	178	162·6	72·4	238	217·4	96·8	298	272·2	121·2
59	53·9	24·0	119	108·7	48·4	179	163·5	72·8	239	218·3	97·2	299	273·2	121·6
60	54·8	24·4	120	109·6	48·8	180	164·4	73·2	240	219·3	97·6	300	274·1	122·0
Dist.	Dep.	Lat.	Dist.	Dep.	Lat.	Dist.	Dep.	Lat.	Dist.	Dep.	Lat.	Dist.	Dep.	Lat.

FOR 66 DEGREES.

Dist.	Lat.	Dep.	Dist.	Lat.	Dep.	Dist.	Lat.	Dep.	Dist.	Lat.	Dep.	Dist.	Lat.	Dep.
1	00·9	00·4	61	55·3	25·8	121	109·7	51·1	181	164·0	76·5	241	218·4	101·9
2	01·8	00·8	62	56·2	26·2	122	110·6	51·6	182	164·9	76·9	242	219·3	102·3
3	02·7	01·3	63	57·1	26·6	123	111·5	52·0	183	165·9	77·3	243	220·2	102·7
4	03·6	01·7	64	58·0	27·0	124	112·4	52·4	184	166·8	77·8	244	221·1	103·1
5	04·5	02·1	65	58·9	27·5	125	113·3	52·8	185	167·7	78·2	245	222·0	103·5
6	05·4	02·5	66	59·8	27·9	126	114·2	53·2	186	168·6	78·6	246	223·0	104·0
7	06·3	03·0	67	60·7	28·3	127	115·1	53·7	187	169·5	79·0	247	223·9	104·4
8	07·3	03·4	68	61·6	28·7	128	116·0	54·1	188	170·4	79·5	248	224·8	104·8
9	08·2	03·8	69	62·5	29·2	129	116·9	54·5	189	171·3	79·9	249	225·7	105·2
10	09·1	04·2	70	63·4	29·6	130	117·8	54·9	190	172·2	80·3	250	226·6	105·7
11	10·0	04·6	71	64·3	30·0	131	118·7	55·4	191	173·1	80·7	251	227·5	106·1
12	10·9	05·1	72	65·3	30·4	132	119·6	55·8	192	174·0	81·1	252	228·4	106·5
13	11·8	05·5	73	66·2	30·9	133	120·5	56·2	193	174·9	81·6	253	229·3	106·9
14	12·7	05·9	74	67·1	31·3	134	121·4	56·6	194	175·8	82·0	254	230·2	107·3
15	13·6	06·3	75	68·0	31·7	135	122·4	57·1	195	176·7	82·4	255	231·1	107·8
16	14·5	06·8	76	68·9	32·1	136	123·3	57·5	196	177·6	82·8	256	232·0	108·2
17	15·4	07·2	77	69·8	32·5	137	124·2	57·9	197	178·5	83·3	257	232·9	108·6
18	16·3	07·6	78	70·7	33·0	138	125·1	58·3	198	179·4	83·7	258	233·8	109·0
19	17·2	08·0	79	71·6	33·4	139	126·0	58·7	199	180·4	84·1	259	234·7	109·5
20	18·1	08·5	80	72·5	33·8	140	126·9	59·2	200	181·3	84·5	260	235·6	109·9
21	19·0	08·9	81	73·4	34·2	141	127·8	59·6	201	182·2	84·9	261	236·5	110·3
22	19·9	09·3	82	74·3	34·7	142	128·7	60·0	202	183·1	85·4	262	237·5	110·7
23	20·8	09·7	83	75·2	35·1	143	129·6	60·4	203	184·0	85·8	263	238·4	111·1
24	21·8	10·1	84	76·1	35·5	144	130·5	60·9	204	184·9	86·2	264	239·3	111·6
25	22·7	10·6	85	77·0	35·9	145	131·4	61·3	205	185·8	86·6	265	240·2	112·0
26	23·6	11·0	86	77·9	36·3	146	132·3	61·7	206	186·7	87·1	266	241·1	112·4
27	24·5	11·4	87	78·8	36·8	147	133·2	62·1	207	187·6	87·5	267	242·0	112·8
28	25·4	11·8	88	79·8	37·2	148	134·1	62·5	208	188·5	87·9	268	242·9	113·3
29	26·3	12·3	89	80·7	37·6	149	13[illegible]	63·0	209	189·4	88·3	269	243·8	113·7
30	27·2	12·7	90	81·6	38·0	150	135·9	63·4	210	190·3	88·7	270	244·7	114·1
31	28·1	13·1	91	82·5	38·5	151	136·9	63·8	211	191·2	89·2	271	245·6	114·5
32	29·0	13·5	92	83·4	38·9	152	137·8	64·2	212	192·1	89·6	272	246·5	115·0
33	29·9	13·9	93	84·3	39·3	153	138·7	64·7	213	193·0	90·0	273	247·4	115·4
34	30·8	14·4	94	85·2	39·7	154	139·6	65·1	214	193·9	90·4	274	248·3	115·8
35	31·7	14·8	95	86·1	40·1	155	140·5	65·5	215	194·9	90·9	275	249·2	116·2
36	32·6	15·2	96	87·0	40·6	156	141·4	65·9	216	195·8	91·3	276	250·1	116·6
37	33·5	15·6	97	87·9	41·0	157	142·3	66·4	217	196·7	91·7	277	251·0	117·1
38	34·4	16·1	98	88·8	41·4	158	143·2	66·8	218	197·6	92·1	278	252·0	117·5
39	35·3	16·5	99	89·7	41·8	159	144·1	67·2	219	198·5	92·6	279	252·9	117·9
40	36·3	16·9	100	90·6	42·3	160	145·0	67·6	220	199·4	93·0	280	253·8	118·3
41	37·2	17·3	101	91·5	42·7	161	145·9	68·0	221	200·3	93·4	281	254·7	118·8
42	38·1	17·7	102	92·4	43·1	162	146·8	68·5	222	201·2	93·8	282	255·6	119·2
43	39·0	18·2	103	93·3	43·5	163	147·7	68·9	223	202·1	94·2	283	256·5	119·6
44	39·9	18·6	104	94·3	44·0	164	148·6	69·3	224	203·0	94·7	284	257·4	120·0
45	40·8	19·0	105	95·2	44·4	165	149·5	69·7	225	203·9	95·1	285	258·3	120·4
46	41·7	19·4	106	96·1	44·8	166	150·4	70·2	226	204·8	95·5	286	259·2	120·9
47	42·6	19·9	107	97·0	45·2	167	151·4	70·6	227	205·7	95·9	287	260·1	121·3
48	43·5	20·3	108	97·9	45·6	168	152·3	71·0	228	206·6	96·4	288	261·0	121·7
49	44·4	20·7	109	98·8	46·1	169	153·2	71·4	229	207·5	96·8	289	261·9	122·1
50	45·3	21·1	110	99·7	46·5	170	154·1	71·8	230	208·5	97·2	290	262·8	122·6
51	46·2	21·6	111	100·6	46·9	171	155·0	72·3	231	209·4	97·6	291	263·7	123·0
52	47·1	22·0	112	101·5	47·3	172	155·9	72·7	232	210·3	98·0	292	264·6	123·4
53	48·0	22·4	113	102·4	47·8	173	156·8	73·1	233	211·2	98·5	293	265·5	123·8
54	48·9	22·8	114	103·3	48·2	174	157·7	73·5	234	212·1	98·9	294	266·5	124·2
55	49·8	23·2	115	104·2	48·6	175	158·6	74·0	235	213·0	99·3	295	267·4	124·7
56	50·8	23·7	116	105·1	49·0	176	159·5	74·4	236	213·9	99·7	296	268·3	125·1
57	51·7	24·1	117	106·0	49·4	177	160·4	74·8	237	214·8	100·2	297	269·2	125·5
58	52·6	24·5	118	106·9	49·9	178	161·3	75·2	238	215·7	100·6	298	270·1	125·9
59	53·5	24·9	119	107·9	50·3	179	162·2	75·6	239	216·6	101·0	299	271·0	126·4
60	54·4	25·4	120	108·8	50·7	180	163·1	76·1	240	217·5	101·4	300	271·9	126·8
Dist.	Dep.	Lat.	Dist.	Dep.	Lat.	Dist.	Dep.	Lat.	Dist.	Dep.	Lat.	Dist.	Dep.	Lat.

FOR 65 DEGREES.

Dist.	Lat.	Dep.	Dist.	Lat.	Dep.	Dist.	Lat.	Dep.	Dist.	Lat.	Dep.	Dist.	Lat.	Dep.
1	00·9	00·4	61	54·8	26·7	121	108·8	53·0	181	162·7	79·3	241	216·6	105·6
2	01·8	00·9	62	55·7	27·2	122	109·7	53·5	182	163·6	79·8	242	217·5	106·1
3	02·7	01·3	63	56·6	27·6	123	110·6	53·9	183	164·5	80·2	243	218·4	106·5
4	03·6	01·8	64	57·5	28·1	124	111·5	54·4	184	165·4	80·7	244	219·3	107·0
5	04·5	02·2	65	58·4	28·5	125	112·3	54·8	185	166·3	81·1	245	220·2	107·4
6	05·4	02·6	66	59·3	28·9	126	113·2	55·2	186	167·2	81·5	246	221·1	107·8
7	06·3	03·1	67	60·2	29·4	127	114·1	55·7	187	168·1	82·0	247	222·0	108·3
8	07·2	03·5	68	61·1	29·8	128	115·0	56·1	188	169·0	82·4	248	222·9	108·7
9	08·1	03·9	69	62·0	30·2	129	115·9	56·5	189	169·9	82·9	249	223·8	109·2
10	09·0	04·4	70	62·9	30·7	130	116·8	57·0	190	170·8	83·3	250	224·7	109·6
11	09·9	04·8	71	63·8	31·1	131	117·7	57·4	191	171·7	83·7	251	225·6	110·0
12	10·8	05·3	72	64·7	31·6	132	118·6	57·9	192	172·6	84·2	252	226·5	110·5
13	11·7	05·7	73	65·6	32·0	133	119·5	58·3	193	173·5	84·6	253	227·4	110·9
14	12·6	06·1	74	66·5	32·4	134	120·4	58·7	194	174·4	85·0	254	228·3	111·3
15	13·5	06·6	75	67·4	32·9	135	121·3	59·2	195	175·3	85·5	255	229·2	111·8
16	14·4	07·0	76	68·3	33·3	136	122·2	59·6	196	176·2	85·9	256	230·1	112·2
17	15·3	07·5	77	69·2	33·8	137	123·1	60·1	197	177·1	86·4	257	231·0	112·7
18	16·2	07·9	78	70·1	34·2	138	124·0	60·5	198	178·0	86·8	258	231·9	113·1
19	17·1	08·3	79	71·0	34·6	139	124·9	60·9	199	178·9	87·2	259	232·8	113·5
20	18·0	08·8	80	71·9	35·1	140	125·8	61·4	200	179·8	87·7	260	233·7	114·0
21	18·9	09·2	81	72·8	35·5	141	126·7	61·8	201	180·7	88·1	261	234·6	114·4
22	19·8	09·6	82	73·7	35·9	142	127·6	62·2	202	181·6	88·6	262	235·5	114·9
23	20·7	10·1	83	74·6	36·4	143	128·5	62·7	203	182·5	89·0	263	236·4	115·3
24	21·6	10·5	84	75·5	36·8	144	129·4	63·1	204	183·4	89·4	264	237·3	115·7
25	22·5	11·0	85	76·4	37·3	145	130·3	63·6	205	184·3	89·9	265	238·2	116·2
26	23·4	11·4	86	77·3	37·7	146	131·2	64·0	206	185·2	90·3	266	239·1	116·6
27	24·3	11·8	87	78·2	38·1	147	132·1	64·4	207	186·1	90·7	267	240·0	117·0
28	25·2	12·3	88	79·1	38·6	148	133·0	64·9	208	186·9	91·2	268	240·9	117·5
29	26·1	12·7	89	80·0	39·0	149	133·[illegible]	[illegible]·3	209	187·8	91·6	269	241·8	117·9
30	27·0	13·2	90	80·9	39·5	150	134·8	65·8	210	188·7	92·1	270	242·7	118·4
31	27·9	13·6	91	81·8	39·9	151	135·7	66·2	211	189·6	92·5	271	243·6	118·8
32	28·8	14·0	92	82·7	40·3	152	136·6	66·6	212	190·5	92·9	272	244·5	119·2
33	29·7	14·5	93	83·6	40·8	153	137·5	67·1	213	191·4	93·4	273	245·4	119·7
34	30·6	14·9	94	84·5	41·2	154	138·4	67·5	214	192·3	93·8	274	246·3	120·1
35	31·5	15·3	95	85·4	41·6	155	139·3	67·9	215	193·2	94·2	275	247·2	120·6
36	32·4	15·8	96	86·3	42·1	156	140·2	68·4	216	194·1	94·7	276	248·1	121·0
37	33·3	16·2	97	87·2	42·5	157	141·1	68·8	217	195·0	95·1	277	249·0	121·4
38	34·2	16·7	98	88·1	43·0	158	142·0	69·3	218	195·9	95·6	278	249·9	121·9
39	35·1	17·1	99	89·0	43·4	159	142·9	69·7	219	196·8	96·0	279	250·8	122·3
40	36·0	17·5	100	89·9	43·8	160	143·8	70·1	220	197·7	96·4	280	251·7	122·7
41	36·9	18·0	101	90·8	44·3	161	144·7	70·6	221	198·6	96·9	281	252·6	123·2
42	37·7	18·4	102	91·7	44·7	162	145·6	71·0	222	199·5	97·3	282	253·5	123·6
43	38·6	18·8	103	92·6	45·2	163	146·5	71·5	223	200·4	97·8	283	254·4	124·1
44	39·5	19·3	104	93·5	45·6	164	147·4	71·9	224	201·3	98·2	284	255·3	124·5
45	40·4	19·7	105	94·4	46·0	165	148·3	72·3	225	202·2	98·6	285	256·2	124·9
46	41·3	20·2	106	95·3	46·5	166	149·2	72·8	226	203·1	99·1	286	257·1	125·4
47	42·2	20·6	107	96·2	46·9	167	150·1	73·2	227	204·0	99·5	287	258·0	125·8
48	43·1	21·0	108	97·1	47·3	168	151·0	73·6	228	204·9	99·9	288	258·9	126·3
49	44·0	21·5	109	98·0	47·8	169	151·9	74·1	229	205·8	100·4	289	259·8	126·7
50	44·9	21·9	110	98·9	48·2	170	152·8	74·5	230	206·7	100·8	290	260·7	127·1
51	45·8	22·4	111	99·8	48·7	171	153·7	75·0	231	207·6	101·3	291	261·5	127·6
52	46·7	22·8	112	100·7	49·1	172	154·6	75·4	232	208·5	101·7	292	262·4	128·0
53	47·6	23·2	113	101·6	49·5	173	155·5	75·8	233	209·4	102·1	293	263·3	128·4
54	48·5	23·7	114	102·5	50·0	174	156·4	76·3	234	210·3	102·6	294	264·2	128·9
55	49·4	24·1	115	103·4	50·4	175	157·3	76·7	235	211·2	103·0	295	265·1	129·3
56	50·3	24·5	116	104·3	50·9	176	158·2	77·2	236	212·1	103·5	296	266·0	129·8
57	51·2	25·0	117	105·2	51·3	177	159·1	77·6	237	213·0	103·9	297	266·9	130·2
58	52·1	25·4	118	106·1	51·7	178	160·0	78·0	238	213·9	104·3	298	267·8	130·6
59	53·0	25·9	119	107·0	52·2	179	160·9	78·5	239	214·8	104·8	299	268·7	131·1
60	53·9	26·3	120	107·9	52·6	180	161·8	78·9	240	215·7	105·2	300	269·6	131·5
Dist.	Dep.	Lat.	Dist.	Dep.	Lat.	Dist.	Dep.	Lat.	Dist.	Dep.	Lat.	Dist.	Dep.	Lat.

FOR 64 DEGREES.

DIFFERENCE OF LATITUDE AND DEPARTURE FOR 27 DEGREES.

Dist.	Lat.	Dep.	Dist.	Lat.	Dep.	Dist.	Lat.	Dep.	Dist.	Lat.	Dep.	Dist.	Lat.	Dep.
1	00·9	00·5	61	54·4	27·7	121	107·8	54·9	181	161·3	82·2	241	214·7	109·4
2	01·8	00·9	62	55·2	28·1	122	108·7	55·4	182	162·2	82·6	242	215·6	109·9
3	02·7	01·4	63	56·1	28·6	123	109·6	55·8	183	163·1	83·1	243	216·5	110·3
4	03·6	01·8	64	57·0	29·1	124	110·5	56·3	184	163·9	83·5	244	217·4	110·8
5	04·5	02·3	65	57·9	29·5	125	111·4	56·7	185	164·8	84·0	245	218·3	111·2
6	05·3	02·7	66	58·8	30·0	126	112·3	57·2	186	165·7	84·4	246	219·2	111·7
7	06·2	03·2	67	59·7	30·4	127	113·2	57·7	187	166·6	84·9	247	220·1	112·1
8	07·1	03·6	68	60·6	30·9	128	114·0	58·1	188	167·5	85·4	248	221·0	112·6
9	08·0	04·1	69	61·5	31·3	129	114·9	58·6	189	168·4	85·8	249	221·9	113·0
10	08·9	04·5	70	62·4	31·8	130	115·8	59·0	190	169·3	86·3	250	222·8	113·5
11	09·8	05·0	71	63·3	32·2	131	116·7	59·5	191	170·2	86·7	251	223·6	114·0
12	10·7	05·4	72	64·2	32·7	132	117·6	59·9	192	171·1	87·2	252	224·5	114·4
13	11·6	05·9	73	65·0	33·1	133	118·5	60·4	193	172·0	87·6	253	225·4	114·9
14	12·5	06·4	74	65·9	33·6	134	119·4	60·8	194	172·9	88·1	254	226·3	115·3
15	13·4	06·8	75	66·8	34·0	135	120·3	61·3	195	173·7	88·5	255	227·2	115·8
16	14·3	07·3	76	67·7	34·5	136	121·2	61·7	196	174·6	89·0	256	228·1	116·2
17	15·1	07·7	77	68·6	35·0	137	122·1	62·2	197	175·5	89·4	257	229·0	116·7
18	16·0	08·2	78	69·5	35·4	138	123·0	62·7	198	176·4	89·9	258	229·9	117·1
19	16·9	08·6	79	70·4	35·9	139	123·8	63·1	199	177·3	90·3	259	230·8	117·6
20	17·8	09·1	80	71·3	36·3	140	124·7	63·6	200	178·2	90·8	260	231·7	118·0
21	18·7	09·5	81	72·2	36·8	141	125·6	64·0	201	179·1	91·3	261	232·6	118·5
22	19·6	10·0	82	73·1	37·2	142	126·5	64·5	202	180·0	91·7	262	233·4	118·9
23	20·5	10·4	83	74·0	37·7	143	127·4	64·9	203	180·9	92·2	263	234·3	119·4
24	21·4	10·9	84	74·8	38·1	144	128·3	65·4	204	181·8	92·6	264	235·2	119·9
25	22·3	11·3	85	75·7	38·6	145	129·2	65·8	205	182·7	93·1	265	236·1	120·3
26	23·2	11·8	86	76·6	39·0	146	130·1	66·3	206	183·5	93·5	266	237·0	120·8
27	24·1	12·3	87	77·5	39·5	147	131·0	66·7	207	184·4	94·0	267	237·9	121·2
28	24·9	12·7	88	78·4	40·0	148	131·9	67·2	208	185·3	94·4	268	238·8	121·7
29	25·8	13·2	89	79·3	40·4	149	132·8	67·6	209	186·2	94·9	269	239·7	122·1
30	26·7	13·6	90	80·2	40·9	150	133·7	68·1	210	187·1	95·3	270	240·6	122·6
31	27·6	14·1	91	81·1	41·3	151	134·5	68·6	211	188·0	95·8	271	241·5	123·0
32	28·5	14·5	92	82·0	41·8	152	135·4	69·0	212	188·9	96·2	272	242·4	123·5
33	29·4	15·0	93	82·9	42·2	153	136·3	69·5	213	189·8	96·7	273	243·2	123·9
34	30·3	15·4	94	83·8	42·7	154	137·2	69·9	214	190·7	97·2	274	244·1	124·4
35	31·2	15·9	95	84·6	43·1	155	138·1	70·4	215	191·6	97·6	275	245·0	124·8
36	32·1	16·3	96	85·5	43·6	156	139·0	70·8	216	192·5	98·1	276	245·9	125·3
37	33·0	16·8	97	86·4	44·0	157	139·9	71·3	217	193·3	98·5	277	246·8	125·8
38	33·9	17·3	98	87·3	44·5	158	140·8	71·7	218	194·2	99·0	278	247·7	126·2
39	34·7	17·7	99	88·2	44·9	159	141·7	72·2	219	195·1	99·4	279	248·6	126·7
40	35·6	18·2	100	89·1	45·4	160	142·6	72·6	220	196·0	99·9	280	249·5	127·1
41	36·5	18·6	101	90·0	45·9	161	143·5	73·1	221	196·9	100·3	281	250·4	127·6
42	37·4	19·1	102	90·9	46·3	162	144·3	73·5	222	197·8	100·8	282	251·3	128·0
43	38·3	19·5	103	91·8	46·8	163	145·2	74·0	223	198·7	101·2	283	252·2	128·5
44	39·2	20·0	104	92·7	47·2	164	146·1	74·5	224	199·6	101·7	284	253·0	128·9
45	40·1	20·4	105	93·6	47·7	165	147·0	74·9	225	200·5	102·1	285	253·9	129·4
46	41·0	20·9	106	94·4	48·1	166	147·9	75·4	226	201·4	102·6	286	254·8	129·8
47	41·9	21·3	107	95·3	48·6	167	148·8	75·8	227	202·3	103·1	287	255·7	130·3
48	42·8	21·8	108	96·2	49·0	168	149·7	76·3	228	203·1	103·5	288	256·6	130·7
49	43·7	22·2	109	97·1	49·5	169	150·6	76·7	229	204·0	104·0	289	257·5	131·2
50	44·6	22·7	110	98·0	49·9	170	151·5	77·2	230	204·9	104·4	290	258·4	131·7
51	45·4	23·2	111	98·9	50·4	171	152·4	77·6	231	205·8	104·9	291	259·3	132·1
52	46·3	23·6	112	99·8	50·8	172	153·3	78·1	232	206·7	105·3	292	260·2	132·6
53	47·2	24·1	113	100·7	51·3	173	154·1	78·5	233	207·6	105·8	293	261·1	133·0
54	48·1	24·5	114	101·6	51·8	174	155·0	79·0	234	208·5	106·2	294	262·0	133·5
55	49·0	25·0	115	102·5	52·2	175	155·9	79·4	235	209·4	106·7	295	262·8	133·9
56	49·9	25·4	116	103·4	52·7	176	156·8	79·9	236	210·3	107·1	296	263·7	134·4
57	50·8	25·9	117	104·2	53·1	177	157·7	80·4	237	211·2	107·6	297	264·6	134·8
58	51·7	26·3	118	105·1	53·6	178	158·6	80·8	238	212·1	108·0	298	265·5	135·3
59	52·6	26·8	119	106·0	54·0	179	159·5	81·3	239	213·0	108·5	299	266·4	135·7
60	53·5	27·2	120	106·9	54·5	180	160·4	81·7	240	213·8	109·0	300	267·3	136·2
Dist.	Dep.	Lat.	Dist.	Dep.	Lat.	Dist.	Dep.	Lat.	Dist.	Dep.	Lat.	Dist.	Dep.	Lat.

FOR 63 DEGREES.

Dist.	Lat.	Dep.	Dist.	Lat.	Dep.	Dist.	Lat.	Dep.	Dist.	Lat.	Dep.	Dist.	Lat.	Dep.
1	00.9	00.5	61	53.9	28.6	121	106.8	56.8	181	159.8	85.0	241	212.8	113.1
2	01.8	00.9	62	54.7	29.1	122	107.7	57.3	182	160.7	85.4	242	213.7	113.6
3	02.6	01.4	63	55.6	29.6	123	108.6	57.7	183	161.6	85.9	243	214.6	114.1
4	03.5	01.9	64	56.5	30.0	124	109.5	58.2	184	162.5	86.4	244	215.4	114.6
5	04.4	02.3	65	57.4	30.5	125	110.4	58.7	185	163.3	86.9	245	216.3	115.0
6	05.3	02.8	66	58.3	31.0	126	111.3	59.2	186	164.2	87.3	246	217.2	115.5
7	06.2	03.3	67	59.2	31.5	127	112.1	59.6	187	165.1	87.8	247	218.1	116.0
8	07.1	03.8	68	60.0	31.9	128	113.0	60.1	188	166.0	88.3	248	219.0	116.4
9	07.9	04.2	69	60.9	32.4	129	113.9	60.6	189	166.9	88.7	249	219.9	116.9
10	08.8	04.7	70	61.8	32.9	130	114.8	61.0	190	167.8	89.2	250	220.7	117.4
11	09.7	05.2	71	62.7	33.3	131	115.7	61.5	191	168.6	89.7	251	221.6	117.9
12	10.6	05.6	72	63.6	33.8	132	116.5	62.0	192	169.5	90.1	252	222.5	118.3
13	11.5	06.1	73	64.5	34.3	133	117.4	62.4	193	170.4	90.6	253	223.4	118.8
14	12.4	06.6	74	65.3	34.7	134	118.3	62.9	194	171.3	91.1	254	224.3	119.2
15	13.2	07.0	75	66.2	35.2	135	119.2	63.4	195	172.2	91.5	255	225.2	119.7
16	14.1	07.5	76	67.1	35.7	136	120.1	63.8	196	173.1	92.0	256	226.0	120.2
17	15.0	08.0	77	68.0	36.1	137	121.0	64.3	197	173.9	92.5	257	226.9	120.7
18	15.9	08.5	78	68.9	36.6	138	121.8	64.8	198	174.8	93.0	258	227.8	121.1
19	16.8	08.9	79	69.8	37.1	139	122.7	65.3	199	175.7	93.4	259	228.7	121.6
20	17.7	09.4	80	70.6	37.6	140	123.6	65.7	200	176.6	93.9	260	229.6	122.1
21	18.5	09.9	81	71.5	38.0	141	124.5	66.2	201	177.5	94.4	261	230.4	122.5
22	19.4	10.3	82	72.4	38.5	142	125.4	66.7	202	178.4	94.8	262	231.3	123.0
23	20.3	10.8	83	73.3	39.0	143	126.3	67.1	203	179.2	95.3	263	232.2	123.5
24	21.2	11.3	84	74.2	39.4	144	127.1	67.6	204	180.1	95.8	264	233.1	123.9
25	22.1	11.7	85	75.1	39.9	145	128.0	68.1	205	181.0	96.2	265	234.0	124.4
26	23.0	12.2	86	75.9	40.4	146	128.9	68.5	206	181.9	96.7	266	234.9	124.9
27	23.8	12.7	87	76.8	40.8	147	129.8	69.0	207	182.8	97.2	267	235.7	125.3
28	24.7	13.1	88	77.7	41.3	148	130.7	69.5	208	183.7	97.7	268	236.6	125.8
29	25.6	13.6	89	78.6	41.8	149	131.6	70.0	209	184.5	98.1	269	237.5	126.3
30	26.5	14.1	90	79.5	42.3	150	132.4	70.4	210	185.4	98.6	270	238.4	126.8
31	27.4	14.6	91	80.3	42.7	151	133.3	70.9	211	186.3	99.1	271	239.3	127.2
32	28.3	15.0	92	81.2	43.2	152	134.2	71.4	212	187.2	99.5	272	240.2	127.7
33	29.1	15.5	93	82.1	43.7	153	135.1	71.8	213	188.1	100.0	273	241.0	128.2
34	30.0	16.0	94	83.0	44.1	154	136.0	72.3	214	189.0	100.5	274	241.9	128.6
35	30.9	16.4	95	83.9	44.6	155	136.9	72.8	215	189.8	100.9	275	242.8	129.1
36	31.8	16.9	96	84.8	45.1	156	137.7	73.2	216	190.7	101.4	276	243.7	129.6
37	32.7	17.4	97	85.6	45.5	157	138.6	73.7	217	191.6	101.9	277	244.6	130.0
38	33.6	17.8	98	86.5	46.0	158	139.5	74.2	218	192.5	102.3	278	245.5	130.5
39	34.4	18.3	99	87.4	46.5	159	140.4	74.6	219	193.4	102.8	279	246.3	131.0
40	35.3	18.8	100	88.3	46.9	160	141.3	75.1	220	194.2	103.3	280	247.2	131.5
41	36.2	19.2	101	89.2	47.4	161	142.2	75.6	221	195.1	103.8	281	248.1	131.9
42	37.1	19.7	102	90.1	47.9	162	143.0	76.1	222	196.0	104.2	282	249.0	132.4
43	38.0	20.2	103	90.9	48.4	163	143.9	76.5	223	196.9	104.7	283	249.9	132.9
44	38.8	20.7	104	91.8	48.8	164	144.8	77.0	224	197.8	105.2	284	250.8	133.3
45	39.7	21.1	105	92.7	49.3	165	145.7	77.5	225	198.7	105.6	285	251.6	133.8
46	40.6	21.6	106	93.6	49.8	166	146.6	77.9	226	199.5	106.1	286	252.5	134.3
47	41.5	22.1	107	94.5	50.2	167	147.5	78.4	227	200.4	106.6	287	253.4	134.7
48	42.4	22.5	108	95.4	50.7	168	148.3	78.9	228	201.3	107.0	288	254.3	135.2
49	43.3	23.0	109	96.2	51.2	169	149.2	79.3	229	202.2	107.5	289	255.2	135.7
50	44.1	23.5	110	97.1	51.6	170	150.1	79.8	230	203.1	108.0	290	256.1	136.1
51	45.0	23.9	111	98.0	52.1	171	151.0	80.3	231	204.0	108.4	291	256.9	136.6
52	45.9	24.4	112	98.9	52.6	172	151.9	80.7	232	204.8	108.9	292	257.8	137.1
53	46.8	24.9	113	99.8	53.1	173	152.7	81.2	233	205.7	109.4	293	258.7	137.6
54	47.7	25.4	114	100.7	53.5	174	153.6	81.7	234	206.6	109.9	294	259.6	138.0
55	48.6	25.8	115	101.5	54.0	175	154.5	82.2	235	207.5	110.3	295	260.5	138.5
56	49.4	26.3	116	102.4	54.5	176	155.4	82.6	236	208.4	110.8	296	261.3	139.0
57	50.3	26.8	117	103.3	54.9	177	156.3	83.1	237	209.3	111.3	297	262.2	139.4
58	51.2	27.2	118	104.2	55.4	178	157.2	83.6	238	210.1	111.7	298	263.1	139.9
59	52.1	27.7	119	105.1	55.9	179	158.0	84.0	239	211.0	112.2	299	264.0	140.4
60	53.0	28.2	120	106.0	56.3	180	158.9	84.5	240	211.9	112.7	300	264.9	140.8
Dist.	Dep.	Lat.	Dist.	Dep.	Lat.	Dist.	Dep.	Lat.	Dist.	Dep.	Lat.	Dist.	Dep.	Lat.

FOR 62 DEGREES.

Dist.	Lat.	Dep.	Dist.	Lat.	Dep.	Dist.	Lat.	Dep.	Dist.	Lat.	Dep.	Dist.	Lat.	Dep.
1	00 9	00·5	61	53·4	29·6	121	105·8	58·7	181	158·3	87·8	241	210·8	116·8
2	01·7	01·0	62	54·2	30·1	122	106 7	59·1	182	159·2	88·2	242	211·7	117·3
3	02·6	01·5	63	55·1	30·5	123	107 6	59·6	183	160·1	88·7	243	212·5	117·8
4	03·5	01·9	64	56·0	31·0	124	108·5	60·1	184	160·9	89·2	244	213·4	118·3
5	04·4	02·4	65	56·9	31·5	125	109 3	60·6	185	161·8	89·7	245	214·3	118·8
6	05·2	02·9	66	57·7	32·0	126	110·2	61·1	186	162·7	90·2	246	215·2	119·3
7	06·1	03·4	67	58 6	32·5	127	111·1	61·6	187	163 6	90·7	247	216·0	119·7
8	07·0	03·9	68	59·5	33·0	128	112·0	62·1	188	164·4	91·1	248	216·9	120·2
9	07·9	04·4	69	60·3	33·5	129	112·8	62·5	189	165·3	91·6	249	217·8	120·7
10	08·7	04·8	70	61·2	33·9	130	113·7	63·0	190	166·2	92·1	250	218·7	121·2
11	09·6	05·3	71	62·1	34·4	131	114·6	63·5	191	167·1	92·6	251	219·5	121·7
12	10·5	05·8	72	63·0	34·9	132	115·4	64·0	192	167·9	93·1	252	220·4	122·2
13	11·4	06·3	73	63·8	35·4	133	116·3	64·5	193	168·8	93·6	253	221·3	122·7
14	12·2	06·8	74	64·7	35·9	134	117·2	65·0	194	169·7	94·1	254	222·2	123·1
15	13·1	07·3	75	65·6	36·4	135	118·1	65·4	195	170·6	94·5	255	223·0	123·6
16	14.0	07·8	76	66·5	36 8	136	118·9	65·9	196	171·4	95·0	256	223·9	124·1
17	14·9	08·2	77	67·3	37·3	137	119·8	66·4	197	172·3	95·5	257	224·8	124·6
18	15·7	08·7	78	68·2	37·8	138	120·7	66·9	198	173·2	96·0	258	225·7	125·1
19	16·6	09·2	79	69·1	38 3	139	121·6	67·4	199	174·0	96 5	259	226·5	125·6
20	17·5	09·7	80	70·0	38·8	140	122·4	67·9	200	174·9	97·0	260	227·4	126·1
21	18·4	10·2	81	70·8	39·3	141	123·3	68·4	201	175·8	97·4	261	228·3	126·5
22	19·2	10·7	82	71·7	39·8	142	124·2	68·8	202	176·7	97·9	262	229·2	127·0
23	20·1	11·2	83	72·6	40·2	143	125·1	69·3	203	177·5	98·4	263	230·0	127·5
24	21·0	11·6	84	73·5	40·7	144	125·9	69·8	204	178·4	98·9	264	230·9	128·0
25	21·9	12·1	85	74·3	41·2	145	126 8	70·3	205	179·3	99·4	265	231·8	128·5
26	22·7	12·6	86	75·2	41·7	146	127·7	70·8	206	180·2	99·9	266	232·6	129·0
27	23·6	13·4	87	76 1	42·2	147	128·6	71·3	207	181·0	100·4	267	233·5	129·4
28	24·5	13·6	88	77·0	42·7	148	129·4	71·8	208	181·9	100·8	268	234·4	129·9
29	25·4	14·1	89	77·8	43·1	149	130·3	71·2	209	182·8	101·3	269	235·3	130·4
30	26·2	14·5	90	78·7	43·6	150	131·2	72·7	210	183·7	101·8	270	236·1	130·9
31	27·1	15·0	91	79·6	44·1	151	132·1	73·2	211	184·5	102·3	271	237·0	131·4
32	28·0	15·5	92	80·5	44·6	152	132·9	73·7	212	185·4	102·8	272	237·9	131·9
33	28·9	16·0	93	81·3	45·1	153	133·8	74·2	213	186·3	103·3	273	238·8	132·4
34	29·7	16·5	94	82·2	45·6	154	134·7	74·7	214	187·2	103·7	274	239 6	132·8
35	30·6	17·0	95	83·1	46·1	155	135 6	75·1	215	188·0	104·2	275	240·5	133·3
36	31·5	17·5	96	84·0	46 5	156	136·4	75·6	216	188·9	104·7	276	241·4	133·8
37	32·4	17·9	97	84·8	47·0	157	137·3	76·1	217	189·8	105·2	277	242·3	134·3
38	33·2	18·4	98	85·7	47·5	158	138·2	76·6	218	190·7	105·7	278	243·1	134·8
39	34·1	18·9	99	86·6	48·0	159	139·1	77·1	219	191·5	106·2	279	244·0	135·3
40	35·0	19·4	100	87·5	48·5	160	139·9	77·6	220	192·4	106·7	280	244·9	135·7
41	35·9	19·9	101	88·3	49 0	161	140·8	78·1	221	193·3	107·1	281	245·8	136·2
42	36·7	20·4	102	89·2	49·5	162	141·7	78·5	222	194 2	107·6	282	246·6	136·7
43	37·6	20 8	103	90·1	49·9	163	142·6	79·0	223	195·0	108·1	283	247·5	137·2
44	38·5	21·3	104	91·0	50·4	164	143·4	79·5	224	195·9	108·6	284	248·4	137·7
45	39·4	21·8	105	91.8	50·9	165	144·3	80·0	225	196·8	109·1	285	249·3	138·2
46	40 2	22·3	106	92 7	51·4	166	145·2	80·5	226	197·7	109·6	286	250·1	138·7
47	41·1	22·8	107	93 6	51·9	167	146·1	81·0	227	198·5	110·1	287	251·0	139·1
48	42·0	23 3	108	94 5	52·4	168	146·9	81·4	228	199·4	110·5	288	251·9	139·6
49	42·9	23·8	109	95·3	52·8	169	147·8	81·9	229	200·3	111·0	289	252·8	140·1
50	43·7	24·2	110	96·2	53·3	170	148·7	82·4	230	201·2	111·5	290	253 6	140·6
51	44·6	24·7	111	97·1	53·8	171	149·6	82 9	231	202·0	112·0	291	254·5	141·1
52	45·5	25·2	112	98·0	54·3	172	150·4	83·4	232	202·9	112·5	292	255·4	141·6
53	46·4	25·7	113	98·8	54 8	173	151·3	83·9	233	203·8	113·0	293	256·3	142·0
54	47·2	26·2	114	99·7	55·3	174	152·2	84·4	234	204·7	113·4	294	257·1	142·5
55	48·1	26·7	115	100·6	55·8	175	153·1	84 8	235	205·5	113·9	295	258·0	143·0
56	49·0	27·1	116	101·5	56 2	176	153·9	85·3	236	206·4	114·4	296	258·9	143·5
57	49·9	27·6	117	102·3	56·7	177	154·8	85·8	237	207·3	114·9	297	259·8	144·0
58	50·7	28·1	118	103·2	57·2	178	155·7	86·3	238	208·2	115·4	298	260·6	144·5
59	51·6	28·6	119	104·1	57·7	179	156·6	86·8	239	209·0	115·9	299	261·5	145·0
60	52·5	29·1	120	105·0	58·2	180	157·4	87·3	240	209 9	116·4	300	262·4	145·4
Dist.	Dep.	Lat.	Dist.	Dep.	Lat.	Dist.	Dep.	Lat.	Dist.	Dep.	Lat.	Dist.	Dep.	Lat.

FOR 61 DEGREES.

Dist.	Lat.	Dep.	Dist.	Lat.	Dep.	Dist.	Lat.	Dep.	Dist.	Lat.	Dep.	Dist.	Lat.	Dep.
1	00·9	00·5	61	52·8	30·5	121	104·8	60·5	181	156·8	90·5	241	208·7	120·5
2	01·7	01·0	62	53·7	31·0	122	105·7	61·0	182	157·6	91·0	242	209·6	121·0
3	02·6	01·5	63	54·6	31·5	123	106·5	61·5	183	158·5	91·5	243	210·4	121·5
4	03·5	02·0	64	55·4	32·0	124	107·4	62·0	184	159·3	92·0	244	211·3	122·0
5	04·3	02·5	65	56·3	32·5	125	108·3	62·5	185	160·2	92·5	245	212·2	122·5
6	05·2	03·0	66	57·2	33·0	126	109·1	63·0	186	161·1	93·0	246	213·0	123·0
7	06·1	03·5	67	58·0	33·5	127	110·0	63·5	187	161·9	93·5	247	213·9	123·5
8	06·9	04·0	68	58·9	34·0	128	110·9	64·0	188	162·8	94·0	248	214·8	124·0
9	07·8	04·5	69	59·8	34·5	129	111·7	64·5	189	163·7	94·5	249	215·6	124·5
10	08·7	05·0	70	60·6	35·0	130	112·6	65·0	190	164·5	95·0	250	216·5	125·0
11	09·5	05·5	71	61·5	35·5	131	113·4	65·5	191	165·4	95·5	251	217·4	125·5
12	10·4	06·0	72	62·4	36·0	132	114·3	66·0	192	166·3	96·0	252	218·2	126·0
13	11·3	06·5	73	63·2	36·5	133	115·2	66·5	193	167·1	96·5	253	219·1	126·5
14	12·1	07·0	74	64·1	37·0	134	116·0	67·0	194	168·0	97·0	254	220·0	127 0
15	13·0	07·5	75	65·0	37·5	135	116·9	67·5	195	168·9	97·5	255	220·8	127·5
16	13·9	08·0	76	65·8	38·0	136	117·8	68·0	196	169·7	98·0	256	221·7	128·0
17	14·7	08·5	77	66·7	38·5	137	118·6	68·5	197	170·6	98·5	257	222·6	128·5
18	15·6	09·0	78	67·5	39·0	138	119·5	69·0	198	171·5	99·0	258	223·4	129·0
19	16·5	09·5	79	68·4	39·5	139	120·4	69·5	199	172·3	99·5	259	224·3	129·5
20	17·3	10·0	80	69·3	40·0	140	121·2	70·0	200	173·2	100·0	260	225·2	130·0
21	18·2	10·5	81	70·1	40·5	141	122·1	70·5	201	174·1	100·5	261	226·0	130·5
22	19·1	11·0	82	71·0	41·0	142	123·0	71·0	202	174·9	101·0	262	226·9	131·0
23	19·9	11·5	83	71·9	41·5	143	123·8	71·5	203	175·8	101·5	263	227·8	131·5
24	20·8	12·0	84	72·7	42·0	144	124·7	72·0	204	176·7	102·0	264	228·6	132·0
25	21·7	12·5	85	73·6	42·5	145	125·6	72·5	205	177·5	102·5	265	229·5	132·5
26	22·5	13·0	86	74·5	43·0	146	126·4	73·0	206	178·4	103·0	266	230·4	133·0
27	23·4	13·5	87	75·3	43·5	147	127·3	73·5	207	179·3	103·5	267	231·2	133·5
28	24·2	14·0	88	76·2	44·0	148	128·2	74·0	208	180·1	104·0	268	232·1	134·0
29	25·1	14·5	89	77·1	44·5	149	129·0	74·5	209	181·0	104·5	269	233·0	134·5
30	26·0	15·0	90	77·9	45·0	150	129·9	75·0	210	181·9	105·0	270	233·8	135·0
31	26·8	15·5	91	78·8	45·5	151	130·8	75·5	211	182·7	105·5	271	234·7	135·5
32	27·7	16·0	92	79·7	46·0	152	131·6	76·0	212	183·6	106·0	272	235·6	136·0
33	28·6	16·5	93	80·5	46·5	153	132·5	76·5	213	184·5	106·5	273	236·4	136·5
34	29·4	17·0	94	81·4	47·0	154	133·4	77·0	214	185·3	107·0	274	237·3	137·0
35	30·3	17·5	95	82·3	47·5	155	134·2	77·5	215	186·2	107·5	275	238·2	137 5
36	31·2	18·0	96	83·1	48·0	156	135·1	78·0	216	187·1	108·0	276	239·0	138·0
37	32·0	18·5	97	84·0	48·5	157	136·0	78·5	217	187·9	108·5	277	239·9	138·5
38	32·9	19·0	98	84·9	49·0	158	136·8	79·0	218	188·8	109·0	278	240·8	139·0
39	33·8	19·5	99	85·7	49·5	159	137·7	79·5	219	189·7	109·5	279	241·6	139·5
40	34·6	20·0	100	86·6	50·0	160	138·6	80·0	220	190·5	110·0	280	242·5	140·0
41	35·5	20·5	101	87·5	50·5	161	139·4	80·5	221	191·4	110·5	281	243·4	140·5
42	36·4	21·0	102	88·3	51·0	162	140·3	81·0	222	192·3	111·0	282	244·2	141·0
43	37·2	21·5	103	89·2	51·5	163	141·2	81·5	223	193·1	111·5	283	245·1	141·5
44	38·1	22·0	104	90·1	52·0	164	142·0	82·0	224	194·0	112·0	284	246·0	142·0
45	39·0	22·5	105	90·9	52·5	165	142·9	82·5	225	194·9	112·5	285	246·8	142·5
46	39·8	23·0	106	91·8	53·0	166	143·8	83·0	226	195·7	113·0	286	247·7	143·0
47	40·7	23·5	107	92·7	53·5	167	144·6	83·5	227	196·6	113·5	287	248·5	143 5
48	41·6	24·0	108	93·5	54·0	168	145·5	84·0	228	197·5	114·0	288	249·4	144·0
49	42·4	24·5	109	94·4	54·5	169	146·4	84·5	229	198·3	114·5	289	250·3	144·5
50	43·3	25·0	110	95·3	55·0	170	147·2	85·0	230	199·2	115·0	290	251·1	145·0
51	44·2	25·5	111	96·1	55·5	171	148·1	85·5	231	200·1	115·5	291	252·0	145·5
52	45·0	26·0	112	97·0	56·0	172	149·0	86·0	232	200·9	116·0	292	252·9	146·0
53	45·9	26·5	113	97·9	56·5	173	149·8	86·5	233	201·8	116·5	293	253·7	146·5
54	46·8	27·0	114	98·7	57·0	174	150·7	87·0	234	202·6	117·0	294	254·6	147·0
55	47·6	27·5	115	99·6	57·5	175	151·6	87·5	235	203·5	117·5	295	255·5	147·5
56	48·5	28·0	116	100·5	58·0	176	152·4	88·0	236	204·4	118·0	296	256·3	148·0
57	49·4	28·5	117	101·3	58·5	177	153·3	88·5	237	205·2	118 5	297	257·2	148·5
58	50·2	29·0	118	102·2	59·0	178	154·2	89·0	238	206·1	119 0	298	258·1	149·0
59	51·1	29·5	119	103·1	59·5	179	155·0	89·5	239	207·0	119·5	299	258·9	149·5
60	52·0	30·0	120	103·9	60·0	180	155·9	90·0	240	207·8	120 0	300	259·8	150·0
Dist.	Dep.	Lat.	Dist.	Dep.	Lat.	Dist.	Dep.	Lat.	Dist.	Dep.	Lat.	Dist.	Dep.	Lat.

FOR 60 DEGREES.

Dist.	Lat.	Dep	Dist.	Lat.	Dep	Dist.	Lat.	Dep.	Dist.	Lat.	Dep.	Dist.	Lat.	Dep.
1	00·9	00·5	61	52·3	31·4	121	103·7	62·3	181	155·1	93·2	241	206·6	124·1
2	01·7	01·0	62	53·1	31·9	122	104·6	62·8	182	156·0	93·7	242	207·4	124·6
3	02·6	01·5	63	54·0	32·4	123	105·4	63·3	183	156·9	94·3	243	208·3	125·2
4	03·4	02·1	64	54·9	33·0	124	106·3	63·9	184	157·7	94·8	244	209·1	125·7
5	04·3	02·6	65	55·7	33·5	125	107·1	64·4	185	158·6	95·3	245	210·0	126·2
6	05·1	03·1	66	56·6	34·0	126	108·0	64·9	186	159·4	95·8	246	210·9	126·7
7	06·0	03·6	67	57·4	34·5	127	108·9	65·4	187	160·3	96·3	247	211·7	127·2
8	06·9	04·1	68	58·3	35·0	128	109·7	65·9	188	161·1	96·8	248	212·6	127·7
9	07·7	04·6	69	59·1	35·5	129	110·6	66·4	189	162·0	97·3	249	213·4	128·2
10	08·6	05·2	70	60·0	36·1	130	111·4	67·0	190	162·9	97·9	250	214·3	128·8
11	09·4	05·7	71	60·9	36·6	131	112·3	67·5	191	163·7	98·4	251	215·1	129·3
12	10·3	06·2	72	61·7	37·1	132	113·1	68·0	192	164·6	98·9	252	216·0	129·8
13	11·1	06·7	73	62·6	37·6	133	114·0	68·5	193	165·4	99·4	253	216·9	130·3
14	12·0	07·2	74	63·4	38·1	134	114·9	69·0	194	166·3	99·9	254	217·7	130·8
15	12·9	07·7	75	64·3	38·6	135	115·7	69·5	195	167·1	100·4	255	218·6	131·3
16	13·7	08·2	76	65·1	39·1	136	116·6	70·0	196	168·0	100·9	256	219·4	131·8
17	14·6	08·8	77	66·0	39·7	137	117·4	70·6	197	168·9	101·5	257	220·3	132·4
18	15·4	09·3	78	66·9	40·2	138	118·3	71·1	198	169·7	102·0	258	221·1	132·9
19	16·3	09·8	79	67·7	40·7	139	119·1	71·6	199	170·6	102·5	259	222·0	133·4
20	17·1	10·3	80	68·6	41·2	140	120·0	72·1	200	171·4	103·0	260	222·9	133·9
21	18·0	10·8	81	69·4	41·7	141	120·9	72·6	201	172·3	103·5	261	223·7	134·4
22	18·9	11·3	82	70·3	42·2	142	121·7	73·1	202	173·1	104·0	262	224·6	134·9
23	19·7	11·8	83	71·1	42·7	143	122·6	73·7	203	174·0	104·6	263	225·4	135·5
24	20·6	12·4	84	72·0	43·3	144	123·4	74·2	204	174·9	105·1	264	226·3	136·0
25	21·4	12·9	85	72·9	43·8	145	124·3	74·7	205	175·7	105·6	265	227·1	136·5
26	22·3	13·4	86	73·7	44·3	146	125·1	75·2	206	176·6	106·1	266	228·0	137·0
27	23·1	13·9	87	74·6	44·8	147	126·0	75·7	207	177·4	106·6	267	228·9	137·5
28	24·0	14·4	88	75·4	45·3	148	126·9	76·2	208	178·3	107·1	268	229·7	138·0
29	24·9	14·9	89	76·3	45·8	149	127·7	76·7	209	179·1	107·6	269	230·6	138·5
30	25·7	15·5	90	77·1	46·4	150	128·6	77·3	210	180·0	108·2	270	231·4	139·1
31	26·6	16·0	91	78·0	46·6	151	129·4	77·8	211	180·9	108·7	271	232·3	139·6
32	27·4	16·5	92	78·9	47·4	152	130·3	78·3	212	181·7	109·2	272	233·1	140·1
33	28·3	17·0	93	79·7	47·9	153	131·1	78·8	213	182·6	109·7	273	234·0	140·6
34	29·1	17·5	94	80·6	48·4	154	132·0	79·3	214	183·4	110·2	274	234·9	141·1
35	30·0	18·0	95	81·4	48·9	155	132·9	79·8	215	184·3	110·7	275	235·7	141·6
36	30·9	18·5	96	82·3	49·4	156	133·7	80·3	216	185·1	111·2	276	236·6	142·2
37	31·7	19·1	97	83·1	50·0	157	134·6	80·9	217	186·0	111·8	277	237·4	142·7
38	32·6	19·6	98	84·0	50·5	158	135·4	81·4	218	186·9	112·3	278	238·3	143·2
39	33·4	20·1	99	84·9	51·0	159	136·3	81·9	219	187·7	112·8	279	239·1	143·7
40	34·3	20·6	100	85·7	51·5	160	137·1	82·4	220	188·6	113·3	280	240·0	144·2
41	35·1	21·1	101	86·6	52·0	161	138·0	82·9	221	189·4	113·8	281	240·9	144·7
42	36·0	21·6	102	87·4	52·5	162	138·9	83·4	222	190·3	114·3	282	241·7	145·2
43	36·9	22·1	103	88·3	53·0	163	139·7	84·0	223	191·1	114·9	283	242·6	145·8
44	37·7	22·7	104	89·1	53·6	164	140·6	84·5	224	192·0	115·4	284	243·4	146·3
45	38·6	23·2	105	90·0	54·1	165	141·4	85·0	225	192·9	115·9	285	244·3	146·8
46	39·4	23·7	106	90·9	54·6	166	142·3	85·5	226	193·7	116·4	286	245·1	147·3
47	40·3	24·2	107	91·7	55·1	167	143·1	86·0	227	194·6	116·9	287	246·0	147·8
48	41·1	24·7	108	92·6	55·6	168	144·0	86·5	228	195·4	117·4	288	246·9	148·3
49	42·0	25·2	109	93·4	56·1	169	144·9	87·0	229	196·3	117·9	289	247·7	148·8
50	42·9	25·8	110	94·3	56·7	170	145·7	87·6	230	197·1	118·5	290	248·6	149·4
51	43·7	26·3	111	95·1	57·2	171	146·6	88·1	231	198·0	119·0	291	249·4	149·9
52	44·6	26·8	112	96·0	57·7	172	147·4	88·6	232	198·9	119·5	292	250·3	150·4
53	45·4	27·3	113	96·9	58·2	173	148·3	89·1	233	199·7	120·0	293	251·1	150·9
54	46·3	27·8	114	97·7	58·7	174	149·1	89·6	234	200·6	120·5	294	252·0	151·4
55	47·1	28·3	115	98·6	59·2	175	150·0	90·1	235	201·4	121·0	295	252·9	151·9
56	48·0	28·8	116	99·4	59·7	176	150·9	90·6	236	202·3	121·5	296	253·7	152·5
57	48·9	29·4	117	100·3	60·3	177	151·7	91·2	237	203·1	122·1	297	254·6	153·0
58	49·7	29·9	118	101·1	60·8	178	152·6	91·7	238	204·0	122·6	298	255·4	153·5
59	50·6	30·4	119	102·0	61·3	179	153·4	92·2	239	204·9	123·1	299	256·3	154·0
60	51·4	30·9	120	102·9	61·8	180	154·3	92·7	240	205·7	123·6	300	257·1	154·5
Dist.	Dep.	Lat.	Dist.	Dep.	Lat.	Dist.	Dep.	Lat.	Dist.	Dep.	Lat.	Dist.	Dep.	Lat.

FOR 59 DEGREES.

Dist.	Lat.	Dep.	Dist.	Lat.	Dep.	Dist.	Lat.	Dep.	Dist.	Lat.	Dep.	Dist.	Lat.	Dep.
1	00·8	00·5	61	51·7	32·3	121	102·6	64·1	181	153·5	95·9	241	204·4	127·7
2	01·7	01·1	62	52·6	32·9	122	103·5	64·7	182	154·3	96·4	242	205·2	128·2
3	02·5	01·6	63	53·4	33·4	123	104·3	65·2	183	155·2	97·0	243	206·1	128·8
4	03·4	02·1	64	54·3	33·9	124	105·2	65·7	184	156·0	97·5	244	206·9	129·3
5	04·2	02·6	65	55·1	34·4	125	106·0	66·2	185	156·9	98·0	245	207·8	129·8
6	05·1	03·2	66	56·0	35·0	126	106·9	66·8	186	157·7	98·6	246	208·6	130·4
7	05·9	03·7	67	56·8	35·5	127	107·7	67·3	187	158·6	99·1	247	209·5	130·9
8	06·8	04·2	68	57·7	36·0	128	108·6	67·8	188	159·4	99·6	248	210·3	131·4
9	07·6	04·8	69	58·5	36·6	129	109·4	68·4	189	160·3	100·2	249	211·2	131·9
10	08·5	05·3	70	59·4	37·1	130	110·2	68·9	190	161·1	100·7	250	212·0	132·5
11	09·3	05·8	71	60·2	37·6	131	111·1	69·4	191	162·0	101·2	251	212·9	133·0
12	10·2	06·4	72	61·1	38·2	132	111·9	69·9	192	162·8	101·7	252	213·7	133·5
13	11·0	06·9	73	61·9	38·7	133	112·8	70·5	193	163·7	102·3	253	214·6	134·1
14	11·9	07·4	74	62·8	39·2	134	113·6	71·0	194	164·5	102·8	254	215·4	134·6
15	12·7	07·9	75	63·6	39·7	135	114·5	71·5	195	165·4	103·3	255	216·3	135·1
16	13·6	08·5	76	64·5	40·3	136	115·3	72·1	196	166·2	103·9	256	217·1	135·7
17	14·4	09·0	77	65·3	40·8	137	116·2	72·6	197	167·1	104·4	257	217·9	136·2
18	15·3	09·5	78	66·1	41·3	138	117·0	73·1	198	167·9	104·9	258	218·8	136·7
19	16·1	10·1	79	67·0	41·9	139	117·9	73·7	199	168·8	105·5	259	219·6	137·2
20	17·0	10·6	80	67·8	42·4	140	118·7	74·2	200	169·6	106·0	260	220·5	137·8
21	17·8	11·1	81	68·7	42·9	141	119·6	74·7	201	170·5	106·5	261	221·3	138·3
22	18·7	11·7	82	69·5	43·5	142	120·4	75·2	202	171·3	107·0	262	222·2	138·8
23	19·5	12·2	83	70·4	44·0	143	121·3	75·8	203	172·2	107·6	263	223·0	139·4
24	20·4	12·7	84	71·2	44·5	144	122·1	76·3	204	173·0	108·1	264	223·9	139·9
25	21·2	13·2	85	72·1	45·0	145	123·0	76·8	205	173·8	108·6	265	224·7	140·4
26	22·0	13·8	86	72·9	45·6	146	123·8	77·4	206	174·7	109·2	266	225·6	141·0
27	22·9	14·3	87	73·8	46·1	147	124·7	77·9	207	175·5	109·7	267	226·4	141·5
28	23·7	14·8	88	74·6	46·6	148	125·5	78·4	208	176·4	110·2	268	227·3	142·0
29	24·6	15·4	89	75·5	47·2	149	126·4	79·0	209	177·2	110·8	269	228·1	142·5
30	25·4	15·9	90	76·3	47·7	150	127·2	79·5	210	178·1	111·3	270	229·0	143·1
31	26·3	16·4	91	77·2	48·2	151	128·1	80·0	211	178·9	111·8	271	229·8	143·6
32	27·1	17·0	92	78·0	48·8	152	128·9	80·5	212	179·8	112·3	272	230·7	144·1
33	28·0	17·5	93	78·9	49·3	153	129·8	81·1	213	180·6	112·9	273	231·5	144·7
34	28·8	18·0	94	79·7	49·8	154	130·6	81·6	214	181·5	113·4	274	232·4	145·2
35	29·7	18·5	95	80·6	50·3	155	131·4	82·1	215	182·3	113·9	275	233·2	145·7
36	30·5	19·1	96	81·4	50·9	156	132·3	82·7	216	183·2	114·5	276	234·1	146·3
37	31·4	19·6	97	82·3	51·4	157	133·1	83·2	217	184·0	115·0	277	234·9	146·8
38	32·2	20·1	98	83·1	51·9	158	134·0	83·7	218	184·9	115·5	278	235·8	147·3
39	33·1	20·7	99	84·0	52·5	159	134·8	84·3	219	185·7	116·1	279	236·6	147·8
40	33·9	21·2	100	84·8	53·0	160	135·7	84·8	220	186·6	116·6	280	237·5	148·4
41	34·8	21·7	101	85·7	53·5	161	136·5	85·3	221	187·4	117·1	281	238·3	148·9
42	35·6	22·3	102	86·5	54·1	162	137·4	85·8	222	188·3	117·6	282	239·1	149·4
43	36·5	22·8	103	87·3	54·6	163	138·2	86·4	223	189·1	118·2	283	240·0	150·0
44	37·3	23·3	104	88·2	55·1	164	139·1	86·9	224	190·0	118·7	284	240·8	150·5
45	38·2	23·8	105	89·0	55·6	165	139·9	87·4	225	190·8	119·2	285	241·7	151·0
46	39·0	24·4	106	89·9	56·2	166	140·8	88·0	226	191·7	119·8	286	242·5	151·6
47	39·9	24·9	107	90·7	56·7	167	141·6	88·5	227	192·5	120·3	287	243·4	152·1
48	40·7	25·4	108	91·6	57·2	168	142·5	89·0	228	193·4	120·8	288	244·2	152·6
49	41·6	26·0	109	92·4	57·8	169	143·3	89·6	229	194·2	121·4	289	245·1	153·1
50	42·4	26·5	110	93·3	58·3	170	144·2	90·1	230	195·1	121·9	290	245·9	153·7
51	43·3	27·0	111	94·1	58·8	171	145·0	90·6	231	195·9	122·4	291	246·8	154·2
52	44·1	27·6	112	95·0	59·4	172	145·9	91·1	232	196·7	122·9	292	247·6	154·7
53	44·9	28·1	113	95·8	59·9	173	146·7	91·7	233	197·6	123·5	293	248·5	155·3
54	45·8	28·6	114	96·7	60·4	174	147·6	92·2	234	198·4	124·0	294	249·3	155·8
55	46·6	29·1	115	97·5	60·9	175	148·4	92·7	235	199·3	124·5	295	250·2	156·3
56	47·5	[illegible]	116	98·4	61·5	176	149·3	93·3	236	200·1	125·1	296	251·0	156·9
57	48·3	30·2	117	99·2	62·0	177	150·1	93·8	237	201·0	125·6	297	251·9	157·4
58	49·2	[illegible]	118	100·1	62·5	178	151·0	94·3	238	201·8	126·1	298	252·7	157·9
59	50·0	31·3	119	100·9	63·1	179	151·8	94·9	239	202·7	126·7	299	253·6	158·4
60	50·9	[illegible]	120	101·8	63·6	180	152·6	95·4	240	203·5	127·2	300	254·4	159·0
Dist.	Dep.	Lat.	Dist.	Dep.	Lat.	Dist.	Dep.	Lat.	Dist.	Dep.	Lat.	Dist.	Dep.	Lat.

FOR 58 DEGREES.

Dist.	Lat.	Dep.	Dist.	Lat.	Dep.	Dist.	Lat.	Dep.	Dist.	Lat.	Dep.	Dist.	Lat.	Dep.
1	00·8	00·5	61	51·2	33·2	121	101·5	65·9	181	151·8	98·6	241	202·1	131·3
2	01·7	01·1	62	52·0	33·8	122	102·3	66·4	182	152·6	99·1	242	203·0	131·8
3	02·5	01·6	63	52·8	34·3	123	103·2	67·0	183	153·5	99·7	243	203·8	132·3
4	03·4	02·2	64	53·7	34·9	124	104·0	67·5	184	154·3	100·2	244	204·6	132·9
5	04·2	02·7	65	54·5	35·4	125	104·8	68·1	185	155·2	100·8	245	205·5	133·4
6	05·0	03·3	66	55·4	35·9	126	105·7	68·6	186	156·0	101·3	246	206·3	134·0
7	05·9	03·8	67	56·2	36·5	127	106·5	69·2	187	156·8	101·8	247	207·2	134·5
8	06·7	04·4	68	57·0	37·0	128	107·3	69·7	188	157·7	102·4	248	208·0	135·1
9	07·5	04·9	69	57·9	37·6	129	108·2	70·3	189	158·5	102·9	249	208·8	135·6
10	08·4	05·4	70	58·7	38·1	130	109·0	70·8	190	159·3	103·5	250	209·7	136·2
11	09·2	06·0	71	59·5	38·7	131	109·9	71·3	191	160·2	104·0	251	210·5	136·7
12	10·1	06·5	72	60·4	39·2	132	110·7	71·9	192	161·0	104·6	252	211·3	137·2
13	10·9	07·1	73	61·2	39·8	133	111·5	72·4	193	161·9	105·1	253	212·2	137·8
14	11·7	07·6	74	62·1	40·3	134	112·4	73·0	194	162·7	105·7	254	213·0	138·3
15	12·6	08·2	75	62·9	40·8	135	113·2	73·5	195	163·5	106·2	255	213·9	138·9
16	13·4	08·7	76	63·7	41·4	136	114·1	74·1	196	164·4	106·7	256	214·7	139·4
17	14·3	09·3	77	64·6	41·9	137	114·9	74·6	197	165·2	107·3	257	215·5	140·0
18	15·1	09·8	78	65·4	42·5	138	115·7	75·2	198	166·1	107·8	258	216·4	140·5
19	15·9	10·3	79	66·3	43·0	139	116·6	75·7	199	166·9	108·4	259	217·2	141·1
20	16·8	10·9	80	67·1	43·6	140	117·4	76·2	200	167·7	108·9	260	218·1	141·6
21	17·6	11·4	81	67·9	44·1	141	118·3	76·8	201	168·6	109·5	261	218·9	142·2
22	18·5	12·0	82	68·8	44·7	142	119·1	77·3	202	169·4	110·0	262	219·7	142·7
23	19·3	12·5	83	69·6	45·2	143	119·9	77·9	203	170·3	110·6	263	220·6	143·2
24	20·1	13·1	84	70·4	45·7	144	120·8	78·4	204	171·1	111·1	264	221·4	143·8
25	21·0	13·6	85	71·3	46·3	145	121·6	79·0	205	171·9	111·7	265	222·2	144·3
26	21·8	14·2	86	72·1	46·8	146	122·4	79·5	206	172·8	112·2	266	223·1	144·9
27	22·6	14·7	87	73·0	47·4	147	123·3	80·1	207	173·6	112·7	267	223·9	145·4
28	23·5	15·2	88	73·8	47·9	148	124·1	80·6	208	174·4	113·3	268	224·8	146·0
29	24·3	15·8	89	74·6	48·5	149	125·0	81·2	209	175·3	113·8	269	225·6	146·5
30	25·2	16·3	90	75·5	49·0	150	125·8	81·7	210	176·1	114·4	270	226·4	147·1
31	26·0	16·9	91	76·3	49·6	151	126·6	82·2	211	177·0	114·9	271	227·3	147·6
32	26·8	17·4	92	77·2	50·1	152	127·5	82·8	212	177·8	115·5	272	228·1	148·1
33	27·7	18·0	93	78·0	50·7	153	128·3	83·3	213	178·6	116·0	273	229·0	148·7
34	28·5	18·5	94	78·8	51·2	154	129·2	83·9	214	179·5	116·6	274	229·8	149·2
35	29·4	19·1	95	79·7	51·7	155	130·0	84·4	215	180·3	117·1	275	230·6	149·8
36	30·2	19·6	96	80·5	52·3	156	130·8	85·0	216	181·2	117·6	276	231·5	150·3
37	31·0	20·2	97	81·4	52·8	157	131·7	85·5	217	182·0	118·2	277	232·3	150·9
38	31·9	20·7	98	82·2	53·4	158	132·5	86·1	218	182·8	118·7	278	233·2	151·4
39	32·7	21·2	99	83·0	53·9	159	133·3	86·6	219	183·7	119·3	279	234·0	152·0
40	33·5	21·8	100	83·9	54·5	160	134·2	87·1	220	184·5	119·8	280	234·8	152·5
41	34·4	22·3	101	84·7	55·0	161	135·0	87·7	221	185·3	120·4	281	235·7	153·0
42	35·2	22·9	102	85·5	55·6	162	135·9	88·2	222	186·2	120·9	282	236·5	153·6
43	36·1	23·4	103	86·4	56·1	163	136·7	88·8	223	187·0	121·5	283	237·3	154·1
44	36·9	24·0	104	87·2	56·6	164	137·5	89·3	224	187·9	122·0	284	238·2	154·7
45	37·7	24·5	105	88·1	57·2	165	138·4	89·9	225	188·7	122·5	285	239·0	155·2
46	38·6	25·1	106	88·9	57·7	166	139·2	90·4	226	189·5	123·1	286	239·9	155·8
47	39·4	25·6	107	89·7	58·3	167	140·1	91·0	227	190·4	123·6	287	240·7	156·3
48	40·3	26·1	108	90·6	58·8	168	140·9	91·5	228	191·2	124·2	288	241·5	156·9
49	41·1	26·7	109	91·4	59·4	169	141·7	92·0	229	192·1	124·7	289	242·4	157·4
50	41·9	27·2	110	92·3	59·9	170	142·6	92·6	230	192·9	125·3	290	243·2	157·9
51	42·8	27·8	111	93·1	60·5	171	143·4	93·1	231	193·7	125·8	291	244·1	158·5
52	43·6	28·3	112	93·9	61·0	172	144·3	93·7	232	194·6	126·4	292	244·9	159·0
53	44·4	28·9	113	94·8	61·5	173	145·1	94·2	233	195·4	126·9	293	245·7	159·6
54	45·3	29·4	114	95·6	62·1	174	145·9	94·8	234	196·2	127·4	294	246·6	160·1
55	46·1	30·0	115	96·4	62·6	175	146·8	95·3	235	197·1	128·0	295	247·4	160·7
56	47·0	30·5	116	97·3	63·2	176	147·6	95·9	236	197·9	128·5	296	248·2	161·2
57	47·8	31·0	117	98·1	63·7	177	148·4	96·4	237	198·8	129·1	297	249·1	161·8
58	48·6	31·6	118	99·0	64·3	178	149·3	96·9	238	199·6	129·6	298	249·9	162·3
59	49·5	32·1	119	99·8	64·8	179	150·1	97·5	239	200·4	130·2	299	250·8	162·8
60	50·3	32·7	120	100·6	65·4	180	151·0	98·0	240	201·3	130·7	300	251·6	163·4
Dist.	Dep.	Lat.	Dist.	Dep.	Lat.	Dist.	Dep.	Lat.	Dist.	Dep.	Lat.	Dist.	Dep.	Lat.

FOR 57 DEGREES.

Dist.	Lat.	Dep.	Dist.	Lat.	Dep.	Dist.	Lat.	Dep.	Dist.	Lat.	Dep.	Dist.	Lat.	Dep.
1	00·8	00·6	61	50·6	34·1	121	100·3	67·7	181	150·1	101·2	241	199·8	134·8
2	01·7	01·1	62	51·4	34·7	122	101·1	68·2	182	150·9	101·8	242	200·6	135·3
3	02·5	01·7	63	52·2	35·2	123	102·0	68·8	183	151·7	102·3	243	201·5	135·9
4	03·3	02·2	64	53·1	35·8	124	102·8	69·3	184	152·5	102·9	244	202·3	136·4
5	04·1	02·8	65	53·9	36·3	125	103·6	69·9	185	153·4	103·5	245	203·1	137·0
6	05·0	03·4	66	54·7	36·9	126	104·5	70·5	186	154·2	104·0	246	203·9	137·6
7	05·8	03·9	67	55·5	37·5	127	105·3	71·0	187	155·0	104·6	247	204·8	138·1
8	06·6	04·5	68	56·4	38·0	128	106·1	71·6	188	155·9	105·1	248	205·6	138·7
9	07·5	05·0	69	57·2	38·6	129	106·9	72·1	189	156·7	105·7	249	206·4	139·2
10	08·3	05·6	70	58·0	39·1	130	107·8	72·7	190	157·5	106·2	250	207·3	139·8
11	09·1	06·2	71	58·9	39·7	131	108·6	73·3	191	158·3	106·8	251	208·1	140·4
12	09·9	06·7	72	59·7	40·3	132	109·4	73·8	192	159·2	107·4	252	208·9	140·9
13	10·8	07·3	73	60·5	40·8	133	110·3	74·4	193	160·0	107·9	253	209·7	141·5
14	11·6	07·8	74	61·3	41·4	134	111·1	74·9	194	160·8	108·5	254	210·6	142·0
15	12·4	08·4	75	62·2	41·9	135	111·9	75·5	195	161·7	109·0	255	211·4	142·6
16	13·3	08·9	76	63·0	42·5	136	112·7	76·1	196	162·5	109·6	256	212·2	143·2
17	14·1	09·5	77	63·8	43·1	137	113·6	76·6	197	163·3	110·2	257	213·1	143·7
18	14·9	10·1	78	64·7	43·6	138	114·4	77·2	198	164·1	110·7	258	213·9	144·3
19	15·8	10·6	79	65·5	44·2	139	115·2	77·7	199	165·0	111·3	259	214·7	144·8
20	16·6	11·2	80	66·3	44·7	140	116·1	78·3	200	165·8	111·8	260	215·5	145·4
21	17·4	11·7	81	67·2	45·3	141	116·9	78·8	201	166·6	112·4	261	216·4	145·9
22	18·2	12·3	82	68·0	45·9	142	117·7	79·4	202	167·5	113·0	262	217·2	146·5
23	19·1	12·9	83	68·8	46·4	143	118·6	80·0	203	168·3	113·5	263	218·0	147·1
24	19·9	13·4	84	69·6	47·0	144	119·4	80·5	204	169·1	114·1	264	218·9	147·6
25	20·7	14·0	85	70·5	47·5	145	120·2	81·1	205	170·0	114·6	265	219·7	148·2
26	21·6	14·5	86	71·3	48·1	146	121·0	81·6	206	170·8	115·2	266	220·5	148·7
27	22·4	15·1	87	72·1	48·6	147	121·9	82·2	207	171·6	115·8	267	221·4	149·3
28	23·2	15·7	88	73·0	49·2	148	122·7	82·8	208	172·4	116·3	268	222·2	149·9
29	24·0	16·2	89	73·8	49·8	149	123·5	83·3	209	173·3	116·9	269	223·0	150·4
30	24·9	16·8	90	74·6	50·3	150	124·4	83·9	210	174·1	117·4	270	223·8	151·0
31	25·7	17·3	91	75·4	50·9	151	125·2	84·4	211	174·9	118·0	271	224·7	151·5
32	26·5	17·9	92	76·3	51·4	152	126·0	85·0	212	175·8	118·5	272	225·5	152·1
33	27·4	18·5	93	77·1	52·0	153	126·8	85·6	213	176·6	119·1	273	226·3	152·7
34	28·2	19·0	94	77·9	52·6	154	127·7	86·1	214	177·4	119·7	274	227·2	153·2
35	29·0	19·6	95	78·8	53·1	155	128·5	86·7	215	178·2	120·2	275	228·0	153·8
36	29·8	20·1	96	79·6	53·7	156	129·3	87·2	216	179·1	120·8	276	228·8	154·3
37	30·7	20·7	97	80·4	54·2	157	130·2	87·8	217	179·9	121·3	277	229·6	154·9
38	31·5	21·2	98	81·2	54·8	158	131·0	88·4	218	180·7	121·9	278	230·5	155·5
39	32·3	21·8	99	82·1	55·4	159	131·8	88·9	219	181·6	122·5	279	231·3	156·0
40	33·2	22·4	100	82·9	55·9	160	132·6	89·5	220	182·4	123·0	280	232·1	156·6
41	34·0	22·9	101	83·7	56·5	161	133·5	90·0	221	183·2	123·6	281	233·0	157·1
42	34·8	23·5	102	84·6	57·0	162	134·3	90·6	222	184·0	124·1	282	233·8	157·7
43	35·6	24·0	103	85·4	57·6	163	135·1	91·1	223	184·9	124·7	283	234·6	158·3
44	36·5	24·6	104	86·2	58·2	164	136·0	91·7	224	185·7	125·3	284	235·4	158·8
45	37·3	25·2	105	87·0	58·7	165	136·8	92·3	225	186·5	125·8	285	236·3	159·4
46	38·1	25·7	106	87·9	59·3	166	137·6	92·8	226	187·4	126·4	286	237·1	159·9
47	39·0	26·3	107	88·7	59·8	167	138·4	93·4	227	188·2	126·9	287	237·9	160·5
48	39·8	26·8	108	89·5	60·4	168	139·3	93·9	228	189·0	127·5	288	238·8	161·0
49	40·6	27·4	109	90·4	61·0	169	140·1	94·5	229	189·8	128·1	289	239·6	161·6
50	41·5	28·0	110	91·2	61·5	170	140·9	95·1	230	190·7	128·6	290	240·4	162·2
51	42·3	28·5	111	92·0	62·1	171	141·8	95·6	231	191·5	129·2	291	241·2	162·7
52	43·1	29·1	112	92·9	62·6	172	142·6	96·2	232	192·3	129·7	292	242·1	163·3
53	43·9	29·6	113	93·7	63·2	173	143·4	96·7	233	193·2	130·3	293	242·9	163·8
54	44·8	30·2	114	94·5	63·7	174	144·3	97·3	234	194·0	130·9	294	243·7	164·4
55	45·6	30·8	115	95·3	64·3	175	145·1	97·9	235	194·8	131·4	295	244·6	165·0
56	46·4	31·3	116	96·2	64·9	176	145·9	98·4	236	195·7	132·0	296	245·4	165·5
57	47·3	31·9	117	97·0	65·4	177	146·7	99·0	237	196·5	132·5	297	246·2	166·1
58	48·1	32·4	118	97·8	66·0	178	147·6	99·5	238	197·3	133·1	298	247·1	166·6
59	48·9	33·0	119	98·7	66·5	179	148·4	100·1	239	198·1	133·6	299	247·9	167·2
60	49·7	33·6	120	99·5	67·1	180	149·2	100·7	240	199·0	134·2	300	248·7	167·8
Dist.	Dep.	Lat.	Dist.	Dep.	Lat.	Dist.	Dep.	Lat.	Dist.	Dep.	Lat.	Dist.	Dep.	Lat.

Dist.	Lat.	Dep.	Dist.	Lat.	Dep.	Dist.	Lat.	Dep.	Dist.	Lat.	Dep.	Dist.	Lat.	Dep.
1	00·8	00·6	61	50·0	35·0	121	99·1	69·4	181	148·3	103·8	241	197·4	138·2
2	01·6	01·1	62	50·8	35·6	122	99·9	70·0	182	149·1	104·4	242	198·2	138·8
3	02·5	01·7	63	51·6	36·1	123	100·8	70·5	183	149·9	105·0	243	199·1	139·4
4	03·3	02·3	64	52·4	36·7	124	101·6	71·1	184	150·7	105·5	244	199·9	140·0
5	04·1	02·9	65	53·2	37·3	125	102·4	71·7	185	151·5	106·1	245	200·7	140·5
6	04·9	03·4	66	54·1	37·9	126	103·2	72·3	186	152·4	106·7	246	201·5	141·1
7	05·7	04·0	67	54·9	38·4	127	104·0	72·8	187	153·2	107·3	247	202·3	141·7
8	06·6	04·6	68	55·7	39·0	128	104·9	73·4	188	154·0	107·8	248	203·1	142·2
9	07·4	05·2	69	56·5	39·6	129	105·7	74·0	189	154·8	108·4	249	204·0	142·8
10	08·2	05·7	70	57·3	40·2	130	106·5	74·6	190	155·6	109·0	250	204·8	143·4
11	09·0	06·3	71	58·2	40·7	131	107·3	75·1	191	156·5	109·6	251	205·6	144·0
12	09·8	06·9	72	59·0	41·3	132	108·1	75·7	192	157·3	110·1	252	206·4	144·5
13	10·6	07·5	73	59·8	41·9	133	108·9	76·3	193	158·1	110·7	253	207·2	145·1
14	11·5	08·0	74	60·6	42·4	134	109·8	76·9	194	158·9	111·3	254	208·1	145·7
15	12·3	08·6	75	61·4	43·0	135	110·6	77·4	195	159·7	111·8	255	208·9	146·3
16	13·1	09·2	76	62·3	43·6	136	111·4	78·0	196	160·6	112·4	256	209·7	146·8
17	13·9	09·8	77	63·1	44·2	137	112·2	78·6	197	161·4	113·0	257	210·5	147·4
18	14·7	10·3	78	63·9	44·7	138	113·0	79·2	198	162·2	113·6	258	211·3	148·0
19	15·6	10·9	79	64·7	45·3	139	113·9	79·7	199	163·0	114·1	259	212·2	148·6
20	16·4	11·5	80	65·5	45·9	140	114·7	80·3	200	163·8	114·7	260	213·0	149·1
21	17·2	12·0	81	66·4	46·5	141	115·5	80·9	201	164·6	115·3	261	213·8	149·7
22	18·0	12·6	82	67·2	47·0	142	116·3	81·4	202	165·5	115·9	262	214·6	150·3
23	18·8	13·2	83	68·0	47·6	143	117·1	82·0	203	166·3	116·4	263	215·4	150·9
24	19·7	13·8	84	68·8	48·2	144	118·0	82·6	204	167·1	117·0	264	216·3	151·4
25	20·5	14·3	85	69·6	48·8	145	118·8	83·2	205	167·9	117·6	265	217·1	152·0
26	21·3	14·9	86	70·4	49·3	146	119·6	83·7	206	168·7	118·2	266	217·9	152·6
27	22·1	15·5	87	71·3	49·9	147	120·4	84·3	207	169·6	118·7	267	218·7	153·1
28	22·9	16·1	88	72·1	50·5	148	121·2	84·9	208	170·4	119·3	268	219·5	153·7
29	23·8	16·6	89	72·9	51·0	149	122·1	85·5	209	171·2	119·9	269	220·4	154·3
30	24·6	17·2	90	73·7	51·6	150	122·9	86·0	210	172·0	120·5	270	221·2	154·9
31	25·4	17·8	91	74·5	52·2	151	123·7	86·6	211	172·8	121·0	271	222·0	155·4
32	26·2	18·4	92	75·4	52·8	152	124·5	87·2	212	173·7	121·6	272	222·8	156·0
33	27·0	18·9	93	76·2	53·3	153	125·3	87·8	213	174·5	122·2	273	223·6	156·6
34	27·9	19·5	94	77·0	53·9	154	126·1	88·3	214	175·3	122·7	274	224·4	157·2
35	28·7	20·1	95	77·8	54·5	155	127·0	88·9	215	176·1	123·3	275	225·3	157·7
36	29·5	20·6	96	78·6	55·1	156	127·8	89·5	216	176·9	123·9	276	226·1	158·3
37	30·3	21·2	97	79·5	55·6	157	128·6	90·1	217	177·8	124·5	277	226·9	158·9
38	31·1	21·8	98	80·3	56·2	158	129·4	90·6	218	178·6	125·0	278	227·7	159·5
39	31·9	22·4	99	81·1	56·8	159	130·2	91·2	219	179·4	125·6	279	228·5	160·0
40	32·8	22·9	100	81·9	57·4	160	131·1	91·8	220	180·2	126·2	280	229·4	160·6
41	33·6	23·5	101	82·7	57·9	161	131·9	92·3	221	181·0	126·8	281	230·2	161·2
42	34·4	24·1	102	83·6	58·5	162	132·7	92·9	222	181·9	127·3	282	231·0	161·7
43	35·2	24·7	103	84·4	59·1	163	133·5	93·5	223	182·7	127·9	283	231·8	162·3
44	36·0	25·2	104	85·2	59·7	164	134·3	94·1	224	183·5	128·5	284	232·6	162·9
45	36·9	25·8	105	86·0	60·2	165	135·2	94·6	225	184·3	129·1	285	233·5	163·5
46	37·7	26·4	106	86·8	60·8	166	136·0	95·2	226	185·1	129·6	286	234·3	164·0
47	38·5	27·0	107	87·6	61·4	167	136·8	95·8	227	185·9	130·2	287	235·1	164·6
48	39·3	27·5	108	88·5	61·9	168	137·6	96·4	228	186·8	130·8	288	235·9	165·2
49	40·1	28·1	109	89·3	62·5	169	138·4	96·9	229	187·6	131·3	289	236·7	165·8
50	41·0	28·7	110	90·1	63·1	170	139·3	97·5	230	188·4	131·9	290	237·6	166·3
51	41·8	29·3	111	90·9	63·7	171	140·1	98·1	231	189·2	132·5	291	238·4	166·9
52	42·6	29·8	112	91·7	64·2	172	140·9	98·7	232	190·0	133·1	292	239·2	167·5
53	43·4	30·4	113	92·6	64·8	173	141·7	99·2	233	190·9	133·6	293	240·0	168·1
54	44·2	31·0	114	93·4	65·4	174	142·5	99·8	234	191·7	134·2	294	240·8	168·6
55	45·1	31·5	115	94·2	66·0	175	143·4	100·4	235	192·5	134·8	295	241·6	169·2
56	45·9	32·1	116	95·0	66·5	176	144·2	100·9	236	193·3	135·4	296	242·5	169·8
57	46·7	32·7	117	95·8	67·1	177	145·0	101·5	237	194·1	135·9	297	243·3	170·4
58	47·5	33·3	118	96·7	67·7	178	145·8	102·1	238	195·0	136·5	298	244·1	170·9
59	48·3	33·8	119	97·5	68·3	179	146·6	102·7	239	195·8	137·1	299	244·9	171·5
60	49·1	34·4	120	98·3	68·8	180	147·4	103·2	240	196·6	137·7	300	245·7	172·1
Dist.	Dep.	Lat.	Dist.	Dep.	Lat.	Dist.	Dep.	Lat.	Dist.	Dep.	Lat.	Dist.	Dep.	Lat.

FOR 55 DEGREES.

Dist.	Lat.	Dep.	Dist.	Lat.	Dep.	Dist.	Lat.	Dep.	Dist.	Lat.	Dep.	Dist.	Lat.	Dep.
1	00·8	00·6	61	49·4	35·9	121	97·9	71·1	181	146·4	106·4	241	195·0	141·7
2	01·6	01·2	62	50·2	36·4	122	98·7	71·7	182	147·2	107·0	242	195·8	142·2
3	02·4	01·8	63	51·0	37·0	123	99·5	72·3	183	148·1	107·6	243	196·6	142·8
4	03·2	02·4	64	51·8	37·6	124	100·3	72·9	184	148·9	108·2	244	197·4	143·4
5	04·0	02·9	65	52·6	38·2	125	101·1	73·5	185	149·7	108·7	245	198·2	144·0
6	04·9	03·5	66	53·4	38·8	126	101·9	74·1	186	150·5	109·3	246	199·0	144·6
7	05·7	04·1	67	54·2	39·4	127	102·7	74·6	187	151·3	109·9	247	199·8	145·2
8	06·5	04·7	68	55·0	40·0	128	103·6	75·2	188	152·1	110·5	248	200·6	145·8
9	07·3	05·3	69	55·8	40·6	129	104·4	75·8	189	152·9	111·1	249	201·4	146·4
10	08·1	05·9	70	56·6	41·1	130	105·2	76·4	190	153·7	111·7	250	202·3	146·9
11	08·9	06·5	71	57·4	41·7	131	106·0	77·0	191	154·5	112·3	251	203·1	147·5
12	09·7	07·1	72	58·2	42·3	132	106·8	77·6	192	155·3	112·9	252	203·9	148·1
13	10·5	07·6	73	59·1	42·9	133	107·6	78·2	193	156·1	113·4	253	204·7	148·7
14	11·3	08·2	74	59·9	43·5	134	108·4	78·8	194	156·9	114·0	254	205·5	149·3
15	12·1	08·8	75	60·7	44·1	135	109·2	79·4	195	157·8	114·6	255	206·3	149·9
16	12·9	09·4	76	61·5	44·7	136	110·0	79·9	196	158·6	115·2	256	207·1	150·5
17	13·8	10·0	77	62·3	45·3	137	110·8	80·5	197	159·4	115·8	257	207·9	151·1
18	14·6	10·6	78	63·1	45·8	138	111·6	81·1	198	160·2	116·4	258	208·7	151·6
19	15·4	11·2	79	63·9	46·4	139	112·5	81·7	199	161·0	117·0	259	209·5	152·2
20	16·2	11·8	80	64·7	47·0	140	113·3	82·3	200	161·8	117·6	260	210·3	152·8
21	17·0	12·3	81	65·5	47·6	141	114·1	82·9	201	162·6	118·1	261	211·2	153·4
22	17·8	12·9	82	66·3	48·2	142	114·9	83·5	202	163·4	118·7	262	212·0	154·0
23	18·6	13·5	83	67·1	48·8	143	115·7	84·1	203	164·2	119·3	263	212·8	154·6
24	19·4	14·1	84	68·0	49·4	144	116·5	84·6	204	165·0	119·9	264	213·6	155·2
25	20·2	14·7	85	68·8	50·0	145	117·3	85·2	205	165·8	120·5	265	214·4	155·8
26	21·0	15·3	86	69·6	50·5	146	118·1	85·8	206	166·7	121·1	266	215·2	156·4
27	21·8	15·9	87	70·4	51·1	147	118·9	86·4	207	167·5	121·7	267	216·0	156·9
28	22·7	16·5	88	71·2	51·7	148	119·7	87·0	208	168·3	122·3	268	216·8	157·5
29	23·5	17·0	89	72·0	52·3	149	120·5	87·6	209	169·1	122·8	269	217·6	158·1
30	24·3	17·6	90	72·8	52·9	150	121·4	88·2	210	169·9	123·4	270	218·4	158·7
31	25·1	18·2	91	73·6	53·5	151	122·2	88·8	211	170·7	124·0	271	219·2	159·3
32	25·9	18·8	92	74·4	54·1	152	123·0	89·3	212	171·5	124·6	272	220·1	159·9
33	26·7	19·4	93	75·2	54·7	153	123·8	89·9	213	172·3	125·2	273	220·9	160·5
34	27·5	20·0	94	76·0	55·3	154	124·6	90·5	214	173·1	125·8	274	221·7	161·1
35	28·3	20·6	95	76·9	55·8	155	125·4	91·1	215	173·9	126·4	275	222·5	161·6
36	29·1	21·2	96	77·7	56·4	156	126·2	91·7	216	174·7	127·0	276	223·3	162·2
37	29·9	21·7	97	78·5	57·0	157	127·0	92·3	217	175·6	127·5	277	224·1	162·8
38	30·7	22·3	98	79·3	57·6	158	127·8	92·9	218	176·4	128·1	278	224·9	163·4
39	31·6	22·9	99	80·1	58·2	159	128·6	93·5	219	177·2	128·7	279	225·7	164·0
40	32·4	23·5	100	80·9	58·8	160	129·4	94·0	220	178·0	129·3	280	226·5	164·6
41	33·2	24·1	101	81·7	59·4	161	130·3	94·6	221	178·8	129·9	281	227·3	165·2
42	34·0	24·7	102	82·5	60·0	162	131·1	95·2	222	179·6	130·5	282	228·1	165·8
43	34·8	25·3	103	83·3	60·5	163	131·9	95·8	223	180·4	131·1	283	229·0	166·3
44	35·6	25·9	104	84·1	61·1	164	132·7	96·4	224	181·2	131·7	284	229·8	166·9
45	36·4	26·5	105	84·9	61·7	165	133·5	97·0	225	182·0	132·3	285	230·6	167·5
46	37·2	27·0	106	85·8	62·3	166	134·3	97·6	226	182·8	132·8	286	231·4	168·1
47	38·0	27·6	107	86·6	62·9	167	135·1	98·2	227	183·6	133·4	287	232·2	168·7
48	38·8	28·2	108	87·4	63·5	168	135·9	98·7	228	184·5	134·0	288	233·0	169·3
49	39·6	28·8	109	88·2	64·1	169	136·7	99·3	229	185·3	134·6	289	233·8	169·9
50	40·5	29·4	110	89·0	64·7	170	137·5	99·9	230	186·1	135·2	290	234·6	170·5
51	41·3	30·0	111	89·8	65·2	171	138·3	100·5	231	186·9	135·8	291	235·4	171·0
52	42·1	30·6	112	90·6	65·8	172	139·2	101·1	232	187·7	136·4	292	236·2	171·6
53	42·9	31·2	113	91·4	66·4	173	140·0	101·7	233	188·5	137·0	293	237·0	172·2
54	43·7	31·7	114	92·2	67·0	174	140·8	102·3	234	189·3	137·5	294	237·9	172·8
55	44·5	32·3	115	93·0	67·6	175	141·6	102·9	235	190·1	138·1	295	238·7	173·4
56	45·3	32·9	116	93·8	68·2	176	142·4	103·5	236	190·9	138·7	296	239·5	174·0
57	46·1	33·5	117	94·7	68·8	177	143·2	104·0	237	191·7	139·3	297	240·3	174·6
58	46·9	34·1	118	95·5	69·4	178	144·0	104·6	238	192·5	139·9	298	241·1	175·2
59	47·7	34·7	119	96·3	69·9	179	144·8	105·2	239	193·4	140·5	299	241·9	175·7
60	48·5	35·3	120	97·1	70·5	180	145·6	105·8	240	194·2	141·1	300	242·7	176·3
Dist.	Dep.	Lat.	Dist.	Dep.	Lat.	Dist.	Dep.	Lat.	Dist.	Dep.	Lat.	Dist	Dep.	Lat.

FOR 54 DEGREES.

Dist.	Lat.	Dep.	Dist.	Lat.	Dep.	Dist.	Lat.	Dep.	Dist.	Lat.	Dep.	Dist.	Lat.	Dep.
1	00·8	00·6	61	48·7	36·7	121	96·6	72·8	181	144·6	108·9	241	192·5	145·0
2	01·6	01·2	62	49·5	37·3	122	97·4	73·4	182	145·4	109·5	242	193·3	145·6
3	02·4	01·8	63	50·3	37·9	123	98·2	74·0	183	146·2	110·1	243	194·1	146·2
4	03·2	02·4	64	51·1	38·5	124	99·0	74·6	184	146·9	110·7	244	194·9	146·8
5	04·0	03·0	65	51·9	39·1	125	99·8	75·2	185	147·7	111·3	245	195·7	147·4
6	04·8	03·6	66	52·7	39·7	126	100·6	75·8	186	148·5	111·9	246	196·5	148·0
7	05·6	04·2	67	53·5	40·3	127	101·4	76·4	187	149·3	112·5	247	197·3	148·6
8	06·4	04·8	68	54·3	40·9	128	102·2	77·0	188	150·1	113·1	248	198·1	149·3
9	07·2	05·4	69	55·1	41·5	129	103·0	77·6	189	150·9	113·7	249	198·9	149·9
10	08·0	06·0	70	55·9	42·1	130	103·8	78·2	190	151·7	114·3	250	199·7	150·5
11	08·8	06·6	71	56·7	42·7	131	104·6	78·8	191	152·5	114·9	251	200·5	151·1
12	09·6	07·2	72	57·5	43·3	132	105·4	79·4	192	153·3	115·5	252	201·3	151·7
13	10·4	07·8	73	58·3	43·9	133	106·2	80·0	193	154·1	116·2	253	202·1	152·3
14	11·2	08·4	74	59·1	44·5	134	107·0	80·6	194	154·9	116·8	254	202·9	152·9
15	12·0	09·0	75	59·9	45·1	135	107·8	81·2	195	155·7	117·4	255	203·7	153·5
16	12·8	09·6	76	60·7	45·7	136	108·6	81·8	196	156·5	118·0	256	204·5	154·1
17	13·6	10·2	77	61·5	46·3	137	109·4	82·4	197	157·3	118·6	257	205·2	154·7
18	14·4	10·8	78	62·3	46·9	138	110·2	83·1	198	158·1	119·2	258	206·0	155·3
19	15·2	11·4	79	63·1	47·5	139	111·0	83·7	199	158·9	119·8	259	206·8	155·9
20	16·0	12·0	80	63·9	48·1	140	111·8	84·3	200	159·7	120·4	260	207·6	156·5
21	16·8	12·6	81	64·7	48·7	141	112·6	84·9	201	160·5	121·0	261	208·4	157·1
22	17·6	13·2	82	65·5	49·3	142	113·4	85·5	202	161·3	121·6	262	209·2	157·7
23	18·4	13·8	83	66·3	50·0	143	114·2	86·1	203	162·1	122·2	263	210·0	158·3
24	19·2	14·4	84	67·1	50·6	144	115·0	86·7	204	162·9	122·8	264	210·8	158·9
25	20·0	15·0	85	67·9	51·2	145	115·8	87·3	205	163·7	123·4	265	211·6	159·5
26	20·8	15·6	86	68·7	51·8	146	116·6	87·9	206	164·5	124·0	266	212·4	160·1
27	21·6	16·2	87	69·5	52·4	147	117·4	88·5	207	165·3	124·6	267	213·2	160·7
28	22·4	16·9	88	70·3	53·0	148	118·2	89·1	208	166·1	125·2	268	214·0	161·3
29	23·2	17·5	89	71·1	53·6	149	119·0	89·7	209	166·9	125·8	269	214·8	161·9
30	24·0	18·1	90	71·9	54·2	150	119·8	90·3	210	167·7	126·4	270	215·6	162·5
31	24·8	18·7	91	72·7	54·8	151	120·6	90·9	211	168·5	127·0	271	216·4	163·1
32	25·6	19·3	92	73·5	55·4	152	121·4	91·5	212	169·3	127·6	272	217·2	163·7
33	26·4	19·9	93	74·3	56·0	153	122·2	92·1	213	170·1	128·2	273	218·0	164·3
34	27·2	20·5	94	75·1	56·6	154	123·0	92·7	214	170·9	128·8	274	218·8	164·9
35	28·0	21·1	95	75·9	57·2	155	123·8	93·3	215	171·7	129·4	275	219·6	165·5
36	28·8	21·7	96	76·7	57·8	156	124·6	93·9	216	172·5	130·0	276	220·4	166·1
37	29·5	22·3	97	77·5	58·4	157	125·4	94·5	217	173·3	130·6	277	221·2	166·7
38	30·3	22·9	98	78·3	59·0	158	126·2	95·1	218	174·1	131·2	278	222·0	167·3
39	31·1	23·5	99	79·1	59·6	159	127·0	95·7	219	174·9	131·8	279	222·8	167·9
40	31·9	24·1	100	79·9	60·2	160	127·8	96·3	220	175·7	132·4	280	223·6	168·5
41	32·7	24·7	101	80·7	60·8	161	128·6	96·9	221	176·5	133·0	281	224·4	169·1
42	33·5	25·3	102	81·5	61·4	162	129·4	97·5	222	177·3	133·6	282	225·2	169·7
43	34·3	25·9	103	82·3	62·0	163	130·2	98·1	223	178·1	134·2	283	226·0	170·3
44	35·1	26·5	104	83·1	62·6	164	131·0	98·7	224	178·9	134·8	284	226·8	170·9
45	35·9	27·1	105	83·9	63·2	165	131·8	99·3	225	179·7	135·4	285	227·6	171·5
46	36·7	27·7	106	84·7	63·8	166	132·6	99·9	226	180·5	136·0	286	228·4	172·1
47	37·5	28·3	107	85·5	64·4	167	133·4	100·5	227	181·3	136·6	287	229·2	172·7
48	38·3	28·9	108	86·3	65·0	168	134·2	101·1	228	182·1	137·2	288	230·0	173·3
49	39·1	29·5	109	87·1	65·6	169	135·0	101·7	229	182·9	137·8	289	230·8	173·9
50	39·9	30·1	110	87·8	66·2	170	135·8	102·3	230	183·7	138·4	290	231·6	174·5
51	40·7	30·7	111	88·6	66·8	171	136·6	102·9	231	184·5	139·0	291	232·4	175·1
52	41·5	31·3	112	89·4	67·4	172	137·4	103·5	232	185·3	139·6	292	233·2	175·7
53	42·3	31·9	113	90·2	68·0	173	138·2	104·1	233	186·1	140·2	293	234·0	176·3
54	43·1	32·5	114	91·0	68·6	174	139·0	104·7	234	186·9	140·8	294	234·8	176·9
55	43·9	33·1	115	91·8	69·2	175	139·8	105·3	235	187·7	141·4	295	235·6	177·5
56	44·7	33·7	116	92·6	69·8	176	140·6	105·9	236	188·5	142·0	296	236·4	178·1
57	45·5	34·3	117	93·4	70·4	177	141·4	106·5	237	189·3	142·6	297	237·2	178·7
58	46·3	34·9	118	94·2	71·0	178	142·2	107·1	238	190·1	143·2	298	238·0	179·3
59	47·1	35·5	119	95·0	71·6	179	143·0	107·7	239	190·9	143·8	299	238·8	179·9
60	47·9	36·1	120	95·8	72·2	180	143·8	108·3	240	191·7	144·4	300	239·6	180·5
Dist.	Dep.	Lat.	Dist.	Dep.	Lat.	Dist.	Dep.	Lat.	Dist.	Dep.	Lat.	Dist.	Dep.	Lat.

FOR 53 DEGREES.

Dist.	Lat.	Dep.	Dist.	Lat.	Dep.	Dist.	Lat.	Dep.	Dist.	Lat.	Dep.	Dist.	Lat.	Dep.
1	00·8	00·6	61	48·1	37·6	121	95·3	74·5	181	142·6	111·4	241	189·9	148·4
2	01·6	01·2	62	48·9	38·2	122	96·1	75·1	182	143·4	112·1	242	190·7	149·0
3	02·4	01·8	63	49·6	38·8	123	96·9	75·7	183	144·2	112·7	243	191·5	149·6
4	03·2	02·5	64	50·4	39·4	124	97·7	76·3	184	145·0	113·3	244	192·3	150·2
5	03·9	03·1	65	51·2	40·0	125	98·5	77·0	185	145·8	113·9	245	193·1	150·8
6	04·7	03·7	66	52·0	40·6	126	99·3	77·6	186	146·6	114·5	246	193·9	151·5
7	05·5	04·3	67	52·8	41·2	127	100·1	78·2	187	147·4	115·1	247	194·6	152·1
8	06·3	04·9	68	53·6	41·9	128	100·9	78·8	188	148·1	115·7	248	195·4	152·7
9	07·1	05·5	69	54·4	42·5	129	101·7	79·4	189	148·9	116·4	249	196·2	153·3
10	07·9	06·2	70	55·2	43·1	130	102·4	80·0	190	149·7	117·0	250	197·0	153·9
11	08·7	06·8	71	55·9	43·7	131	103·2	80·7	191	150·5	117·6	251	197·8	154·5
12	09·5	07·4	72	56·7	44·3	132	104·0	81·3	192	151·3	118·2	252	198·6	155·1
13	10·2	08·0	73	57·5	44·9	133	104·8	81·9	193	152·1	118·8	253	199·4	155·8
14	11·0	08·6	74	58·3	45·6	134	105·6	82·5	194	152·9	119·4	254	200·2	156·4
15	11·8	09·2	75	59·1	46·2	135	106·4	83·1	195	153·7	120·1	255	200·9	157·0
16	12·6	09·9	76	59·9	46·8	136	107·2	83·7	196	154·5	120·7	256	201·7	157·6
17	13·4	10·5	77	60·7	47·4	137	108·0	84·3	197	155·2	121·3	257	202·5	158·2
18	14·2	11·1	78	61·5	48·0	138	108·7	85·0	198	156·0	121·9	258	203·3	158·8
19	15·0	11·7	79	62·3	48·6	139	109·5	85·6	199	156·8	122·5	259	204·1	159·5
20	15·8	12·3	80	63·0	49·3	140	110·3	86·2	200	157·6	123·1	260	204·9	160·1
21	16·5	12·9	81	63·8	49·9	141	111·1	86·8	201	158·4	123·7	261	205·7	160·7
22	17·3	13·5	82	64·6	50·5	142	111·9	87·4	202	159·2	124·4	262	206·5	161·3
23	18·1	14·2	83	65·4	51·1	143	112·7	88·0	203	160·0	125·0	263	207·2	161·9
24	18·9	14·8	84	66·2	51·7	144	113·5	88·7	204	160·8	125·6	264	208·0	162·5
25	19·7	15·4	85	67·0	52·3	145	114·3	89·3	205	161·5	126·2	265	208·8	163·2
26	20·5	16·0	86	67·8	52·9	146	115·0	89·9	206	162·3	126·8	266	209·6	163·8
27	21·3	16·6	87	68·6	53·6	147	115·8	90·5	207	163·1	127·4	267	210·4	164·4
28	22·1	17·2	88	69·3	54·2	148	116·6	91·1	208	163·9	128·1	268	211·2	165·0
29	22·9	17·9	89	70·1	54·8	149	117·4	91·7	209	164·7	128·7	269	212·0	165·6
30	23·6	18·5	90	70·9	55·4	150	118·2	92·3	210	165·5	129·3	270	212·8	166·2
31	24·4	19·1	91	71·7	56·0	151	119·0	93·0	211	166·3	129·9	271	213·6	166·8
32	25·2	19·7	92	72·5	56·6	152	119·8	93·6	212	167·1	130·5	272	214·3	167·5
33	26·0	20·3	93	73·3	57·3	153	120·6	94·2	213	167·8	131·1	273	215·1	168·1
34	26·8	20·9	94	74·1	57·9	154	121·4	94·8	214	168·6	131·8	274	215·9	168·7
35	27·6	21·5	95	74·9	58·5	155	122·1	95·4	215	169·4	132·4	275	216·7	169·3
36	28·4	22·2	96	75·6	59·1	156	122·9	96·0	216	170·2	133·0	276	217·5	169·9
37	29·2	22·8	97	76·4	59·7	157	123·7	96·7	217	171·0	133·6	277	218·3	170·5
38	29·9	23·4	98	77·2	60·3	158	124·5	97·3	218	171·8	134·2	278	219·1	171·2
39	30·7	24·0	99	78·0	61·0	159	125·3	97·9	219	172·6	134·8	279	219·9	171·8
40	31·5	24·6	100	78·8	61·6	160	126·1	98·5	220	173·4	135·4	280	220·6	172·4
41	32·3	25·2	101	79·6	62·2	161	126·9	99·1	221	174·2	136·1	281	221·4	173·0
42	33·1	25·9	102	80·4	62·8	162	127·7	99·7	222	174·9	136·7	282	222·2	173·6
43	33·9	26·5	103	81·2	63·4	163	128·4	100·4	223	175·7	137·3	283	223·0	174·2
44	34·7	27·1	104	82·0	64·0	164	129·2	101·0	224	176·5	137·9	284	223·8	174·8
45	35·5	27·7	105	82·7	64·6	165	130·0	101·6	225	177·3	138·5	285	224·6	175·5
46	36·2	28·3	106	83·5	65·3	166	130·8	102·2	226	178·1	139·1	286	225·4	176·1
47	37·0	28·9	107	84·3	65·9	167	131·6	102·8	227	178·9	139·8	287	226·2	176·7
48	37·8	29·6	108	85·1	66·5	168	132·4	103·4	228	179·7	140·4	288	226·9	177·3
49	38·6	30·2	109	85·9	67·1	169	133·2	104·0	229	180·5	141·0	289	227·7	177·9
50	39·4	30·8	110	86·7	67·7	170	134·0	104·7	230	181·2	141·6	290	228·5	178·5
51	40·2	31·4	111	87·5	68·3	171	134·7	105·3	231	182·0	142·2	291	229·3	179·2
52	41·0	32·0	112	88·3	69·0	172	135·5	105·9	232	182·8	142·8	292	230·1	179·8
53	41·8	32·6	113	89·0	69·6	173	136·3	106·5	233	183·6	143·4	293	230·9	180·4
54	42·6	33·2	114	89·8	70·2	174	137·1	107·1	234	184·4	144·1	294	231·7	181·0
55	43·3	33·9	115	90·6	70·8	175	137·9	107·7	235	185·2	144·7	295	232·5	181·6
56	44·1	34·5	116	91·4	71·4	176	138·7	108·4	236	186·0	145·3	296	233·3	182·2
57	44·9	35·1	117	92·2	72·0	177	139·5	109·0	237	186·8	145·9	297	234·0	182·9
58	45·7	35·7	118	93·0	72·6	178	140·3	109·6	238	187·5	146·5	298	234·8	183·5
59	46·5	36·3	119	93·8	73·3	179	141·1	110·2	239	188·3	147·1	299	235·6	184·1
60	47·3	36·9	120	94·6	73·9	180	141·8	110·8	240	189·1	147·8	300	236·4	184·7
Dist.	Dep.	Lat.	Dist.	Dep.	Lat.	Dist.	Dep.	Lat.	Dist.	Dep.	Lat.	Dist.	Dep.	Lat.

FOR 52 DEGREES.

DIFFERENCE OF LATITUDE AND DEPARTURE FOR 39 DEGREES.

Dist.	Lat.	Dep.	Dist.	Lat.	Dep.	Dist.	Lat.	Dep.	Dist.	Lat.	Dep.	Dist.	Lat.	Dep.
1	00·8	00·6	61	47·4	38·4	121	94·0	76·1	181	140·7	113·9	241	187·3	151·7
2	01·6	01·3	62	48·2	39·0	122	94·8	76·8	182	141·4	114·5	242	188·1	152·3
3	02·3	01·9	63	49·0	39·6	123	95·6	77·4	183	142·2	115·2	243	188·8	152·9
4	03·1	02·5	64	49·7	40·3	124	96·4	78·0	184	143·0	115·8	244	189·6	153·6
5	03·9	03·1	65	50·5	40·9	125	97·1	78·7	185	143·8	116·4	245	190·4	154·2
6	04·7	03·8	66	51·3	41·5	126	97·9	79·3	186	144·5	117·1	246	191·2	154·8
7	05·4	04·4	67	52·1	42·2	127	98·7	79·9	187	145·3	117·7	247	192·0	155·4
8	06·2	05·0	68	52·8	42·8	128	99·5	80·6	188	146·1	118·3	248	192·7	156·1
9	07·0	05·7	69	53·6	43·4	129	100·3	81·2	189	146·9	118·9	249	193·5	156·7
10	07·8	06·3	70	54·4	44·1	130	101·0	81·8	190	147·7	119·6	250	194·3	157·3
11	08·5	06·9	71	55·2	44·7	131	101·8	82·4	191	148·4	120·2	251	195·1	158·0
12	09·3	07·6	72	56·0	45·3	132	102·6	83·1	192	149·2	120·8	252	195·8	158·6
13	10·1	08·2	73	56·7	45·9	133	103·4	83·7	193	150·0	121·5	253	196·6	159·2
14	10·9	08·8	74	57·5	46·6	134	104·1	84·3	194	150·8	122·1	254	197·4	159·8
15	11·7	09·4	75	58·3	47·2	135	104·9	85·0	195	151·5	122·7	255	198·2	160·5
16	12·4	10·1	76	59·1	47·8	136	105·7	85·6	196	152·3	123·3	256	198·9	161·1
17	13·2	10·7	77	59·8	48·5	137	106·5	86·2	197	153·1	124·0	257	199·7	161·7
18	14·0	11·3	78	60·6	49·1	138	107·2	86·8	198	153·9	124·6	258	200·5	162·4
19	14·8	12·0	79	61·4	49·7	139	108·0	87·5	199	154·7	125·2	259	201·3	163·0
20	15·5	12·6	80	62·2	50·3	140	108·8	88·1	200	155·4	125·9	260	202·1	163·6
21	16·3	13·2	81	62·9	51·0	141	109·6	88·7	201	156·2	126·5	261	202·8	164·3
22	17·1	13·8	82	63·7	51·6	142	110·4	89·4	202	157·0	127·1	262	203·6	164·9
23	17·9	14·5	83	64·5	52·2	143	111·1	90·0	203	157·8	127·8	263	204·4	165·5
24	18·7	15·1	84	65·3	52·9	144	111·9	90·6	204	158·5	128·4	264	205·2	166·1
25	19·4	15·7	85	66·1	53·5	145	112·7	91·3	205	159·3	129·0	265	205·9	166·8
26	20·2	16·4	86	66·8	54·1	146	113·5	91·9	206	160·1	129·6	266	206·7	167·4
27	21·0	17·0	87	67·6	54·8	147	114·2	92·5	207	160·9	130·3	267	207·5	168·0
28	21·8	17·6	88	68·4	55·4	148	115·0	93·1	208	161·6	130·9	268	208·3	168·7
29	22·5	18·3	89	69·2	56·0	149	115·8	93·8	209	162·4	131·5	269	209·1	169·3
30	23·3	18·9	90	69·9	56·6	150	116·6	94·4	210	163·2	132·2	270	209·8	169·9
31	24·1	19·5	91	70·7	57·3	151	117·3	95·0	211	164·0	132·8	271	210·6	170·5
32	24·9	20·1	92	71·5	57·9	152	118·1	95·7	212	164·8	133·4	272	211·4	171·2
33	25·6	20·8	93	72·3	58·5	153	118·9	96·3	213	165·5	134·0	273	212·2	171·8
34	26·4	21·4	94	73·1	59·2	154	119·7	96·9	214	166·3	134·7	274	212·9	172·4
35	27·2	22·0	95	73·8	59·8	155	120·5	97·5	215	167·1	135·3	275	213·7	173·1
36	28·0	22·7	96	74·6	60·4	156	121·2	98·2	216	167·9	135·9	276	214·5	173·7
37	28·8	23·3	97	75·4	61·0	157	122·0	98·8	217	168·6	136·6	277	215·3	174·3
38	29·5	23·9	98	76·2	61·7	158	122·8	99·4	218	169·4	137·2	278	216·0	175·0
39	30·3	24·5	99	76·9	62·3	159	123·6	100·1	219	170·2	137·8	279	216·8	175·6
40	31·1	25·2	100	77·7	62·9	160	124·3	100·7	220	171·0	138·5	280	217·6	176·2
41	31·9	25·8	101	78·5	63·6	161	125·1	101·3	221	171·7	139·1	281	218·4	176·8
42	32·6	26·4	102	79·3	64·2	162	125·9	101·9	222	172·5	139·7	282	219·2	177·5
43	33·4	27·1	103	80·0	64·8	163	126·7	102·6	223	173·3	140·3	283	219·9	178·1
44	34·2	27·7	104	80·8	65·4	164	127·5	103·2	224	174·1	141·0	284	220·7	178·7
45	35·0	28·3	105	81·6	66·1	165	128·2	103·8	225	174·9	141·6	285	221·5	179·4
46	35·7	28·9	106	82·4	66·7	166	129·0	104·5	226	175·6	142·2	286	222·3	180·0
47	36·5	29·6	107	83·2	67·3	167	129·8	105·1	227	176·4	142·9	287	223·0	180·6
48	37·3	30·2	108	83·9	68·0	168	130·6	105·7	228	177·2	143·5	288	223·8	181·2
49	38·1	30·8	109	84·7	68·6	169	131·3	106·4	229	178·0	144·1	289	224·6	181·9
50	38·9	31·5	110	85·5	69·2	170	132·1	107·0	230	178·7	144·7	290	225·4	182·5
51	39·6	32·1	111	86·3	69·9	171	132·9	107·6	231	179·5	145·4	291	226·1	183·1
52	40·4	32·7	112	87·0	70·5	172	133·7	108·2	232	180·3	146·0	292	226·9	183·8
53	41·2	33·4	113	87·8	71·1	173	134·4	108·9	233	181·1	146·6	293	227·7	184·4
54	42·0	34·0	114	88·6	71·7	174	135·2	109·5	234	181·9	147·3	294	228·5	185·0
55	42·7	34·6	115	89·4	72·4	175	136·0	110·1	235	182·6	147·9	295	229·3	185·6
56	43·5	35·2	116	90·1	73·0	176	136·8	110·8	236	183·4	148·5	296	230·0	187·3
57	44·3	35·9	117	90·9	73·6	177	137·6	111·4	237	184·2	149·1	297	230·8	186·9
58	45·1	36·5	118	91·7	74·3	178	138·3	112·0	238	185·0	149·8	298	231·6	187·5
59	45·9	37·1	119	92·5	74·9	179	139·1	112·6	239	185·7	150·4	299	232·4	188·2
60	46·6	37·8	120	93·3	75·5	180	139·9	113·3	240	186·5	151·0	300	233·1	188·8
Dist.	Dep.	Lat.	Dist.	Dep.	Lat.	Dist.	Dep.	Lat.	Dist.	Dep.	Lat.	Dist.	Dep.	Lat.

FOR 51 DEGREES.

Dist.	Lat.	Dep.	Dist.	Lat.	Dep.	Dist.	Lat.	Dep.	Dist.	Lat.	Dep.	Dist.	Lat.	Dep.
1	00·8	00·6	61	46·7	39·2	121	92·7	77·8	181	138·7	116·3	241	184·6	154·9
2	01·5	01·3	62	47·5	39·9	122	93·5	78·4	182	139·4	117·0	242	185·4	155·6
3	02·3	01·9	63	48·3	40·5	123	94·2	79·1	183	140·2	117·6	243	186·1	156·2
4	03·1	02·6	64	49·0	41·1	124	95·0	79·7	184	141·0	118·3	244	186·9	156·8
5	03·8	03·2	65	49·8	41·8	125	95·8	80·3	185	141·7	118·9	245	187·7	157·5
6	04·6	03·9	66	50·6	42·4	126	96·5	81·0	186	142·5	119·6	246	188·4	158·1
7	05·4	04·5	67	51·3	43·1	127	97·3	81·6	187	143·3	120·2	247	189·2	158·8
8	06·1	05·1	68	52·1	43·7	128	98·1	82·3	188	144·0	120·8	248	190·0	159·4
9	06·9	05·8	69	52·9	44·4	129	98·8	82·9	189	144·8	121·5	249	190·7	160·1
10	07·7	06·4	70	53·6	45·0	130	99·6	83·6	190	145·5	122·1	250	191·5	160·7
11	08·4	07·1	71	54·4	45·6	131	100·4	84·2	191	146·3	122·8	251	192·3	161·3
12	09·2	07·7	72	55·2	46·3	132	101·1	84·8	192	147·1	123·4	252	193·0	162·0
13	10.0	08·4	73	55·9	46·9	133	101.9	85·5	193	147·8	124·1	253	193·8	162·6
14	10·7	09 0	74	56·7	47·6	134	102·6	86·1	194	148·6	124·7	254	194·6	163·3
15	11·5	09·6	75	57·5	48·2	135	103·4	86·8	195	149·4	125·3	255	195·3	163·9
16	12·3	10·3	76	58·2	48·9	136	104·2	87·4	196	150·1	126·0	256	196·1	164·6
17	13·0	10·9	77	59·0	49·5	137	104·9	88·1	197	150·9	126·6	257	196·9	165·2
18	13·8	11·6	78	59·8	50·1	138	105·7	88·7	198	151·7	127·3	258	197·6	165·8
19	14·6	12·2	79	60·5	50·8	139	106·5	89·3	199	152·4	127·9	259	198·4	166·5
20	15·3	12·9	80	61·3	51·4	140	107·2	90·0	200	153·2	128·6	260	199·2	167·1
21	16·1	13·5	81	62·0	52·1	141	108·0	90·6	201	154·0	129·2	261	199·9	167·8
22	16·9	14·1	82	62·8	52·7	142	108·8	91·3	202	154·7	129·8	262	200·7	168·4
23	17·6	14·8	83	63·6	53·4	143	109·5	91·9	203	155·5	130·5	263	201·5	169·1
24	18·4	15·4	84	64·3	54·0	144	110·3	92·6	204	156·3	131·1	264	202·2	169·7
25	19·2	16·1	85	65·1	54·6	145	111·1	93·2	205	157·0	131·8	265	203·0	170·3
26	19·9	16·7	86	65·9	55·3	146	111·8	93·8	206	157·8	132·4	266	203·8	171·0
27	20·7	17·4	87	66·6	55·9	147	112·6	94·5	207	158·6	133·1	267	204·5	171·6
28	21·4	18·0	88	67·4	56·6	148	113·4	95·1	208	159·3	133·7	268	205·3	172·3
29	22·2	18·6	89	68·2	57·2	149	114·1	95·8	209	160·1	134·3	269	206·1	172·9
30	23·0	19·3	90	68·9	57·9	150	114·9	96·4	210	160·9	135·0	270	206·8	173·6
31	23·7	19·9	91	69·7	58·5	151	115·7	97·1	211	161·6	135·6	271	207·6	174·2
32	24·5	20·6	92	70·5	59·1	152	116·4	97·7	212	162·4	136·3	272	208·4	174·8
33	25·3	21·2	93	71·2	59·8	153	117·2	98·3	213	163·2	136·9	273	209·1	175·5
34	26·0	21·9	94	72·0	60·4	154	118·0	99·0	214	163·9	137·6	274	209·9	176·1
35	26·8	22·5	95	72·8	61·1	155	118.7	99·6	215	164·7	138·2	275	210·7	176·8
36	27·6	23·1	96	73·5	61·7	156	119·5	100·3	216	165·5	138·8	276	211·4	177·4
37	28·3	23·8	97	74·3	62·4	157	120·3	100·9	217	166·2	139·5	277	212·2	178·1
38	29·1	24·4	98	75·1	63·0	158	121·0	101·6	218	167·0	140.1	278	213·0	178·7
39	29·9	25·1	99	75·8	63·6	159	121·8	102·2	219	167·8	140·8	279	213·7	179·3
40	30·6	25·7	100	76·6	64·3	160	122·6	102·8	220	168·5	141·4	280	214·5	180·0
41	31·4	26·4	101	77·4	64 9	161	123·3	103·5	221	169·3	142·1	281	215·3	180·6
42	32·2	27·0	102	78·1	65·6	162	124·1	104·1	222	170·1	142·7	282	216·0	181·3
43	32·9	27·6	103	78·9	66·2	163	124·9	104·8	223	170·8	143·3	283	216·8	181·9
44	33·7	28·3	104	79·7	66·8	164	125·6	105·4	224	171·6	144·0	284	217·6	182·6
45	34·5	28·9	105	80·4	67·5	165	126·4	106·1	225	172·4	144·6	285	218·3	183·2
46	35·2	29·6	106	81·2	68·1	166	127·2	106·7	226	173·1	145·3	286	219·1	183·8
47	36·0	30·2	107	82·0	68·8	167	127·9	107·3	227	173·9	145·9	287	219·9	184·5
48	36·8	30·9	108	82·7	69·4	168	128·7	108·0	228	174·7	146·6	288	220·6	185·1
49	37·5	31·5	109	83·5	70·1	169	129·5	108·6	229	175·4	147·2	289	221·4	185·8
50	38·3	32·1	110	84·3	70·7	170	130·2	109·3	230	176·2	147·8	290	222·2	186 4
51	39·1	32·8	111	85·0	71·3	171	131·0	109·9	231	177·0	148·5	291	222·9	187·1
52	39·8	33·4	112	85·8	72·0	172	131·8	110·6	232	177·7	149·1	292	223·7	187·7
53	40·6	34·1	113	86·6	72·6	173	132·5	111·2	233	178·5	149·8	293	224·5	188·3
54	41·4	34·7	114	87·3	73·3	174	133·3	111·8	234	179·3	150·4	294	225·2	189·0
55	42·1	35·4	115	88·1	73·9	175	134·1	112·5	235	180·0	151·1	295	226·0	189·6
56	42·9	36·0	116	88·9	74·6	176	134·8	113·1	236	180·8	151·7	296	226·7	190·3
57	43·7	36·6	117	89·6	75·2	177	135·6	113·8	237	181·6	152·3	297	227·5	190·9
58	44·4	37·3	118	90·4	75·8	178	136·4	114·4	238	182·3	153·0	298	228·3	191·6
59	45·2	37·9	119	91·2	76·5	179	137·1	115·1	239	183·1	153·6	299	229·0	192·2
60	46·0	38·6	120	91·9	77·1	180	137·9	115·7	240	183·9	154·3	300	229·8	192·8
Dist.	Dep.	Lat.	Dist.	Dep.	Lat.	Dist.	Dep.	Lat.	Dist.	Dep.	Lat.	Dist.	Dep.	Lat.

FOR 50 DEGREES.

Dist.	Lat.	Dep.	Dist.	Lat.	Dep.	Dist.	Lat.	Dep.	Dist.	Lat.	Dep.	Dist.	Lat.	Dep.
1	00 8	00·7	61	46·0	40·0	121	91·3	79·4	181	136·6	118·7	241	181·9	158·1
2	01·5	01·3	62	46·8	40·7	122	92·1	80·0	182	137·4	119·4	242	182·6	158·8
3	02·3	02·0	63	47·5	41·3	123	92·8	80·7	183	138·1	120·1	243	183·4	159·4
4	03·0	02·6	64	48·3	42·0	124	93·6	81·4	184	138·9	120·7	244	184·1	160·1
5	03·8	03·3	65	49·1	42·6	125	94·3	82·0	185	139·6	121·4	245	184·9	160·7
6	04·5	03·9	66	49·8	43·3	126	95·1	82·7	186	140·4	122·0	246	185·7	161·4
7	05·3	04·6	67	50·6	44·0	127	95·8	83 3	187	141·1	122·7	247	186·4	162·0
8	06·0	05·2	68	51·3	44·6	128	96·6	84·0	188	141·9	123·3	248	187·2	162·7
9	06·8	05·9	69	52·1	45·3	129	97·4	84 6	189	142·6	124·0	249	187·9	163·4
10	07·5	06·6	70	52·8	45·9	130	98·1	85 3	190	143·4	124·7	250	188·7	164·0
11	08·3	07·2	71	53·6	46 6	131	98·9	85·9	191	144·1	125·3	251	189·4	164·7
12	09·1	07·9	72	54·3	47·2	132	99·6	86·6	192	144·9	126·0	252	190·2	165·3
13	09·8	08·5	73	55·1	47·9	133	100·4	87·3	193	145·7	126·6	253	190·9	166·0
14	10·6	09·2	74	55·8	48·5	134	101·1	87·9	194	146·4	127·3	254	191·7	166·6
15	11·3	09·8	75	56·6	49·2	135	101·9	88·6	195	147·2	127·9	255	192·5	167·3
16	12·1	10·5	76	57·4	49·9	136	102·6	89·2	196	147·9	128·6	256	193·2	168·0
17	12·8	11·2	77	58·1	50·5	137	103·4	89·9	197	148·7	129·2	257	194 0	168·6
18	13·6	11·8	78	58 9	51·2	138	104·1	90·5	198	149·4	129·9	258	194·7	169·3
19	14·3	12·5	79	59·6	51·8	139	104·9	91·2	199	150 2	130·6	259	195·5	169·9
20	15·1	13·1	80	60·4	52·5	140	105·7	91·8	200	150·9	131·2	260	196·2	170·6
21	15·8	13·8	81	61·1	53·1	141	106·4	92·5	201	151·7	131·9	261	197·0	171·2
22	16·6	14·4	82	61·9	53·8	142	107·2	93·2	202	152·5	132·5	262	197·7	171 9
23	17·4	15·1	83	62·6	54 5	143	107·9	93·8	203	153·2	133·2	263	198·5	172·5
24	18·1	15·7	84	63·4	55·1	144	108 7	94·5	204	154·0	133·8	264	199·2	173·2
25	18·9	16·4	85	64·2	55·8	145	109·4	95·1	205	154·7	134·5	265	200·0	173 9
26	19 6	17·1	86	64·9	56·4	146	110·2	95·8	206	155·5	135·1	266	200·8	174·5
27	20·4	17·7	87	65·7	57·1	147	110·9	96·4	207	156·2	135·8	267	201·5	175·2
28	21·1	18·4	88	66·4	57·7	148	111·7	97·1	208	157·0	136·5	268	202·3	175·8
29	21·9	19·0	89	67·2	58·4	149	112·5	97·8	209	157 7	137·1	269	203·0	176·5
30	22·6	19·7	90	67·9	59·0	150	113·2	98·4	210	158·5	137·8	270	203·8	177·1
31	23·4	20·3	91	68 7	59·7	151	114·0	99·1	211	159·2	138·4	271	204·5	177·8
32	24·2	21·0	92	69·4	60·4	152	114·7	99·7	212	160·0	139·1	272	205·3	178·4
33	24·9	21·6	93	70·2	61·0	153	115·5	100·4	213	160·8	139·7	273	206·0	179·1
34	25·7	22·3	94	70·9	61·7	154	116·2	101·0	214	161·5	140·4	274	206·8	179·8
35	26·4	23·0	95	71·7	62·3	155	117·0	101·7	215	162·3	141·1	275	207·5	180·4
36	27·2	23·6	96	72·5	63·0	156	117·7	102·3	216	163·0	141·7	276	208·3	181·1
37	27·9	24·3	97	73·2	63 6	157	118·5	103·0	217	163 8	142·4	277	209·1	181·7
38	28·7	24·9	98	74·0	64·3	158	119·2	103·7	218	164·5	143·0	278	209·8	182·4
39	29·4	25·6	99	74·7	64·9	159	120·0	104·3	219	165·3	143·7	279	210·6	183·0
40	30·2	26·2	100	75·5	65·6	160	120·8	105·0	220	166·0	144·3	280	211·3	183·7
41	30·9	26·9	101	76·2	66·3	161	121·5	105·6	221	166·8	145·0	281	212·1	184·4
42	31·7	27·6	102	77·0	66·9	162	122·3	106·3	222	167·5	145·6	282	212·8	185·0
43	32·5	28·2	103	77·7	67·6	163	123·0	106·9	223	168·3	146·3	283	213·6	185·7
44	33·2	28·9	104	78·5	68·2	164	123·8	107·6	224	169·1	147·0	284	214·3	186·3
45	34·0	29·5	105	79·2	68·9	165	124·5	108·2	225	169·8	147·6	285	215·1	187·0
46	34·7	30·2	106	80 0	69·5	166	125·3	108·9	226	170·6	148·3	286	215·8	187·6
47	35·5	30·8	107	80·8	70·2	167	126·0	109·6	227	171·3	148·9	287	216·6	188·3
48	36·2	31·5	108	81·5	70·9	168	126·8	110·2	228	172·1	149·6	288	217·4	188·9
49	37·0	32·1	109	82·3	71·5	169	127·5	110·9	229	172·8	150·2	289	218·1	189·6
50	37·7	32·8	110	83·0	72·2	170	128·3	111·5	230	173·6	150 9	290	218·9	190·3
51	38·5	33·5	111	83·8	72·8	171	129·1	112·2	231	174·3	151·5	291	219·6	190·9
52	39·2	34·1	112	84·5	73·5	172	129·8	112·8	232	175·1	152·2	292	220·4	191·6
53	40·0	34·8	113	85·3	74·1	173	130·6	113·5	233	175·8	152·9	293	221·1	192·2
54	40·8	35·4	114	86·0	74·8	174	131·3	114·2	234	176·6	153·5	294	221·9	192·9
55	41·5	36·1	115	86·8	75·4	175	132·1	114·8	235	177·4	154·2	295	222·6	193·5
56	42·3	36·7	116	87·5	76·1	176	132·8	115·5	236	178·1	154·8	296	223·4	194·2
57	43·0	37·4	117	88·3	76·8	177	133·6	116·1	237	178·9	155·5	297	224·1	194·8
58	43·8	38 1	118	89·1	77·4	178	134·3	116·8	238	179·6	156·1	298	224 9	195·5
59	44·5	38·7	119	89·8	78·1	179	135·1	117·4	239	180·4	156·8	299	225·7	196·2
60	45·3	39·4	120	90·6	78·7	180	135·8	118·1	240	181·1	157·5	300	226·4	196·8
Dist.	Dep.	Lat.	Dist.	Dep.	Lat.	Dist.	Dep.	Lat.	Dist.	Dep.	Lat.	Dist.	Dep.	Lat.

FOR 49 DEGREES.

Dist.	Lat.	Dep.	Dist.	Lat.	Dep.	Dist.	Lat.	Dep.	Dist.	Lat.	Dep.	Dist.	Lat.	Dep.
1	00·7	00·7	61	45·3	40·8	121	89·9	81·0	181	134·5	121·1	241	179·1	161·3
2	01·5	01·3	62	46·1	41·5	122	90·7	81·6	182	135·3	121·8	242	179·8	161·9
3	02·2	02·0	63	46·8	42·2	123	91·4	82·3	183	136·0	122·5	243	180·6	162·6
4	03·0	02·7	64	47·6	42·8	124	92·1	83·0	184	136·7	123·1	244	181·3	163·3
5	03·7	03·3	65	48·3	43·5	125	92·9	83·6	185	137·5	123·8	245	182·1	163·9
6	04·5	04·0	66	49·0	44·2	126	93·6	84·3	186	138·2	124·5	246	182·8	164·6
7	05·2	04·7	67	49·8	44·8	127	94·4	85·0	187	139·0	125·1	247	183·6	165·3
8	05·9	05·4	68	50·5	45·5	128	95·1	85·6	188	139·7	125·8	248	184·3	165·9
9	06·7	06·0	69	51·3	46·2	129	95·9	86·3	189	140·5	126·5	249	185·0	166·6
10	07·4	06·7	70	52·0	46·8	130	96·6	87·0	190	141·2	127·1	250	185·8	167·3
11	08·2	07·4	71	52·8	47·5	131	97·4	87·7	191	141·9	127·8	251	186·5	168·0
12	08·9	08·0	72	53·5	48·2	132	98·1	88·3	192	142·7	128·5	252	187·3	168·6
13	09·7	08·7	73	54·2	48·8	133	98·8	89·0	193	143·4	129·1	253	188·0	169·3
14	10·4	09·4	74	55·0	49·5	134	99·6	89·7	194	144·2	129·8	254	188·8	170·0
15	11·1	10·0	75	55·7	50·2	135	100·3	90·3	195	144·9	130·5	255	189·5	170·6
16	11·9	10·7	76	56·5	50·9	136	101·1	91·0	196	145·7	131·1	256	190·2	171·3
17	12·6	11·4	77	57·2	51·5	137	101·8	91·7	197	146·4	131·8	257	191·0	172·0
18	13·4	12·0	78	58·0	52·2	138	102·6	92·3	198	147·1	132·5	258	191·7	172·6
19	14·1	12·7	79	58·7	52·9	139	103·3	93·0	199	147·9	133·2	259	192·5	173·3
20	14·9	13·4	80	59·5	53·5	140	104·0	93·7	200	148·6	133·8	260	193·2	174·0
21	15·6	14·1	81	60·2	54·2	141	104·8	94·3	201	149·4	134·5	261	194·0	174·6
22	16·3	14·7	82	60·9	54·9	142	105·5	95·0	202	150·1	135·2	262	194·7	175·3
23	17·1	15·4	83	61·7	55·5	143	106·3	95·7	203	150·9	135·8	263	195·4	176·0
24	17·8	16·1	84	62·4	56·2	144	107·0	96·4	204	151·6	136·5	264	196·2	176·7
25	18·6	16·7	85	63·2	56·9	145	107·8	97·0	205	152·3	137·2	265	196·9	177·3
26	19·3	17·4	86	63·9	57·5	146	108·5	97·7	206	153·1	137·8	266	197·7	178·0
27	20·1	18·1	87	64·7	58·2	147	109·2	98·4	207	153·8	138·5	267	198·4	178·7
28	20·8	18·7	88	65·4	58·9	148	110·0	99·0	208	154·6	139·2	268	199·2	179·3
29	21·6	19·4	89	66·1	59·6	149	110·7	99·7	209	155·3	139·8	269	199·9	180·0
30	22·3	20·1	90	66·9	60·2	150	111·5	100·4	210	156·1	140·5	270	200·6	180·7
31	23·0	20·7	91	67·6	60·9	151	112·2	101·0	211	156·8	141·2	271	201·4	181·3
32	23·8	21·4	92	68·4	61·6	152	113·0	101·7	212	157·5	141·9	272	202·1	182·0
33	24·5	22·1	93	69·1	62·2	153	113·7	102·4	213	158·3	142·5	273	202·9	182·7
34	25·3	22·8	94	69·9	62·9	154	114·4	103·0	214	159·0	143·2	274	203·6	183·3
35	26·0	23·4	95	70·6	63·6	155	115·2	103·7	215	159·8	143·9	275	204·4	184·0
36	26·8	24·1	96	71·3	64·2	156	115·9	104·4	216	160·5	144·5	276	205·1	184·7
37	27·5	24·8	97	72·1	64·9	157	116·7	105·1	217	161·3	145·2	277	205·9	185·3
38	28·2	25·4	98	72·8	65·6	158	117·4	105·7	218	162·0	145·9	278	206·6	186·0
39	29·0	26·1	99	73·6	66·2	159	118·2	106·4	219	162·7	146·5	279	207·3	186·7
40	29·7	26·8	100	74·3	66·9	160	118·9	107·1	220	163·5	147·2	280	208·1	187·4
41	30·5	27·4	101	75·1	67·6	161	119·6	107·7	221	164·2	147·9	281	208·8	188·0
42	31·2	28·1	102	75·8	68·3	162	120·4	108·4	222	165·0	148·5	282	209·6	188·7
43	32·0	28·8	103	76·5	68·9	163	121·1	109·1	223	165·7	149·2	283	210·3	189·4
44	32·7	29·4	104	77·3	69·6	164	121·9	109·7	224	166·5	149·9	284	211·1	190·0
45	33·4	30·1	105	78·0	70·3	165	122·6	110·4	225	167·2	150·6	285	211·8	190·7
46	34·2	30·8	106	78·8	70·9	166	123·4	111·1	226	168·0	151·2	286	212·5	191·4
47	34·9	31·4	107	79·5	71·6	167	124·1	111·7	227	168·7	151·9	287	213·3	192·0
48	35·7	32·1	108	80·3	72·3	168	124·8	112·4	228	169·4	152·6	288	214·0	192·7
49	36·4	32·8	109	81·0	72·9	169	125·6	113·1	229	170·2	153·2	289	214·8	193·4
50	37·2	33·5	110	81·7	73·6	170	126·3	113·8	230	170·9	153·9	290	215·5	194·0
51	37·9	34·1	111	82·5	74·3	171	127·1	114·4	231	171·7	154·6	291	216·3	194·7
52	38·6	34·8	112	83·2	74·9	172	127·8	115·1	232	172·4	155·2	292	217·0	195·4
53	39·4	35·5	113	84·0	75·6	173	128·6	115·8	233	173·2	155·9	293	217·7	196·1
54	40·1	36·1	114	84·7	76·3	174	129·3	116·4	234	173·9	156·6	294	218·5	196·7
55	40·9	36·8	115	85·5	77·0	175	130·1	117·1	235	174·6	157·2	295	219·2	197·4
56	41·6	37·5	116	86·2	77·6	176	130·8	117·8	236	175·4	157·9	296	220·0	198·1
57	42·4	38·1	117	86·9	78·3	177	131·5	118·4	237	176·1	158·6	297	220·7	198·7
58	43·1	38·8	118	87·7	79·0	178	132·3	119·1	238	176·9	159·3	298	221·5	199·4
59	43·8	39·5	119	88·4	79·6	179	133·0	119·8	239	177·6	159·9	299	222·2	200·1
60	44·6	40·1	120	89·2	80·3	180	133·8	120·4	240	178·4	160·6	300	222·9	200·7
Dist.	Dep.	Lat.	Dist.	Dep.	Lat.	Dist.	Dep.	Lat.	Dist.	Dep.	Lat.	Dist.	Dep.	Lat.

FOR 48 DEGREES.

Dist.	Lat.	Dep.	Dist.	Lat.	Dep.	Dist.	Lat.	Dep.	Dist.	Lat.	Dep.	Dist.	Lat.	Dep.
1	00·7	00·7	61	44·6	41·6	121	88·5	82·5	181	132·4	123·4	241	176·3	164·4
2	01·5	01·4	62	45·3	42·3	122	89·2	83·2	182	133·1	124·1	242	177·0	165·0
3	02·2	02·0	63	46·1	43·0	123	90·0	83·9	183	133·8	124·8	243	177·7	165·7
4	02·9	02·7	64	46·8	43·6	124	90·7	84·6	184	134·6	125·5	244	178·5	166·4
5	03·7	03·4	65	47·5	44·3	125	91·4	85·2	185	135·3	126·2	245	179·2	167·1
6	04·4	04·1	66	48·3	45·0	126	92·2	85·9	186	136·0	126·9	246	179·9	167·8
7	05·1	04·8	67	49·0	45·7	127	92·9	86·6	187	136·8	127·5	247	180·6	168·5
8	05·9	05·5	68	49·7	46·4	128	93·6	87·3	188	137·5	128·2	248	181·4	169·1
9	06·6	06·1	69	50·5	47·1	129	94·3	88·0	189	138·2	128·9	249	182·1	169·8
10	07·3	06·8	70	51·2	47·7	130	95·1	88·7	190	139·0	129·6	250	182·8	170·5
11	08·0	07·5	71	51·9	48·4	131	95·8	89·3	191	139·7	130·3	251	183·6	171·2
12	08·8	08·2	72	52·7	49·1	132	96·5	90·0	192	140·4	130·9	252	184·3	171·9
13	09·5	08·9	73	53·4	49·8	133	97·3	90·7	193	141·2	131·6	253	185·0	172·5
14	10·2	09·5	74	54·1	50·5	134	98·0	91·4	194	141·9	132·3	254	185·8	173·2
15	11·0	10·2	75	54·9	51·1	135	98·7	92·1	195	142·6	133·0	255	186·5	173·9
16	11·7	10·9	76	55·6	51·8	136	99·5	92·8	196	143·3	133·7	256	187·2	174·6
17	12·4	11·6	77	56·3	52·5	137	100·2	93·4	197	144·1	134·4	257	188·0	175·3
18	13·2	12·3	78	57·0	53·2	138	100·9	94·1	198	144·8	135·0	258	188·7	176·0
19	13·9	13·0	79	57·8	53·9	139	101·7	94·8	199	145·5	135·7	259	189·4	176·6
20	14·6	13·6	80	58·5	54·6	140	102·4	95·5	200	146·3	136·4	260	190·2	177·3
21	15·4	14·3	81	59·2	55·2	141	103·1	96·2	201	147·0	137·1	261	190·9	178·0
22	16·1	15·0	82	60·0	55·9	142	103·9	96·8	202	147·7	137·8	262	191·6	178·7
23	16·8	15·7	83	60·7	56·6	143	104·6	97·5	203	148·5	138·4	263	192·3	179·4
24	17·6	16·4	84	61·4	57·3	144	105·3	98·2	204	149·2	139·1	264	193·1	180·0
25	18·3	17·0	85	62·2	58·0	145	106·0	98·9	205	149·9	139·8	265	193·8	180·7
26	19·0	17·7	86	62·9	58·7	146	106·8	99·6	206	150·7	140·5	266	194·5	181·4
27	19·7	18·4	87	63·6	59·3	147	107·5	100·3	207	151·4	141·2	267	195·3	182·1
28	20·5	19·1	88	64·4	60·0	148	108·2	100·9	208	152·1	141·9	268	196·0	182·8
29	21·2	19·8	89	65·1	60·7	149	109·0	101·6	209	152·9	142·5	269	196·7	183·5
30	21·9	20·5	90	65·8	61·4	150	109·7	102·3	210	153·6	143·2	270	197·5	184·1
31	22·7	21·1	91	66·6	62·1	151	110·4	103·0	211	154·3	143·9	271	198·2	184·8
32	23·4	21·8	92	67·3	62·7	152	111·2	103·7	212	155·0	144·6	272	198·9	185·5
33	24·1	22·5	93	68·0	63·4	153	111·9	104·3	213	155·8	145·3	273	199·7	186·2
34	24·9	23·2	94	68·7	64·1	154	112·6	105·0	214	156·5	145·9	274	200·4	186·9
35	25·6	23·9	95	69·5	64·8	155	113·4	105·7	215	157·2	146·6	275	201·1	187·5
36	26·3	24·6	96	70·2	65·5	156	114·1	106·4	216	158·0	147·3	276	201·9	188·2
37	27·1	25·2	97	70·9	66·2	157	114·8	107·1	217	158·7	148·0	277	202·6	188·9
38	27·8	25·9	98	71·7	66·8	158	115·6	107·8	218	159·4	148·7	278	203·3	189·6
39	28·5	26·6	99	72·4	67·5	159	116·3	108·4	219	160·2	149·4	279	204·0	190·3
40	29·3	27·3	100	73·1	68·2	160	117·0	109·1	220	160·9	150·0	280	204·8	191·0
41	30·0	28·0	101	73·9	68·9	161	117·7	109·8	221	161·6	150·7	281	205·5	191·6
42	30·7	28·6	102	74·6	69·6	162	118·5	110·5	222	162·4	151·4	282	206·2	192·3
43	31·4	29·3	103	75·3	70·2	163	119·2	111·2	223	163·1	152·1	283	207·0	193·0
44	32·2	30·0	104	76·1	70·9	164	119·9	111·8	224	163·8	152·8	284	207·7	193·7
45	32·9	30·7	105	76·8	71·6	165	120·7	112·5	225	164·6	153·4	285	208·4	194·4
46	33·6	31·4	106	77·5	72·3	166	121·4	113·2	226	165·3	154·1	286	209·2	195·1
47	34·4	32·1	107	78·3	73·0	167	122·1	113·9	227	166·0	154·8	287	209·9	195·7
48	35·1	32·7	108	79·0	73·7	168	122·9	114·6	228	166·7	155·5	288	210·6	196·4
49	35·8	33·4	109	79·7	74·3	169	123·6	115·3	229	167·5	156·2	289	211·4	197·1
50	36·6	34·1	110	80·4	75·0	170	124·3	115·9	230	168·2	156·9	290	212·1	197·8
51	37·3	34·8	111	81·2	75·7	171	125·1	116·6	231	168·9	157·5	291	212·8	198·5
52	38·0	35·5	112	81·9	76·4	172	125·8	117·3	232	169·7	158·2	292	213·6	199·1
53	38·8	36·1	113	82·6	77·1	173	126·5	118·0	233	170·4	158·9	293	214·3	199·8
54	39·5	36·8	114	83·4	77·7	174	127·3	118·7	234	171·1	159·6	294	215·0	200·5
55	40·2	37·5	115	84·1	78·4	175	128·0	119·3	235	171·9	160·3	295	215·7	201·2
56	41·0	38·2	116	84·8	79·1	176	128·7	120·0	236	172·6	161·0	296	216·5	201·9
57	41·7	38·9	117	85·6	79·8	177	129·4	120·7	237	173·3	161·6	297	217·2	202·6
58	42·4	39·6	118	86·3	80·5	178	130·2	121·4	238	174·1	162·3	298	217·9	203·2
59	43·1	40·2	119	87·0	81·2	179	130·9	122·1	239	174·8	163·0	299	218·7	203·9
60	43·9	40·9	120	87·8	81·8	180	131·6	122·8	240	175·5	163·7	300	219·4	204·6
Dist.	Dep.	Lat.	Dist.	Dep.	Lat.	Dist.	Dep.	Lat.	Dist.	Dep.	Lat.	Dist.	Dep.	Lat.

FOR 47 DEGREES.

Dist.	Lat.	Dep.	Dist.	Lat.	Dep.	Dist.	Lat.	Dep.	Dist.	Lat.	Dep.	Dist.	Lat.	Dep.
1	00·7	00·7	61	43·9	42·4	121	87·0	84·1	181	130·2	125·7	241	173·4	167·4
2	01·4	01·4	62	44·6	43·1	122	87·8	84·7	182	130·9	126·4	242	174·1	168·1
3	02·2	02·1	63	45·3	43·8	123	88·5	85·4	183	131·6	127·1	243	174·8	168·8
4	02·9	02·8	64	46·0	44·5	124	89·2	86·1	184	132·4	127·8	244	175·5	169·5
5	03·6	03·5	65	46·8	45·2	125	89 9	86·8	185	133·1	128·5	245	176·2	170·2
6	04·3	04·2	66	47·5	45 8	126	90·6	87·5	186	133·8	129·2	246	177·0	170·9
7	05·0	04·9	67	48 2	46·5	127	91·4	88·2	187	134·5	129 9	247	177·7	171·6
8	05·8	05·6	68	48·9	47·2	128	92·1	88·9	188	135·2	130·6	248	178·4	172·3
9	06·5	06·3	69	49·6	47·9	129	92·8	89·6	189	136·0	131·3	249	179·1	173·0
10	07·2	06·9	70	50·4	48·6	130	93·5	90·3	190	136·7	132·0	250	179·8	173·7
11	07·9	07·6	71	51·1	49·3	131	94·2	91·0	191	137·4	132·7	251	180·6	174·4
12	08·6	08·3	72	51·8	50·0	132	95·0	91·7	192	138·1	133·4	252	181·3	175·1
13	09·4	09·0	73	52·5	50·7	133	95·7	92·4	193	138·8	134·1	253	182·0	175·7
14	10·1	09·7	74	53·2	51·4	134	96·4	93·1	194	139·6	134·8	254	182·7	176·4
15	10·8	10·4	75	54·0	52·1	135	97·1	93·8	195	140·3	135·5	255	183·4	177·1
16	11·5	11·1	76	54·7	52 8	136	97·8	94·5	196	141·0	136·2	256	184·2	177·8
17	12·2	11·8	77	55·4	53·5	137	98·5	95·2	197	141·7	136·8	257	184·9	178·5
18	12·9	12·5	78	56·1	54·2	138	99 3	95·9	198	142·4	137·5	258	185·6	179·2
19	13·7	13·2	79	56·8	54·9	139	100·0	96·6	199	143·1	138·2	259	186·3	179·9
20	14·4	13·9	80	57·5	55·6	140	100·7	97·3	200	143·9	138·9	260	187·0	180·6
21	15·1	14·6	81	58·3	56·3	141	101·4	97·9	201	144·6	139·6	261	187·7	181·3
22	15 8	15·3	82	59·0	57·0	142	102·1	98·6	202	145·3	140·3	262	188·5	182·0
23	16·5	16·0	83	59·7	57·7	143	102·9	99·3	203	146·0	141·0	263	189·2	182·7
24	17·3	16·7	84	60·4	58·4	144	103·6	100·0	204	146·7	141·7	264	189·9	183·4
25	18·0	17·4	85	61·1	59·0	145	104·3	100·7	205	147·5	142·4	265	190·6	184·1
26	18·7	18·1	86	61·9	59·7	146	105·0	101·4	206	148·2	143·1	266	191·3	184·8
27	19·4	18·8	87	62·6	60·4	147	105·7	102·1	207	148·9	143·8	267	192·1	185·5
28	20·1	19·5	88	63·3	61·1	148	106·5	102·8	208	149·6	144·5	268	192·8	186·2
29	20·9	20·1	89	64·0	61·8	149	107·2	103·5	209	150 3	145·2	269	193·5	186·9
30	21·6	20 8	90	64·7	62·5	150	107·9	104·2	210	151·1	145·9	270	194·2	187·6
31	22·3	21·5	91	65·5	63·2	151	108·6	104·9	211	151·8	146·6	271	194·9	188·3
32	23·0	22·2	92	66·2	63·9	152	109·3	105·6	212	152·5	147·3	272	195 7	188·9
33	23·7	22·9	93	66·9	64·6	153	110·1	106·3	213	153·2	148·0	273	196·4	189·6
34	24·5	23 6	94	67·6	65 3	154	110·8	107·0	214	153·9	148·7	274	197·1	190·3
35	25·2	24·3	95	68·3	66·0	155	111·5	107·7	215	154·7	149·4	275	197·8	191·0
36	25·9	25 0	96	69·1	66·7	156	112·2	108·4	216	155·4	150·0	276	198·5	191·7
37	26·6	25·7	97	69·8	67·4	157	112·9	109·1	217	156·1	150·7	277	199·3	192·4
38	27·3	26·4	98	70·5	68·1	158	113·7	109·8	218	156·8	151·4	278	200·0	193·1
39	28·1	27·1	99	71·2	68·8	159	114·4	110·5	219	157·5	152·1	279	200·7	193·8
40	28·8	27·8	100	71·9	69·5	160	115·1	111·1	220	158·3	152·8	280	201·4	194·5
41	29·5	28·5	101	72·7	70·2	161	115·8	111·8	221	159 0	153·5	281	202·1	195·2
42	30·2	29·2	102	73·4	70·9	162	116·5	112·5	222	159·7	154·2	282	202·9	195·9
43	30·9	29·9	103	74·1	71·5	163	117·3	113·2	223	160·4	154·9	283	203·6	196·6
44	31·7	30·6	104	74·8	72·2	164	118·0	113·9	224	161·1	155·6	284	204·3	197·3
45	32·4	31·3	105	75·5	72·9	165	118·7	114·6	225	161·9	156·3	285	205·0	198·0
46	33·1	32·0	106	76·3	73·6	166	119·4	115·3	226	162·6	157·0	286	205·7	198·7
47	33·8	32·6	107	77·0	74·3	167	120·1	116·0	227	163·3	157·7	287	206·5	199·4
48	34·5	33·3	108	77·7	75 0	168	120·8	116·7	228	164·0	158·4	288	207·2	200·1
49	35·2	34·0	109	78·4	75·7	169	121·6	117·4	229	164·7	159·1	289	207·9	200·8
50	36·0	34·7	110	79·1	76·4	170	122·3	118·1	230	165·4	159·8	290	208·6	201·5
51	36·7	35·4	111	79·8	77·1	171	123·0	118·8	231	166·2	160·5	291	209·3	202·1
52	37·4	36·1	112	80·6	77·8	172	123·7	119·5	232	166·9	161·2	292	210·0	202·8
53	38·1	36·8	113	81·3	78·5	173	124·4	120·2	233	167·6	161·9	293	210·8	203·5
54	38·8	37·5	114	82·0	79·2	174	125·2	120·9	234	168·3	162·6	294	211·5	204·2
55	39·6	38·2	115	82·7	79·9	175	125·9	121·6	235	169·0	163·2	295	212·2	204·9
56	40·3	38·9	116	83·4	80·6	176	126·6	122·3	236	169·8	163·9	296	212·9	205·6
57	41·0	39·6	117	84·2	81·3	177	127·3	123·0	237	170·5	164·6	297	213·6	206·3
58	41·7	40 3	118	84·9	82·0	178	128·0	123·6	238	171·2	165·3	298	214·4	207·0
59	42·4	41·0	119	85·6	82·7	179	128·8	124·3	239	171·9	166·0	299	215·1	207·7
60	43·2	41·7	120	86·3	83·4	180	129·5	125·0	240	172·6	166·7	300	215·8	208·4
Dist.	Dep.	Lat.	Dist.	Dep.	Lat.	Dist.	Dep.	Lat.	Dist.	Dep.	Lat.	Dist.	Dep.	Lat.

FOR 46 DEGREES.

Dist.	Lat.	Dep.	Dist.	Lat.	Dep.	Dist.	Lat.	Dep.	Dist.	Lat.	Dep.	Dist.	Lat.	Dep.
1	00·7	00·7	61	43·1	43·1	121	85·6	85·6	181	128·0	128·0	241	170·4	170·4
2	01·4	01·4	62	43·8	43·8	122	86·3	86·3	182	128·7	128·7	242	171·1	171·1
3	02·1	02·1	63	44·5	44·5	123	87·0	87·0	183	129·4	129·4	243	171·8	171·8
4	02·8	02·8	64	45·3	45·3	124	87·7	87·7	184	130·1	130·1	244	172·5	172·5
5	03·5	03·5	65	46·0	46·0	125	88·4	88·4	185	130·8	130·8	245	173·2	173·2
6	04·2	04·2	66	46·7	46·7	126	89·1	89·1	186	131·5	131·5	246	173·9	173·9
7	04·9	04·9	67	47·4	47·4	127	89·8	89·8	187	132·2	132·2	247	174·7	174·7
8	05·7	05·7	68	48·1	48·1	128	90·5	90·5	188	132·9	132·9	248	175·4	175·4
9	06·4	06·4	69	48·8	48·8	129	91·2	91·2	189	133·6	133·6	249	176·1	176·1
10	07·1	07·1	70	49·5	49·5	130	91·9	91·9	190	134·3	134·3	250	176·8	176·8
11	07·8	07·8	71	50·2	50·2	131	92·6	92·6	191	135·1	135·1	251	177·5	177·5
12	08·5	08·5	72	50·9	50·9	132	93·3	93·3	192	135·8	135·8	252	178·2	178·2
13	09·2	09·2	73	51·6	51·6	133	94·0	94·0	193	136·5	136·5	253	178·9	178·9
14	09·9	09·9	74	52·3	52·3	134	94·8	94·8	194	137·2	137·2	254	179·6	179·6
15	10·6	10·6	75	53·0	53·0	135	95·5	95·5	195	137·9	137·9	255	180·3	180·3
16	11·3	11·3	76	53·7	53·7	136	96·2	96·2	196	138·6	138·6	256	181·0	181·0
17	12·0	12·0	77	54·4	54·4	137	96·9	96·9	197	139·3	139·3	257	181·7	181·7
18	12·7	12·7	78	55·2	55·2	138	97·6	97·6	198	140·0	140·0	258	182·4	182·4
19	13·4	13·4	79	55·9	55·9	139	98·3	98·3	199	140·7	140·7	259	183·1	183·1
20	14·1	14·1	80	56·6	56·6	140	99·0	99·0	200	141·4	141·4	260	183·8	183·8
21	14·8	14·8	81	57·3	57·3	141	99·7	99·7	201	142·1	142·1	261	184·6	184·6
22	15·6	15·6	82	58·0	58·0	142	100·4	100·4	202	142·8	142·8	262	185·3	185·3
23	16·3	16·3	83	58·7	58·7	143	101·1	101·1	203	143·5	143·5	263	186·0	186·0
24	17·0	17·0	84	59·4	59·4	144	101·8	101·8	204	144·2	144·2	264	186·7	186·7
25	17·7	17·7	85	60·1	60·1	145	102·5	102·5	205	145·0	145·0	265	187·4	187·4
26	18·4	18·4	86	60·8	60·8	146	103·2	103·2	206	145·7	145·7	266	188·1	188·1
27	19·1	19·1	87	61·5	61·5	147	103·9	103·9	207	146·4	146·4	267	188·8	188·8
28	19·8	19·8	88	62·2	62·2	148	104·7	104·7	208	147·1	147·1	268	189·5	189·5
29	20·5	20·5	89	62·9	62·9	149	105·4	105·4	209	147·8	147·8	269	190·2	190·2
30	21·2	21·2	90	63·6	63·6	150	106·1	106·1	210	148·5	148·5	270	190·9	190·9
31	21·9	21·9	91	64·3	64·3	151	106·8	106·8	211	149·2	149·2	271	191·6	191·6
32	22·6	22·6	92	65·1	65·1	152	107·5	107·5	212	149·9	149·9	272	192·3	192·3
33	23·3	23·3	93	65·8	65·8	153	108·2	108·2	213	150·6	150·6	273	193·0	193·0
34	24·0	24·0	94	66·5	66·5	154	108·9	108·9	214	151·3	151·3	274	193·7	193·7
35	24·7	24·7	95	67·2	67·2	155	109·6	109·6	215	152·0	152·0	275	194·5	194·5
36	25·5	25·5	96	67·9	67·9	156	110·3	110·3	216	152·7	152·7	276	195·2	195·2
37	26·2	26·2	97	68·6	68·6	157	111·0	111·0	217	153·4	153·4	277	195·9	195·9
38	26·9	26·9	98	69·3	69·3	158	111·7	111·7	218	154·1	154·1	278	196·6	196·6
39	27·6	27·6	99	70·0	70·0	159	112·4	112·4	219	154·9	154·9	279	197·3	197·3
40.	28·3	28·3	100	70·7	70·7	160	113·1	113·1	220	155·6	155·6	280	198·0	198·0
41	29·0	29·0	101	71·4	71·4	161	113·8	113·8	221	156·3	156·3	281	198·7	198·7
42	29·7	29·7	102	72·1	72·1	162	114·5	114·5	222	157·0	157·0	282	199·4	199·4
43	30·4	30·4	103	72·8	72·8	163	115·3	115·3	223	157·7	157·7	283	200·1	200·1
44	31·1	31·1	104	73·5	73·5	164	116·0	116·0	224	158·4	158·4	284	200·8	200·8
45	31·8	31·8	105	74·2	74·2	165	116·7	116·7	225	159·1	159·1	285	201·5	201·5
46	32·5	32·5	106	75·0	75·0	166	117·4	117·4	226	159·8	159·8	286	202·2	202·2
47	33·2	33·2	107	75·7	75·7	167	118·1	118·1	227	160·5	160·5	287	202·9	202·9
48	33·9	33·9	108	76·4	76·4	168	118·8	118·8	228	161·2	161·2	288	203·6	203·6
49	34·6	34·6	109	77·1	77·1	169	119·5	119·5	229	161·9	161·9	289	204·3	204·3
50	35·4	35·4	110	77·8	77·8	170	120·2	120·2	230	162·6	162·6	290	205·1	205·1
51	36·1	36·1	111	78·5	78·5	171	120·9	120·9	231	163·3	163·3	291	205·8	205·8
52	36·8	36·8	112	79·2	79·2	172	121·6	121·6	232	164·0	164·0	292	206·5	206·5
53	37·5	37·5	113	79·9	79·9	173	122·3	122·3	233	164·8	164·8	293	207·2	207·2
54	38·2	38·2	114	80·6	80·6	174	123·0	123·0	234	165·5	165·5	294	207·9	207·9
55	38·9	38·9	115	81·3	81·3	175	123·7	123·7	235	166·2	166·2	295	208·6	208·6
56	39·6	39·6	116	82·0	82·0	176	124·4	124·4	236	166·9	166·9	296	209·3	209·3
57	40·3	40·3	117	82·7	82·7	177	125·2	125·2	237	167·6	167·6	297	210·0	210·0
58	41·0	41·0	118	83·4	83·4	178	125·9	125·9	238	168·3	168·3	298	210·7	210·7
59	41·7	41·7	119	84·1	84·1	179	126·6	126·6	239	169·0	169·0	299	211·4	211·4
60	42·4	42·4	120	84·9	84·9	180	127·3	127·3	240	169·7	169·7	300	212·1	212·1
Dist.	Dep.	Lat.	Dist.	Dep.	Lat.	Dist.	Dep.	Lat.	Dist.	Dep.	Lat.	Dist.	Dep.	Lat.

Table 6.

′ ″	0° Co-sine.	0° Parts for ″	1° Co-sine.	1° Parts for ″	2° Co-sine.	2° Parts for ″	3° Co-sine.	3° Parts for ″	4° Co-sine.	4° Parts for ″	5° Co-sine.	5° Parts for ″
0	000000	0	999848	0	999391	0	998630	0	997564	0	996195	0
1	00	0	843	0	381	0	614	0	544	0	6169	0
2	00	0	837	0	370	0	599	1	523	1	6144	1
3	00	0	832	0	360	1	584	1	503	1	6118	1
4	999999	0	827	0	350	1	568	1	482	1	6093	2
5	99	0	821	0	339	1	552	1	462	2	6067	2
6	99	0	816	1	328	1	537	2	441	2	6041	3
7	98	0	810	1	318	1	521	2	420	3	6015	3
8	97	1	804	1	307	1	505	2	399	3	5989	3
9	97	1	799	1	296	2	489	2	378	3	5963	4
10	96	1	793	1	285	2	473	3	357	4	5937	4
11	999995	1	999787	1	999274	2	998457	3	997336	4	995911	5
12	94	1	781	1	263	2	441	3	315	4	884	5
13	93	1	774	1	252	2	425	4	293	5	858	6
14	92	1	768	1	240	3	408	4	272	5	832	6
15	91	1	762	2	229	3	392	4	250	5	805	7
16	89	1	756	2	218	3	375	4	229	6	778	7
17	88	1	749	2	206	3	359	4	207	6	752	7
18	86	1	743	2	194	3	342	5	185	6	725	8
19	85	1	736	2	183	4	325	5	163	7	698	8
20	83	1	729	2	171	4	308	5	141	7	671	9
21	999981	1	999722	2	999159	4	998291	6	997119	7	995644	9
22	80	2	716	2	147	4	274	6	7097	8	617	10
23	78	2	709	2	135	4	257	6	7075	8	589	10
24	76	2	701	3	123	5	240	7	7053	8	562	11
25	74	2	694	3	111	5	223	7	7030	9	535	11
26	71	2	687	3	098	5	205	7	7008	9	507	11
27	69	2	680	3	086	5	188	8	6985	10	480	12
28	67	2	672	3	073	5	170	8	6963	10	452	12
29	64	2	665	3	061	6	153	8	6940	11	424	13
30	62	2	657	3	048	6	135	9	6917	11	396	14
31	999959	2	999650	4	999036	7	998117	9	996895	12	995368	15
32	57	2	642	4	9023	7	8099	10	872	13	340	15
33	54	2	634	4	9010	7	8081	10	849	13	312	16
34	51	2	626	4	8997	8	8063	11	825	14	284	16
35	48	2	618	4	8984	8	8045	11	802	14	256	17
36	45	2	610	5	8971	8	8027	11	779	15	227	18
37	42	2	602	5	8957	9	8008	12	756	15	199	18
38	39	2	594	5	8944	9	7990	12	732	16	171	19
39	36	2	585	5	8931	9	7972	12	709	16	142	19
40	32	2	577	6	8917	9	7953	13	685	16	113	20
41	999929	2	999568	6	998904	10	997934	13	996661	17	995084	20
42	925	2	560	6	890	10	916	13	637	17	5056	21
43	922	2	551	6	876	10	897	14	614	17	5027	21
44	918	2	542	6	862	10	878	14	590	18	4998	21
45	914	3	534	6	848	11	859	14	566	18	4969	22
46	911	3	525	6	834	11	840	15	541	19	4939	22
47	907	3	516	7	820	11	821	15	517	19	4910	23
48	903	3	507	7	806	11	802	15	493	20	4881	23
49	898	3	497	7	792	11	782	16	469	20	4851	24
50	894	3	488	7	778	12	763	16	444	20	4822	25
51	999890	3	999479	7	998763	12	997743	16	996420	21	994792	25
52	886	3	469	7	749	12	724	16	395	21	763	26
53	881	3	460	7	734	12	704	17	370	22	733	26
54	877	3	450	8	719	13	684	17	345	22	703	27
55	872	3	441	8	705	13	665	17	320	22	673	27
56	867	3	431	8	690	13	645	18	295	23	643	28
57	863	3	421	8	675	14	625	18	270	23	613	28
58	858	3	411	8	660	14	605	18	245	24	583	29
59	853	3	401	8	645	14	584	19	220	24	552	29
60	848	4	391	9	630	14	564	19	195	24	522	29

′	6°		7°		8°		9°		10°		11°	
″	Co-sine.	Parts for ″	Co-sine.	Parts for ″	Co-sine.	Parts for ″	Co-sine.	Parts for ″	Co-sine.	Parts for ″	Co-sine.	Parts for ″
0	994522	0	992546	0	990268	0	987688	0	984808	0	981627	0
1	491	1	511	1	0228	1	643	1	757	1	572	1
2	461	1	475	1	0187	1	597	1	707	2	516	2
3	430	2	439	2	0146	2	551	2	656	3	460	3
4	400	2	404	2	0106	3	506	3	605	3	405	4
5	369	3	368	3	0065	3	460	4	554	4	349	5
6	338	3	332	4	0024	4	414	5	503	5	293	6
7	307	4	296	4	989983	5	368	6	452	6	237	7
8	276	4	260	5	9942	6	322	7	401	7	181	7
9	245	5	224	6	9900	6	275	8	350	8	124	8
10	214	5	187	6	9859	7	229	8	299	9	068	9
11	994182	6	992151	7	989818	8	987183	9	984247	9	981012	10
12	4151	6	2115	7	776	8	7136	10	4196	10	0955	11
13	4120	7	2078	8	735	9	7090	11	4144	11	0899	12
14	4088	7	2042	9	693	10	7043	12	4092	12	0842	13
15	4056	8	2005	9	651	10	6996	12	4041	13	0785	14
16	4025	8	1968	10	610	11	6950	13	3989	14	0729	15
17	3993	9	1931	10	568	12	6903	13	3937	15	0672	16
18	3961	9	1894	11	526	12	6856	14	3885	16	0615	17
19	3929	10	1857	12	484	13	6809	15	3833	16	0558	18
20	3897	10	1820	12	442	14	6762	15	3781	17	0501	19
21	993865	11	991783	13	989399	14	986714	16	983729	18	980443	20
22	833	11	746	13	357	15	667	17	676	19	0386	21
23	800	12	709	14	315	16	620	18	624	20	0329	22
24	768	13	671	15	272	17	572	19	572	21	0271	23
25	736	13	634	15	230	17	525	19	519	22	0214	24
26	703	14	596	16	187	18	477	20	466	22	0156	25
27	670	14	558	17	145	19	429	21	414	23	0098	26
28	638	15	521	17	102	19	382	22	361	24	0041	27
29	605	15	483	18	059	20	334	23	308	25	979983	28
30	572	16	445	19	016	21	286	24	255	26	9925	29
31	993539	17	991407	20	988973	22	986238	25	983202	27	979867	30
32	506	18	369	21	930	23	6189	26	3149	28	809	31
33	473	19	331	22	887	24	6141	27	3096	29	750	32
34	440	19	292	22	843	25	6093	28	3042	30	692	33
35	406	20	254	23	800	26	6045	29	2989	31	634	34
36	373	21	216	24	756	26	5996	30	2935	32	575	35
37	339	21	177	24	713	27	5948	31	2882	33	517	36
38	306	22	138	25	669	28	5899	32	2828	34	458	37
39	272	22	100	26	626	28	5850	33	2774	35	399	38
40	238	23	061	26	582	29	5801	33	2721	36	341	39
41	993205	23	991022	27	988538	30	985752	34	982667	37	979282	40
42	3171	24	983	28	494	31	704	35	613	38	9223	41
43	3137	24	944	28	450	32	654	35	559	39	9164	42
44	3103	25	905	29	406	32	605	36	505	40	9105	43
45	3069	26	866	30	362	33	556	37	450	41	9046	44
46	3034	26	827	30	317	34	507	38	396	41	8986	45
47	3000	27	787	31	273	35	457	39	342	42	8927	46
48	2966	28	748	32	228	35	408	40	287	43	8867	47
49	2931	28	708	32	184	36	358	41	233	44	8808	48
50	2896	29	669	33	139	37	309	42	178	45	8748	49
51	992862	29	990629	34	988095	38	985259	42	982123	46	978689	50
52	827	30	589	34	8050	38	5209	43	2069	47	629	51
53	792	30	549	35	8005	39	5159	44	2014	48	569	52
54	757	31	510	36	7960	40	5109	45	1959	49	509	54
55	722	31	469	36	7915	41	5059	45	1904	50	449	55
56	687	32	429	37	7870	41	5009	46	1849	50	389	56
57	652	32	389	38	7825	42	4959	47	1793	51	329	57
58	617	33	349	38	7779	43	4909	48	1738	52	268	58
59	582	34	309	39	7734	44	4858	49	1683	53	208	59
60	546	34	268	39	7688	44	4808	50	1627	54	148	60

′ ″	12° Co-sine.	12° Parts for ″	13° Co-sine.	13° Parts for ″	14° Co-sine.	14° Parts for ″	15° Co-sine.	15° Parts for ″	16° Co-sine.	16° Parts for ″	17° Co-sine.	17° Parts for ″
0	978148	0	974370	0	970296	0	965926	0	961262	0	956305	0
1	8087	1	4305	1	0225	1	850	1	1182	1	6220	1
2	8026	2	4239	2	0155	2	775	2	1101	3	6135	3
3	7966	3	4173	3	0084	3	700	4	1021	4	6049	4
4	7905	4	4108	4	0014	5	624	5	0940	5	5964	6
5	7844	5	4042	6	969943	6	548	6	0860	7	5879	7
6	7783	6	3976	7	9872	7	473	8	0779	8	5793	9
7	7722	7	3910	8	9801	8	397	9	0698	9	5707	10
8	7661	8	3844	9	9730	9	321	10	0618	11	5622	11
9	7600	9	3778	10	9659	10	245	11	0537	12	5536	13
10	7539	10	3712	11	9588	12	169	13	0456	14	5450	14
11	977477	11	973645	13	969517	13	965093	14	960375	15	955364	16
12	7416	12	579	14	9445	14	5016	15	0294	16	5278	17
13	7354	13	512	15	9374	15	4940	16	0213	18	5192	19
14	7293	14	446	16	9302	16	4864	18	0131	19	5106	20
15	7231	15	379	17	9231	18	4787	19	0050	20	5020	22
16	7169	16	313	18	9159	19	4711	20	959968	22	4934	23
17	7108	17	246	19	9088	20	4634	21	9887	23	4847	24
18	7046	18	179	20	9016	21	4557	23	9805	24	4761	26
19	6984	19	112	21	8944	22	4481	24	9724	26	4674	27
20	6922	20	045	22	8872	24	4404	26	9642	27	4588	29
21	976859	22	972978	24	968800	25	964327	27	959560	28	954501	30
22	797	23	911	25	728	26	4250	28	9478	30	4414	32
23	735	24	843	26	656	27	4173	29	9396	31	4327	33
24	672	25	776	27	583	28	4095	31	9314	32	4240	35
25	610	26	708	28	511	30	4018	32	9232	34	4153	36
26	547	27	641	29	438	31	3941	33	9150	35	4066	37
27	485	28	573	30	366	32	3863	34	9067	36	3979	39
28	422	29	506	31	293	33	3786	36	8985	38	3892	40
29	359	30	438	32	220	34	3708	37	8902	39	3804	42
30	296	31	370	34	148	36	3631	38	8820	41	3717	44
31	976233	32	972302	35	968075	37	963553	40	958737	43	953629	45
32	6170	33	2234	36	8002	38	3475	42	8654	44	3542	47
33	6107	35	2166	38	7929	40	3397	43	8572	46	3454	48
34	6044	36	2098	39	7856	41	3319	44	8489	47	3366	50
35	5980	37	2029	40	7783	43	3241	46	8406	49	3279	51
36	5917	38	1961	41	7709	44	3163	47	8323	50	3191	53
37	5853	39	1893	42	7636	45	3084	48	8239	51	3103	55
38	5790	40	1824	44	7562	47	3006	49	8156	53	3015	56
39	5726	41	1755	45	7489	48	2928	51	8073	54	2926	58
40	5662	42	1687	46	7415	49	2849	52	7990	55	2838	59
41	975599	43	971618	47	967342	50	962770	53	957906	57	952750	61
42	535	44	1549	48	7268	52	692	55	823	58	2662	62
43	471	45	1480	49	7194	53	613	56	739	59	2573	64
44	407	46	1411	50	7120	54	534	57	655	61	2484	65
45	342	47	1342	52	7046	55	455	59	571	62	2396	67
46	278	49	1273	53	6972	57	376	60	488	64	2307	68
47	214	50	1204	54	6898	58	297	61	404	65	2218	70
48	149	51	1134	55	6823	59	218	63	320	66	2129	71
49	085	52	1065	56	6749	60	139	64	235	68	2040	73
50	020	53	0995	57	6675	62	059	65	151	69	1951	74
51	974956	54	970926	59	966600	63	961980	67	957067	71	951862	76
52	891	55	856	60	6526	64	901	68	6983	72	773	77
53	826	56	786	61	6451	65	821	69	6898	74	684	79
54	761	57	717	62	6376	66	741	71	6814	75	594	80
55	696	58	647	63	6301	68	662	72	6729	77	505	82
56	631	59	577	64	6226	69	582	73	6644	78	415	83
57	566	60	507	66	6151	70	502	75	6560	80	326	85
58	501	61	436	67	6076	72	422	76	6475	81	236	86
59	436	62	366	68	6001	73	342	78	6390	82	146	88
60	370	64	296	69	5926	74	262	79	6305	83	057	89

′	18°		19°		20°		21°		22°		23°	
″	Cosine.	Parts for ″	Cosine.	Parts for ″	Cosine.	Parts for ″	Cosine.	Parts for ″	Cosine.	Parts for ″	Cosine.	Parts for ″
0	951057	0	945519	0	939693	0	933580	0	927184	0	920505	0
1	0967	2	5424	2	9593	2	3476	2	7075	2	0391	2
2	0877	3	5329	3	9494	3	3372	4	6966	4	0277	4
3	0787	5	5234	5	9394	5	3267	5	6857	5	0164	6
4	0696	6	5139	6	9294	7	3163	7	6747	7	0050	8
5	0606	8	5044	8	9194	8	3058	9	6638	9	919936	10
6	0516	9	4949	10	9094	10	2954	11	6529	11	9822	11
7	0425	11	4854	11	8994	12	2849	12	6419	13	9707	13
8	0335	12	4758	13	8894	13	2744	14	6310	15	9593	15
9	0244	14	4663	14	8794	15	2639	16	6200	17	9479	17
10	0154	15	4568	16	8694	17	2534	18	6090	18	9364	19
11	950063	17	944472	18	938593	18	932429	19	925981	20	919250	21
12	949972	18	4376	19	8493	20	2324	21	5871	22	9135	23
13	9881	20	4281	21	8393	22	2219	23	5761	24	9021	25
14	9790	21	4185	22	8292	23	2113	25	5651	26	8906	27
15	9699	23	4089	24	8191	25	2008	26	5541	28	8791	29
16	9608	24	3993	26	8091	27	1902	28	5430	29	8676	31
17	9517	26	3897	27	7990	28	1797	30	5320	31	8561	33
18	9426	27	3801	29	7889	30	1691	32	5210	33	8446	35
19	9334	29	3705	30	7788	32	1586	33	5099	35	8331	37
20	9243	30	3609	32	7687	34	1480	35	4989	37	8216	38
21	949151	32	943512	34	937586	35	931374	37	924878	39	918101	40
22	9060	33	3416	35	7485	37	1268	39	4768	40	7986	42
23	8968	35	3319	37	7383	39	1162	40	4657	42	7870	44
24	8876	36	3223	38	7282	40	1056	42	4546	44	7755	46
25	8784	38	3126	40	7181	42	0950	44	4435	46	7639	48
26	8692	39	3029	42	7079	44	0843	46	4324	48	7523	50
27	8600	41	2932	43	6977	46	0737	48	4213	50	7408	52
28	8508	42	2836	45	6876	47	0631	50	4102	52	7292	54
29	8416	44	2739	47	6774	49	0524	52	3991	54	7176	56
30	8324	46	2642	48	6672	51	0418	53	3880	56	7060	58
31	948231	48	942544	51	936570	53	930311	55	923768	58	916944	60
32	8139	50	2447	52	6468	55	0204	57	3657	60	6828	62
33	8046	51	2350	54	6366	57	0097	59	3545	62	6712	64
34	7954	53	2253	56	6264	58	929991	61	3434	64	6596	66
35	7861	54	2155	57	6162	60	9884	63	3322	65	6479	68
36	7768	56	2058	59	6060	62	9777	65	3210	67	6363	70
37	7676	57	1960	60	5957	63	9969	67	3098	69	6246	72
38	7583	59	1862	62	5855	65	9562	69	2987	71	6130	74
39	7490	61	1764	64	5752	67	9455	71	2875	73	6013	76
40	7397	62	1667	66	5650	69	9348	72	2762	75	5896	78
41	947304	64	941569	67	935547	70	929240	74	922650	77	915780	80
42	7210	65	1471	69	5444	72	9133	76	2538	79	5663	82
43	7117	67	1372	71	5341	74	9025	78	2426	81	5546	84
44	7024	68	1274	72	5238	75	8917	80	2313	83	5429	86
45	6930	70	1176	74	5135	77	8810	81	2201	84	5312	88
46	6837	71	1078	75	5032	79	8702	83	2088	86	5194	90
47	6743	73	0979	77	4929	81	8594	85	1976	88	5077	92
48	6649	75	0881	79	4826	82	8486	87	1863	90	4960	94
49	6556	76	0782	80	4722	84	8378	89	1750	92	4842	96
50	6462	78	0684	82	4619	86	8270	90	1638	94	4725	98
51	946368	79	940585	84	934515	87	928161	92	921525	96	914607	100
52	6274	81	0486	85	4412	89	8053	94	1412	98	4490	102
53	6180	82	0387	87	4308	91	7945	96	1299	100	4372	104
54	6085	84	0288	89	4205	93	7836	98	1185	101	4254	106
55	5991	85	0189	90	4101	95	7728	100	1072	103	4136	108
56	5897	87	0090	92	3997	96	7619	101	0959	105	4018	110
57	5802	88	939991	94	3893	98	7510	103	0846	107	3900	112
58	5708	90	9891	95	3789	100	7402	105	0732	109	3782	114
59	5613	92	9792	97	3685	101	7293	107	0619	110	3664	116
60	5519	93	9693	98	3580	103	7184	109	0505	112	3546	118

′ / ″	24° Co-sine.	24° Parts for ″	25° Co-sine.	25° Parts for ″	26° Co-sine.	26° Parts for ″	27° Co-sine.	27° Parts for ″	28° Co-sine.	28° Parts for ″	29° Co-sine.	29° Parts for ″
0	913546	0	906308	0	898794	0	891007	0	882948	0	874620	0
1	8427	2	6185	2	8667	2	0874	2	2811	2	4479	2
2	8309	4	6062	4	8539	4	0742	4	2674	4	4338	5
3	8190	6	5939	6	8411	6	0610	6	2538	6	4196	7
4	8072	8	5815	8	8283	8	0478	8	2401	9	4055	9
5	2953	10	5692	10	8156	11	0345	11	2264	11	3914	12
6	2834	12	5569	12	8028	13	0213	13	2127	13	3772	14
7	2715	14	5445	14	7900	15	0080	15	1990	16	3631	16
8	2597	16	5322	16	7772	17	889948	17	1853	18	3489	19
9	2478	18	5198	18	7643	19	9815	19	1716	21	3348	21
10	2358	20	5075	21	7515	21	9682	21	1578	23	3206	24
11	912239	22	904951	23	897387	23	889549	23	881441	25	873064	26
12	2120	24	4827	25	7258	26	9416	26	1304	27	2922	29
13	2001	26	4703	27	7130	28	9283	28	1166	30	2780	31
14	1882	28	4579	29	7001	30	9150	30	1028	32	2638	33
15	1762	30	4455	31	6873	32	9017	32	0891	34	2496	36
16	1643	32	4331	33	6744	34	8884	35	0753	37	2354	38
17	1523	34	4207	35	6615	36	8751	37	0615	39	2212	40
18	1403	36	4083	37	6486	38	8617	39	0477	41	2069	43
19	1284	38	3958	39	6358	40	8484	41	0339	43	1927	45
20	1164	40	3834	41	6229	43	8350	44	0201	46	1784	47
21	911044	42	903709	43	896099	45	888217	46	880063	48	871642	49
22	0924	44	3585	45	5970	47	8083	48	879925	52	1499	52
23	0804	46	3460	47	5841	49	7949	50	9787	54	1357	54
24	0684	48	3335	49	5712	52	7815	52	9649	56	1214	56
25	0564	50	3211	51	5582	54	7682	55	9510	58	1071	59
26	0443	52	3086	54	5453	57	7548	58	9372	60	0928	61
27	0323	54	2961	56	5323	58	7413	60	9233	62	0785	64
28	0202	56	2836	58	5194	60	7279	62	9095	64	0642	66
29	0082	58	2711	60	5064	62	7145	64	8956	67	0499	69
30	909961	60	2585	63	4934	65	7011	67	8817	69	0356	71
31	909841	62	902460	65	894805	67	886877	69	878678	71	870212	74
32	9720	64	2335	67	4675	69	6742	71	8539	73	0069	77
33	9599	66	2209	69	4545	71	6608	74	8400	76	869926	79
34	9478	68	2084	71	4415	73	6473	76	8261	78	9782	82
35	9357	70	1958	73	4284	75	6338	78	8122	81	9639	84
36	9236	72	1833	75	4154	78	6204	81	7983	84	9495	87
37	9115	74	1707	77	4024	80	6069	83	7844	86	9351	89
38	8994	76	1581	79	3894	82	5934	85	7704	89	9207	91
39	8873	78	1455	81	3763	84	5799	87	7565	91	9064	94
40	8751	80	1329	84	3633	86	5664	90	7425	93	8920	96
41	908630	82	901203	86	893502	89	885529	92	877286	95	868776	98
42	8508	84	1077	88	3371	91	5394	94	7146	97	8632	101
43	8387	86	0951	90	3241	93	5258	96	7006	100	8487	103
44	8265	88	0825	92	3110	95	5123	98	6867	102	8343	105
45	8143	90	0698	95	2979	97	4988	101	6727	105	8199	108
46	8021	92	0572	97	2848	100	4852	103	6587	107	8054	110
47	7900	94	0445	99	2717	102	4717	105	6447	109	7910	112
48	7778	96	0319	101	2586	104	4581	107	6307	112	7766	115
49	7655	98	0192	103	2455	106	4445	110	6167	114	7621	117
50	7533	100	0065	105	2323	108	4310	112	6026	117	7476	119
51	907411	102	899939	107	892192	111	884174	114	875886	119	867331	122
52	7289	104	9812	109	2061	113	4038	116	5746	122	7187	124
53	7167	106	9685	111	1929	115	3902	119	5605	124	7042	127
54	7044	109	9558	113	1798	117	3766	121	5465	126	6897	129
55	6922	111	9431	116	1666	119	3630	124	5324	129	6752	132
56	6799	113	9304	118	1534	122	3493	126	5183	131	6607	134
57	6676	115	9176	120	1402	124	3357	129	5042	133	6461	137
58	6554	117	9049	122	1271	126	3221	131	4902	136	6316	139
59	6431	119	8922	124	1139	129	3084	133	4761	138	6171	142
60	6308	121	8794	127	1007	131	2948	136	4620	140	6025	144

′	30°		31°		32°		33°		34°		35°	
″	Co-sine.	Parts for ″	Co-sine.	Parts for ″	Co-sine.	Parts for ″	Co-sine.	Parts for ″	Co-sine.	Parts for ″	Co-sine.	Parts for ″
0	866025	0	857167	0	848048	0	838671	0	829038	0	819152	0
1	5880	2	7017	3	7894	3	8512	3	8875	3	8985	3
2	5734	5	6868	5	7740	5	8354	5	8712	6	8818	6
3	5589	7	6718	8	7585	8	8195	8	8549	8	8651	9
4	5443	9	6567	10	7431	10	8036	11	8386	11	8484	11
5	5297	12	6417	13	7277	13	7878	13	8223	14	8317	14
6	5151	15	6267	15	7122	16	7719	16	8060	16	8150	17
7	5006	17	6117	17	6967	18	7560	19	7897	19	7982	20
8	4860	19	5966	20	6813	20	7401	22	7734	22	7815	23
9	4713	22	5816	22	6658	23	7242	24	7571	25	7648	25
10	4567	24	5666	25	6503	26	7083	27	7407	27	7480	28
11	864421	27	855515	27	846348	28	836924	29	827244	30	817313	31
12	4275	29	5364	30	6193	31	6764	32	7081	33	7145	34
13	4128	32	5214	32	6038	33	6605	35	6917	36	6977	36
14	3982	34	5063	35	5883	36	6446	38	6753	38	6809	39
15	3836	37	4912	38	5728	39	6286	40	6590	41	6642	42
16	3689	39	4761	40	5573	41	6127	43	6426	44	6474	44
17	3542	41	4610	43	5417	44	5967	46	6262	47	6306	47
18	3396	44	4459	45	5262	47	5807	48	6098	49	6138	50
19	3249	46	4308	47	5106	49	5648	51	5934	52	5970	53
20	3102	49	4156	50	4951	52	5488	54	5770	55	5801	56
21	862955	51	854005	52	844795	54	835328	56	825606	57	815633	58
22	2808	54	3854	55	4640	57	5168	59	5442	60	5465	61
23	2661	56	3702	57	4484	60	5008	62	5278	63	5296	64
24	2514	59	3551	60	4328	62	4848	65	5113	65	5128	67
25	2366	61	3399	62	4172	65	4688	67	4949	68	4959	70
26	2219	63	3248	65	4016	68	4527	70	4785	71	4791	73
27	2072	66	3096	67	3860	72	4367	72	4620	73	4622	76
28	1924	68	2944	70	3704	74	4207	75	4456	76	4453	79
29	1777	71	2792	73	3548	76	4046	78	4291	79	4284	82
30	1629	74	2640	76	3391	78	3886	81	4126	82	4116	84
31	861482	77	852488	78	843235	81	833725	84	823961	84	813947	87
32	1334	80	2336	81	3079	84	3565	87	3797	87	3778	90
33	1186	82	2184	83	2922	87	3404	90	3632	90	3608	93
34	1038	84	2032	85	2766	90	3243	93	3467	93	3439	95
35	0890	87	1879	88	2609	92	3082	95	3302	96	3270	98
36	0742	89	1727	90	2452	94	2921	98	3136	99	3101	101
37	0594	92	1575	93	2296	97	2760	101	2971	102	2931	104
38	0446	94	1422	96	2139	99	2599	103	2806	105	2762	107
39	0298	97	1269	99	1982	102	2438	106	2641	108	2592	110
40	0149	99	1117	102	1825	105	2277	108	2475	111	2423	113
41	860001	102	850964	105	841668	108	832115	111	822310	114	812253	115
42	859852	103	0811	107	1511	111	1954	114	2144	116	2084	118
43	9704	106	0658	109	1354	113	1793	116	1978	119	1914	121
44	9555	109	0505	111	1196	115	1631	119	1813	122	1744	124
45	9406	112	0352	114	1039	118	1470	121	1647	125	1574	127
46	9258	114	0199	117	0882	121	1308	124	1481	128	1404	130
47	9109	116	0046	119	0724	123	1146	127	1315	131	1234	133
48	8960	118	849893	122	0567	126	0984	129	1149	134	1064	136
49	8811	121	9739	125	0409	128	0823	132	0983	136	0894	139
50	8662	124	9586	128	0251	131	0661	135	0817	139	0723	142
51	858513	126	849433	131	840094	134	830499	138	820651	142	810553	144
52	8364	129	9279	133	839936	136	0337	141	0485	145	0383	147
53	8214	131	9125	135	9778	139	0174	143	0318	147	0212	150
54	8065	134	8972	138	9620	142	0012	146	0152	150	0042	153
55	7916	136	8818	140	9462	144	829850	148	819985	153	809871	156
56	7766	139	8664	143	9·04	147	9688	151	9819	156	9700	159
57	7616	142	8510	145	9146	150	9525	154	9652	158	9530	162
58	7467	145	8356	148	8987	152	9363	156	9486	161	9359	164
59	7317	147	8202	151	8829	155	9200	159	9319	164	9188	167
60	7167	149	8048	153	8671	157	9038	162	9152	166	9017	170

′ ″	36° Co-sine.	36° Parts for ″	37° Co-sine.	37° Parts for ″	38° Co-sine.	38° Parts for ″	39° Co-sine.	39° Parts for ″	40° Co-sine.	40° Parts for ″	41° Co-sine.	41° Parts for ″
0	809017	0	798636	0	788011	0	777146	0	766044	0	754710	0
1	8846	3	8460	3	7832	3	6963	3	5857	3	4519	3
2	8675	6	8285	6	7652	6	6780	6	5670	6	4328	6
3	8504	9	8110	9	7473	9	6596	9	5483	9	4137	10
4	8333	11	7935	12	7294	12	6413	12	5296	13	3946	13
5	8161	14	7759	15	7114	15	6230	15	5109	16	3755	16
6	7990	17	7584	18	6935	18	6046	18	4921	19	3563	19
7	7819	20	7408	20	6756	21	5863	21	4734	22	3372	22
8	7647	23	7233	23	6576	24	5679	24	4547	25	3181	25
9	7475	26	7057	26	6396	27	5496	27	4359	28	2989	28
10	7304	29	6882	29	6217	30	5312	31	4171	31	2798	32
11	807132	32	796706	32	786037	33	775128	34	763984	34	752606	35
12	6960	34	6530	35	5857	36	4945	37	3796	38	2415	38
13	6789	37	6354	38	5677	39	4761	40	3608	41	2223	41
14	6617	40	6178	41	5497	42	4577	43	3420	44	2032	44
15	6445	43	6002	44	5317	45	4393	46	3232	47	1840	48
16	6273	46	5826	47	5137	48	4209	49	3044	50	1648	51
17	6101	49	5650	50	4957	51	4024	52	2856	53	1456	54
18	5928	52	5473	53	4776	54	3840	55	2668	57	1264	57
19	5756	55	5297	56	4596	57	3656	58	2480	60	1072	60
20	5584	57	5121	59	4416	60	3472	61	2292	63	0880	64
21	805411	60	794944	62	784235	63	773287	65	762104	66	750688	67
22	5239	63	4768	65	4055	66	3103	68	1915	69	0496	70
23	5066	66	4591	68	3874	69	2918	71	1727	72	0303	73
24	4894	69	4415	71	3694	72	2734	74	1538	75	0111	86
25	4721	72	4238	74	3513	75	2549	77	1350	78	749919	80
26	4548	75	4061	76	3332	78	2364	80	1161	82	9726	83
27	4376	78	3884	79	3151	81	2179	83	0972	85	9534	86
28	4203	81	3707	82	2970	84	1995	86	0784	88	9341	89
29	4030	84	3530	85	2789	87	1810	89	0595	91	9148	92
30	3857	86	3353	88	2608	90	1625	92	0406	94	8956	96
31	803684	89	793176	92	782427	94	771440	97	760217	98	748763	101
32	3511	92	2999	95	2246	97	1254	100	0028	101	8570	104
33	3338	95	2822	98	2065	100	1069	103	759839	105	8377	107
34	3164	98	2644	101	1883	103	0884	106	9650	108	8184	110
35	2991	101	2467	104	1702	106	0699	109	9461	111	7991	113
36	2818	104	2290	107	1520	109	0513	112	9271	114	7798	117
37	2644	107	2112	110	1339	112	0328	115	9082	117	7605	120
38	2471	110	1935	113	1157	115	0142	118	8893	120	7412	123
39	2297	113	1757	116	0976	118	769957	121	8703	123	7218	126
40	2123	116	1579	119	0794	121	9771	124	8514	127	7025	129
41	801950	118	791401	121	780612	125	769585	127	758324	130	746832	133
42	1776	121	1224	124	0430	128	9400	130	8134	133	6638	136
43	1602	124	1046	127	0249	131	9214	133	7945	136	6445	139
44	1428	127	0868	130	0067	134	9028	136	7755	139	6251	142
45	1254	130	0690	133	779884	137	8842	139	7565	142	6057	145
46	1080	133	0512	136	9702	140	8656	143	7375	146	5864	149
47	0906	136	0333	139	9520	143	8470	146	7185	149	5670	152
48	0731	139	0155	142	9338	146	8284	149	6995	152	5476	155
49	0557	142	789977	145	9156	149	8097	152	6805	155	5282	159
50	0383	145	9798	148	8973	152	7911	155	6615	158	5088	162
51	800208	147	789620	151	778791	155	767725	158	756425	161	744894	166
52	0034	150	9441	154	8608	158	7538	161	6234	165	4700	169
53	799859	153	9263	157	8426	161	7352	164	6044	168	4506	172
54	9685	156	9084	160	8243	164	7165	167	5854	171	4312	175
55	9510	159	8905	163	8060	167	6979	171	5663	174	4117	178
56	9335	162	8727	166	7878	170	6792	174	5472	177	3923	181
57	9160	165	8548	169	7695	173	6605	177	5282	180	3728	184
58	8985	168	8369	172	7512	176	6418	180	5091	184	3534	188
59	8811	171	8190	175	7329	179	6231	183	4900	187	3339	191
60	8636	174	8011	178	7146	182	6044	186	4710	190	3145	194

′ / ″	42° Co-sine.	42° Parts for ″	43° Co-sine.	43° Parts for ″	44° Co-sine.	44° Parts for ″	45° Co-sine.	45° Parts for ″	46° Co-sine.	46° Parts for ″	47° Co-sine.	47° Parts for ″
0	743145	0	731354	0	719340	0	707107	0	694658	0	681998	0
1	2950	3	1155	3	9138	3	6901	3	4449	3	1786	4
2	2755	7	0957	7	8936	7	6695	7	4240	7	1573	7
3	2561	10	0758	10	8733	10	6489	10	4030	11	1360	10
4	2366	13	0560	13	8531	14	6284	14	3821	14	1147	14
5	2171	17	0361	16	8329	17	6078	17	3611	18	0934	18
6	1976	20	0162	20	8126	20	5872	21	3402	21	0721	21
7	1781	23	729963	23	7924	24	5666	24	3192	25	0508	25
8	1586	26	9765	26	7721	27	5459	28	2983	28	0295	28
9	1391	29	9566	29	7519	31	5253	31	2773	32	0081	32
10	1195	33	9367	33	7316	34	5047	34	2563	35	679868	36
11	741000	36	729168	36	717113	38	704841	38	692353	39	679655	39
12	0805	39	8969	39	6911	41	4634	41	2143	42	9441	43
13	0609	42	8770	42	6708	45	4428	45	1933	46	9228	46
14	0414	45	8570	46	6505	48	4221	48	1723	49	9014	50
15	0218	49	8371	50	6302	51	4015	52	1513	52	8801	53
16	0023	52	8172	53	6099	55	3808	55	1303	56	8587	57
17	739827	55	7972	56	5896	58	3601	59	1093	59	8373	60
18	9631	58	7773	60	5693	62	3395	62	0882	63	8160	64
19	9435	62	7573	63	5490	65	3188	66	0672	66	7946	67
20	9239	65	7374	66	5286	68	2981	69	0462	70	7732	71
21	739043	68	727174	70	715083	72	702774	73	690251	73	677518	74
22	8848	71	6974	73	4880	75	2567	76	0041	77	7304	78
23	8651	75	6775	76	4676	79	2360	80	689830	80	7090	81
24	8455	78	6575	80	4473	82	2153	83	9620	84	6876	85
25	8259	81	6375	83	4269	85	1946	86	9409	87	6662	89
26	8063	84	6175	86	4066	88	1739	90	9198	91	6448	92
27	7867	88	5975	90	3862	92	1531	93	8987	94	6233	96
28	7670	91	5775	93	3658	96	1324	97	8776	98	6019	99
29	7474	94	5575	96	3454	99	1117	100	8566	101	5805	103
30	7277	98	5374	100	3250	102	0909	103	8355	105	5590	107
31	737081	103	725174	104	713047	106	700702	107	688144	110	675376	111
32	6884	106	4974	107	2843	109	0494	111	7932	113	5161	115
33	6687	110	4773	110	2639	112	0287	114	7721	117	4947	118
34	6491	113	4573	113	2434	116	0079	118	7510	120	4732	122
35	6294	116	4372	117	2230	119	699871	121	7299	124	4517	125
36	6097	119	4172	120	2026	123	9663	125	7088	127	4302	129
37	5900	123	3971	123	1822	126	9455	128	6876	131	4088	133
38	5703	126	3771	127	1617	130	9248	132	6665	134	3873	136
39	5506	129	3570	130	1413	133	9040	135	6453	138	3658	140
40	5309	132	3369	134	1209	137	8832	139	6242	141	3443	143
41	735112	135	723168	137	711004	140	698623	142	686030	144	673228	147
42	4915	139	2967	141	0799	143	8415	145	5818	148	3013	151
43	4717	142	2766	144	0595	146	8207	149	5607	152	2797	154
44	4520	145	2565	147	0390	150	7999	152	5395	156	2582	158
45	4323	149	2364	150	0185	153	7790	156	5183	159	2367	161
46	4125	152	2163	154	709981	157	7582	159	4971	163	2151	165
47	3927	155	1962	157	9776	160	7374	163	4759	167	1936	169
48	3730	158	1760	161	9571	164	7165	166	4547	170	1721	172
49	3532	162	1559	164	9366	167	6957	170	4335	174	1505	176
50	3334	165	1357	168	9161	171	6748	173	4123	177	1290	179
51	733137	169	721156	171	708956	174	696539	177	683911	181	671074	183
52	2939	172	0954	174	8750	177	6330	180	3698	184	0858	186
53	2741	175	0753	177	8545	181	6122	184	3486	188	0642	190
54	2543	178	0551	181	8340	184	5913	187	3274	191	0427	193
55	2345	182	0349	184	8135	188	5704	191	3061	195	0211	197
56	2147	185	0148	188	7929	191	5495	194	2849	198	669995	201
57	1949	188	719946	191	7724	195	5286	198	2636	202	9779	204
58	1750	191	9744	194	7518	198	5077	201	2424	205	9563	208
59	1552	194	9542	197	7312	202	4868	205	2211	209	9347	211
60	1354	197	9340	201	7107	205	4658	208	1998	212	9131	214

′ ″	48° Co-sine.	48° Parts for ″	49° Co-sine.	49° Parts for ″	50° Co-sine.	50° Parts for ″	51° Co-sine.	51° Parts for ″	52° Co-sine.	52° Parts for ″	53° Co-sine.	53° Parts for ″
0	669131	0	656059	0	642788	0	629320	0	615661	0	601815	0
1	8914	4	5840	4	2565	4	9094	4	5432	4	1583	4
2	8698	7	5620	7	2342	8	8868	8	5208	8	1350	8
3	8482	11	5400	11	2119	11	8642	11	4974	12	1118	12
4	8265	14	5180	15	1896	15	8416	15	4744	16	0885	16
5	8049	18	4961	19	1673	19	8189	19	4515	19	0653	19
6	7833	22	4741	22	1450	22	7963	23	4285	23	0420	23
7	7616	25	4521	26	1226	26	7737	26	4056	27	0188	27
8	7399	29	4301	30	1003	30	7510	30	3826	31	599955	31
9	7183	32	4081	33	0780	34	7284	34	3596	35	9722	35
10	6966	36	3861	37	0557	37	7057	38	3367	38	9489	39
11	666749	39	653641	41	640333	41	626830	42	613137	42	599256	43
12	6532	43	3421	44	0110	45	6604	45	2907	46	9024	47
13	6316	46	3200	48	639886	49	6377	49	2677	50	8791	50
14	6099	50	2980	52	9663	53	6150	53	2447	54	8558	54
15	5882	54	2760	55	9439	56	5923	57	2217	57	8325	58
16	5665	57	2539	59	9215	60	5697	61	1987	61	8092	62
17	5448	61	2319	63	8992	64	5470	64	1757	65	7858	66
18	5230	64	2098	66	8768	68	5243	68	1527	69	7625	70
19	5013	68	1878	70	8544	72	5016	72	1297	73	7392	74
20	4796	72	1657	73	8320	75	4789	76	1067	77	7159	78
21	664579	75	351437	77	638096	78	624561	80	610836	81	596925	82
22	4361	79	1216	81	7872	82	4334	84	0606	85	6692	86
23	4144	82	0995	85	7648	86	4107	88	0376	89	6458	90
24	3926	86	0774	89	7424	90	3880	92	0145	92	6225	94
25	3709	90	0553	93	7200	94	3652	95	609915	96	5991	98
26	3491	93	0332	96	6976	97	3425	99	9684	100	5758	102
27	3273	97	0111	100	6751	101	3197	103	9454	104	5524	106
28	3056	101	649890	103	6527	105	2970	107	9223	108	5290	110
29	2838	105	9669	107	6303	109	2742	111	8992	111	5057	114
30	2620	109	9448	110	6078	112	2515	114	8761	115	4823	117
31	662402	114	649227	115	635854	117	622287	119	608531	119	594589	121
32	2184	118	9006	118	5629	121	2059	123	8300	123	4355	125
33	1966	121	8784	122	5405	124	1831	127	8069	127	4121	129
34	1748	125	8563	126	5180	128	1604	131	7838	131	3887	133
35	1530	128	8341	129	4955	131	1376	134	7607	135	3653	137
36	1312	132	8120	133	4731	134	1148	138	7376	139	3419	141
37	1094	136	7898	137	4506	138	0920	142	7145	143	3185	145
38	0875	139	7677	141	4281	142	0692	146	6914	147	2951	149
39	0657	143	7455	144	4056	146	0464	150	6682	151	2716	153
40	0439	146	7233	148	3831	150	0235	153	6451	154	2482	156
41	660220	150	647012	152	633606	153	620007	157	606220	158	592248	160
42	0002	154	6790	155	3381	157	619779	161	5988	162	2013	164
43	659783	157	6568	159	3156	161	9551	165	5757	166	1779	168
44	9565	161	6346	163	2931	165	9322	169	5526	170	1544	172
45	9346	164	6124	167	2705	169	9094	172	5294	174	1310	176
46	9127	168	5902	171	2480	172	8865	176	5062	178	1075	180
47	8908	172	5680	174	2255	176	8637	180	4831	182	0840	184
48	8690	175	5458	178	2029	180	8408	184	4599	186	0606	188
49	8471	179	5236	181	1804	183	8180	188	4367	190	0371	192
50	8252	183	5013	185	1578	187	7951	191	4136	194	0136	195
51	658033	187	644791	188	631353	191	617722	195	603904	197	589901	199
52	7814	190	4569	192	1127	195	7494	199	3672	201	9666	203
53	7594	194	4346	196	0902	199	7265	203	3440	205	9431	207
54	7375	197	4124	200	0676	202	7036	206	3208	209	9196	211
55	7156	201	3901	204	0450	206	6807	210	2976	213	8961	215
56	6937	204	3679	207	0224	210	6578	214	2744	217	8726	219
57	6717	208	3456	211	629998	214	6349	218	2512	220	8491	223
58	6498	212	3233	215	9772	218	6120	221	2280	224	8256	227
59	6279	215	3010	218	9546	221	5891	225	2047	228	8021	231
60	6059	219	2788	222	9320	225	5661	228	1815	231	7785	234

′ ″	54° Co-sine.	54° Parts for ″	55° Co-sine.	55° Parts for ″	56° Co-sine.	56° Parts for ″	57° Co-sine.	57° Parts for ″	58° Co-sine.	58° Parts for ″	59° Co-sine.	59° Parts for ″
0	587785	0	573576	0	559193	0	544639	0	529919	0	515038	0
1	7550	4	3338	4	8952	4	4395	4	9673	4	4789	4
2	7315	8	3100	8	8710	8	4151	8	9426	8	4539	8
3	7079	12	2861	12	8469	12	3907	12	9179	12	4290	12
4	6844	16	2623	16	8228	16	3663	16	8932	17	4040	17
5	6608	20	2384	20	7987	20	3419	20	8685	21	3791	21
6	6372	24	2146	24	7745	24	3174	24	8438	25	3541	25
7	6137	28	1907	28	7504	28	2930	28	8191	29	3292	29
8	5901	32	1669	32	7262	32	2686	32	7944	33	3042	33
9	5665	36	1430	36	7021	36	2442	37	7697	37	2792	37
10	5429	39	1191	40	6779	40	2197	41	7450	41	2543	42
11	585194	43	570952	44	556537	45	541953	45	527203	45	512293	46
12	4958	47	0714	48	6296	49	1708	49	6956	49	2043	50
13	4722	51	0475	52	6054	53	1464	53	6709	54	1793	54
14	4486	55	0236	56	5812	57	1219	57	6461	58	1543	58
15	4250	59	569997	60	5570	61	0975	61	6214	62	1293	63
16	4014	63	9758	64	5328	65	0730	65	5967	66	1043	67
17	3777	67	9519	68	5086	69	0485	69	5719	70	0793	71
18	3541	71	9280	72	4844	73	0240	73	5472	74	0543	75
19	3305	75	9040	76	4602	77	539996	77	5224	78	0293	79
20	3069	79	8801	80	4360	81	9751	81	4977	82	0043	83
21	582832	83	568562	84	554118	85	539506	86	524729	87	509792	87
22	2596	87	8323	88	3876	89	9261	90	4481	91	9542	91
23	2360	91	8083	92	3634	93	9016	94	4234	95	9292	95
24	2123	95	7844	96	3392	97	8771	98	3986	99	9041	99
25	1886	99	7604	100	3149	101	8526	102	3738	103	8791	104
26	1650	103	7365	104	2907	105	8281	106	3490	107	8541	108
27	1413	107	7125	108	2664	109	8035	110	3242	111	8290	112
28	1176	111	6886	112	2422	113	7790	114	2995	115	8040	117
29	0940	115	6646	116	2180	117	7545	118	2747	119	7789	121
30	0703	118	6406	120	1937	122	7300	122	2499	124	7538	126
31	580466	122	566166	124	551694	126	537054	127	522251	128	507288	130
32	0229	126	5927	128	1452	130	6809	131	2002	132	7037	134
33	579992	130	5687	132	1209	134	6563	135	1754	136	6786	138
34	9755	134	5447	136	0966	138	6318	139	1506	141	6536	142
35	9518	138	5207	140	0724	142	6072	143	1258	145	6285	146
36	9281	142	4967	144	0481	146	5827	148	1010	149	6034	151
37	9044	146	4727	148	0238	150	5581	152	0761	153	5783	155
38	8807	150	4487	152	549995	154	5336	156	0513	158	5532	159
39	8570	154	4247	156	9752	158	5090	160	0265	162	5281	163
40	8332	158	4007	160	9509	162	4844	164	0016	166	5030	168
41	578095	162	563766	164	549266	166	534598	168	519768	170	504779	172
42	7858	166	3526	168	9023	171	4352	172	9519	174	4528	176
43	7620	170	3286	172	8780	175	4107	176	9271	178	4277	180
44	7383	174	3045	176	8536	179	3861	180	9022	182	4025	184
45	7145	178	2805	180	8293	183	3615	184	8773	186	3774	188
46	6908	182	2564	184	8050	187	3369	189	8525	190	3523	193
47	6670	186	2324	188	7807	191	3122	193	8276	195	3271	197
48	6432	190	2083	192	7563	195	2876	197	8027	199	3020	201
49	6195	194	1843	196	7320	199	2630	201	7778	203	2769	205
50	5957	198	1602	200	7076	203	2384	205	7529	207	2517	210
51	575719	202	561361	204	546833	207	532138	209	517280	212	502266	214
52	5481	206	1121	208	6589	211	1891	213	7031	216	2014	218
53	5243	210	0880	212	6346	215	1645	217	6782	220	1762	222
54	5005	214	0639	216	6102	219	1399	221	6533	224	1511	226
55	4767	218	0398	220	5858	223	1152	226	6284	228	1259	230
56	4529	222	0157	224	5615	227	0906	230	6035	233	1007	235
57	4291	226	559916	228	5371	231	0659	234	5786	237	0756	239
58	4053	230	9675	232	5127	235	0413	238	5537	241	0504	243
59	3815	234	9434	236	4883	239	0166	242	5287	245	0252	247
60	3576	237	9193	240	4639	243	529919	246	5038	249	0000	251

′	60°		61°		62°		63°		64°		65°	
″	Co-sine.	Parts for ″	Co-sine.	Parts for ″	Co-sine.	Parts for ″	Co-sine.	Parts for ″	Co-sine.	Parts for ″	Co-sine.	Parts for ″
0	500000	0	484810	0	469472	0	453991	0	438371	0	422618	0
1	499748	4	4555	4	9215	4	3731	4	8110	4	2355	4
2	9496	8	4301	8	8958	8	3472	9	7848	9	2091	9
3	9244	12	4046	13	8701	13	3213	13	7587	13	1827	13
4	8992	17	3792	17	8444	17	2954	17	7325	17	1563	18
5	8740	21	3537	21	8187	21	2694	22	7063	22	1300	22
6	8488	25	3282	25	7930	26	2435	26	6802	26	1036	26
7	8236	30	3028	30	7673	30	2175	30	6540	31	0772	31
8	7983	34	2773	34	7416	34	1916	35	6278	35	0508	35
9	7731	38	2518	38	7158	38	1656	39	6017	39	0244	40
10	7479	42	2263	43	6901	43	1397	43	5755	44	419980	44
11	497226	46	482009	47	466644	47	451137	48	435493	48	419716	48
12	6974	50	1754	51	6387	51	0878	52	5231	52	9452	53
13	6722	54	1499	55	6129	55	0618	56	4969	57	9188	57
14	6469	58	1244	59	5872	60	0358	61	4707	61	8924	62
15	6217	63	0989	63	5615	64	0098	65	4445	66	8660	66
16	5964	67	0734	67	5357	68	449839	69	4183	70	8396	71
17	5711	71	0479	72	5100	72	9579	74	3921	74	8131	75
18	5459	75	0224	76	4842	77	9319	78	3659	79	7867	79
19	5206	79	479968	80	4585	81	9059	82	3397	83	7603	84
20	4953	84	9713	85	4327	85	8799	87	3135	87	7339	88
21	494701	88	479458	89	464069	90	448539	91	432873	92	417074	92
22	4448	92	9203	93	3812	94	8279	95	2610	96	6810	97
23	4195	96	8947	97	3554	98	8019	100	2348	100	6545	101
24	3942	100	8692	101	3296	103	7759	104	2086	105	6281	106
25	3689	105	8436	106	3038	107	7499	108	1823	109	6016	110
26	3436	109	8181	110	2780	111	7239	113	1561	113	5752	114
27	3183	113	7926	115	2523	115	6979	117	1299	118	5487	119
28	2930	117	7670	119	2265	120	6718	121	1036	122	5223	123
29	2677	121	7414	123	2007	124	6458	126	0774	126	4958	128
30	2424	126	7159	128	1749	129	6198	130	0511	131	4693	132
31	492170	131	476903	132	461491	133	445938	134	430249	136	414429	137
32	1917	135	6647	136	1233	138	5677	139	429986	140	4164	141
33	1664	140	6392	141	0974	142	5417	143	9723	145	3899	146
34	1411	144	6136	145	0716	146	5156	147	9461	149	3634	150
35	1157	148	5880	149	0458	151	4896	152	9198	153	3369	154
36	0904	152	5624	154	0200	155	4635	156	8935	158	3104	159
37	0650	156	5368	158	459942	159	4375	160	8672	162	2840	163
38	0397	161	5112	162	9683	164	4114	165	8410	167	2575	168
39	0143	165	4856	166	9425	168	3853	169	8147	171	2310	172
40	489890	169	4600	171	9167	172	3593	174	7884	175	2045	177
41	489636	173	474344	175	458908	177	443332	178	427621	180	411780	181
42	9383	178	4088	179	8650	181	3071	182	7358	184	1514	185
43	9129	182	3832	183	8391	185	2810	187	7095	189	1249	189
44	8875	186	3576	187	8133	189	2550	191	6832	193	0984	194
45	8621	190	3320	192	7874	194	2289	195	6569	197	0719	199
46	8367	195	3063	196	7615	198	2028	199	6306	202	0454	203
47	8114	199	2807	200	7357	202	1767	204	6043	206	0188	207
48	7860	203	2551	204	7098	207	1506	208	5779	210	409923	212
49	7606	207	2294	208	6839	211	1245	212	5516	215	9658	216
50	7352	212	2038	213	6580	215	0984	217	5253	219	9392	221
51	487098	216	471782	217	456322	220	440723	221	424990	224	409127	225
52	6844	220	1525	221	6063	224	0462	226	4726	228	8862	230
53	6590	224	1269	225	5804	228	0200	230	4463	232	8596	234
54	6335	229	1012	230	5545	233	439939	234	4199	237	8331	239
55	6081	233	0755	234	5286	237	9678	239	3936	241	8065	243
56	5827	237	0499	238	5027	241	9417	243	3673	245	7799	247
57	5573	241	0242	242	4768	246	9155	247	3409	250	7534	252
58	5318	245	469985	247	4509	250	8894	251	3146	254	7268	256
59	5064	249	9728	251	4250	254	8633	256	2882	259	7002	260
60	4810	254	9472	256	3991	258	8371	260	2618	263	6737	265

′	66°		67°		68°		69°		70°		71°	
″	Co-sine.	Parts for ″	Co-sine.	Parts for ″	Co-sine.	Parts for ″	Co-sine.	Parts for ″	Co-sine.	Parts for ″	Co-sine.	Parts for ″
0	406737	0	390731	0	374607	0	358368	0	342020	0	325568	0
1	6471	4	0463	4	4337	5	8096	5	1747	5	5293	5
2	6205	9	0196	9	4067	9	7825	9	1473	9	5018	9
3	5939	13	389928	13	3797	14	7553	14	1200	14	4743	14
4	5673	18	9660	18	3528	18	7281	18	0927	18	4468	18
5	5408	22	9392	22	3258	23	7010	23	0653	23	4193	23
6	5142	27	9124	27	2988	27	6738	27	0380	27	3917	27
7	4876	31	8856	31	2718	32	6466	32	0106	32	3642	32
8	4610	36	8588	36	2448	36	6194	36	339833	36	3367	37
9	4344	40	8320	40	2178	41	5923	41	9559	41	3092	41
10	4078	44	8052	45	1908	45	5651	46	9285	46	2816	46
11	403811	49	387784	49	371638	50	355379	50	339012	50	322541	51
12	3545	53	7516	54	1368	54	5107	54	8738	55	2266	55
13	3279	58	7247	58	1098	59	4835	59	8464	59	1990	60
14	3013	62	6979	63	0828	63	4563	63	8191	64	1715	64
15	2747	66	6711	67	0557	68	4291	68	7917	68	1440	69
16	2480	71	6443	72	0287	72	4019	73	7643	73	1164	74
17	2214	76	6174	76	0017	77	3747	77	7369	78	0889	78
18	1948	80	5906	81	369747	81	3475	82	7095	82	0613	83
19	1681	85	5638	85	9477	86	3203	87	6821	87	0337	87
20	1415	89	5369	89	9206	90	2931	91	6548	91	0062	92
21	401149	94	385101	93	368936	95	352658	96	336274	96	319786	96
22	0882	98	4832	98	8665	100	2386	100	6000	100	9511	101
23	0616	103	4564	102	8395	104	2114	105	5726	105	9235	106
24	0349	107	4295	107	8125	108	1842	109	5452	109	8959	110
25	0083	112	4027	111	7854	113	1569	114	5178	114	8684	115
26	399816	116	3758	116	7584	117	1297	118	4903	118	8408	119
27	9549	121	3490	121	7313	122	1025	123	4629	123	8132	124
28	9283	125	3221	125	7043	126	0752	127	4355	127	7856	128
29	9016	129	2952	130	6772	131	0480	132	4081	132	7581	133
30	8749	133	2683	134	6501	135	0207	136	3807	137	7305	138
31	398482	138	382415	139	366231	140	349935	141	333533	142	317029	143
32	8216	142	2146	143	5960	144	9662	145	3258	146	6753	147
33	7949	147	1877	148	5689	149	9390	150	2984	151	6477	152
34	7682	151	1608	152	5418	153	9117	155	2710	155	6201	157
35	7415	156	1339	157	5148	158	8845	159	2436	160	5925	161
36	7148	160	1070	161	4877	162	8572	164	2161	165	5649	166
37	6881	165	0801	166	4606	167	8299	168	1887	169	5373	171
38	6614	169	0532	170	4335	171	8027	173	1612	173	5097	175
39	6347	174	0263	175	4064	176	7754	177	1338	178	4821	180
40	6080	178	379994	179	3793	180	7481	182	1063	183	4545	184
41	395813	182	379725	184	363522	185	347209	186	330789	187	314269	189
42	5546	187	9456	188	3251	189	6936	191	0514	192	3993	193
43	5278	191	9187	193	2980	194	6663	195	0240	197	3716	198
44	5011	196	8918	197	2709	198	6390	200	329965	201	3440	202
45	4744	200	8649	202	2438	203	6117	205	9691	206	3164	207
46	4477	205	8379	206	2167	207	5844	209	9416	210	2888	212
47	4209	209	8110	211	1896	212	5571	214	9141	215	2611	216
48	3942	214	7841	215	1625	216	5298	218	8867	220	2335	221
49	3675	218	7571	220	1353	221	5025	223	8592	224	2059	225
50	3407	223	7302	224	1082	226	4752	228	8317	229	1782	230
51	393140	227	377033	229	360811	230	344479	232	328042	234	311506	235
52	2872	231	6763	233	0540	235	4206	237	7768	238	1229	239
53	2605	236	6494	238	0268	239	3933	241	7493	243	0953	244
54	2337	240	6224	242	359997	244	3660	246	7218	247	0676	248
55	2070	245	5955	247	9725	248	3387	250	6943	252	0400	253
56	1802	249	5685	251	9454	253	3113	255	6668	256	0123	258
57	1534	254	5416	256	9183	257	2840	259	6393	261	309847	262
58	1267	258	5146	260	8911	262	2567	264	6118	265	9570	267
59	0999	263	4876	265	8640	266	2294	268	5843	270	9294	271
60	0731	267	4607	269	8368	271	2020	273	5568	274	9017	276

′	72°		73°		74°		75°		76°		77°	
″	Co-sine.	Parts for ″	Co-sine.	Parts for ″	Co-sine.	Parts for ″	Co-sine.	Parts for ″	Co-sine.	Parts for ″	Co-sine.	Parts for ″
0	309017	0	292372	0	275637	0	258819	0	241922	0	224951	0
1	8740	5	2094	5	5358	5	8538	5	1640	5	4668	5
2	8464	9	1815	9	5078	9	8257	9	1357	9	4384	9
3	8187	14	1537	14	4798	14	7976	14	1075	14	4101	14
4	7910	18	1259	19	4519	19	7695	19	0793	19	3817	19
5	7633	23	0981	23	4239	23	7414	23	0510	24	3534	24
6	7357	28	0702	28	3959	28	7133	28	0228	28	3250	28
7	7080	32	0424	32	3679	33	6852	33	239946	33	2967	33
8	6803	37	0146	37	3400	37	6571	37	9663	38	2683	38
9	6526	42	289867	42	3120	42	6289	42	9381	43	2399	43
10	6249	46	9589	46	2840	47	6008	47	9098	47	2116	47
11	305972	51	289310	51	272560	51	255727	52	238816	52	221832	52
12	5695	55	9032	56	2280	56	5446	56	8534	57	1549	57
13	5418	60	8753	60	2000	61	5165	61	8251	61	1265	62
14	5141	65	8475	65	1720	65	4883	66	7968	66	0981	66
15	4864	69	8196	70	1440	70	4602	70	7686	71	0697	71
16	4587	74	7918	74	1161	75	4321	75	7403	75	0414	76
17	4310	78	7639	79	0881	79	4039	80	7121	80	0130	81
18	4033	83	7361	84	0600	84	3758	84	6838	85	219846	85
19	3756	88	7082	88	0320	89	3477	89	6556	90	9562	90
20	3479	92	6803	93	0040	93	3195	94	6273	94	9279	95
21	303202	97	286525	98	269760	98	252914	98	235990	99	218995	100
22	2924	102	6246	102	9480	103	2632	103	5708	104	8711	104
23	2647	106	5967	107	9200	107	2351	108	5425	109	8427	109
24	2370	111	5688	112	8920	112	2069	113	5142	113	8143	114
25	2093	116	5410	116	8640	117	1788	117	4859	118	7859	119
26	1815	120	5131	121	8359	121	1506	122	4577	123	7575	123
27	1538	125	4852	126	8079	126	1225	127	4294	127	7292	128
28	1261	130	4573	130	7799	131	0943	131	4011	132	7008	133
29	0983	134	4294	135	7519	135	0662	136	3728	137	6724	138
30	0706	139	4015	139	7238	140	0380	141	3445	141	6440	142
31	300428	143	283736	144	266958	145	250098	146	233163	146	216156	147
32	0151	148	3458	149	6678	150	249817	150	2880	151	5872	152
33	299873	153	3179	154	6397	154	9535	155	2597	156	5588	157
34	9596	157	2900	158	6117	159	9253	160	2314	161	5304	161
35	9318	162	2621	163	5837	164	8972	165	2031	165	5019	166
36	9041	167	2342	168	5556	169	8690	169	1748	170	4735	171
37	8763	171	2062	172	5276	173	8408	174	1465	175	4451	176
38	8486	176	1783	177	4995	178	8126	179	1182	179	4167	180
39	8208	181	1504	182	4715	183	7845	183	0899	184	3883	185
40	7930	185	1225	186	4434	187	7563	188	0616	189	3599	190
41	297653	190	280946	191	264154	192	247281	193	230333	194	213315	195
42	7375	195	0667	196	3873	197	6999	198	0050	198	3030	199
43	7097	199	0388	200	3593	201	6717	202	229767	203	2746	204
44	6819	204	0108	205	3312	206	6435	207	9484	208	2462	209
45	6542	208	279829	210	3031	211	6153	212	9200	213	2178	213
46	6264	213	9550	214	2751	215	5871	216	8917	217	1893	218
47	5986	218	9270	219	2470	220	5589	221	8634	222	1609	223
48	5708	222	8991	224	2189	225	5307	225	8351	227	1325	228
49	5430	227	8712	228	1909	230	5025	230	8068	232	1040	232
50	5152	231	8432	233	1628	234	4743	235	7784	236	0756	237
51	294874	236	278153	238	261347	239	244461	240	227501	241	210472	242
52	4596	241	7874	242	1066	244	4179	245	7218	246	0187	247
53	4318	245	7594	247	0785	248	3897	249	6935	250	209903	251
54	4040	250	7315	252	0505	253	3615	254	6651	255	9619	256
55	3762	254	7035	256	0224	258	3333	259	6368	260	9334	261
56	3484	259	6756	261	259943	262	3051	263	6085	265	9050	266
57	3206	264	6476	266	9662	267	2769	268	5801	269	8765	270
58	2928	268	6197	270	9381	272	2486	273	5518	274	8481	275
59	2650	273	5917	275	9100	276	2204	277	5235	279	8196	280
60	2372	277	5637	279	8819	281	1922	282	4951	283	7912	284

′ / ″	78° Co-sine	Parts for ″	79° Co-sine.	Parts for ″	80° Co-sine.	Parts for ″	81° Co-sine.	Parts for ″	82° Co-sine.	Parts for ″	83° Co-sine.	Parts for ″
0	207912	0	190809	0	173648	0	156435	0	139173	0	121869	0
1	7627	5	0523	5	3362	5	6147	5	8885	5	1581	5
2	7343	9	0238	10	3075	10	5860	10	8597	10	1292	10
3	7058	14	189952	14	2789	14	5573	14	8309	14	1003	14
4	6773	19	9667	19	2502	19	5285	19	8021	19	0714	19
5	6489	24	9381	24	2216	24	4998	24	7733	24	0426	24
6	6204	28	9095	29	1929	29	4710	29	7445	29	0137	29
7	5920	33	8810	33	1643	33	4423	33	7156	34	119848	34
8	5635	38	8524	38	1356	38	4136	38	6868	38	9559	39
9	5350	43	8239	43	1069	43	3848	43	6580	43	9270	43
10	5066	47	7953	48	0783	48	3561	48	6292	48	8982	48
11	204781	52	187667	52	170496	52	153273	53	136004	53	118693	53
12	4496	57	7381	57	0210	57	2986	57	5716	58	8404	58
13	4211	62	7096	62	169923	62	2698	62	5427	62	8115	63
14	3927	66	6810	67	9636	67	2411	67	5139	67	7826	67
15	3642	71	6524	71	9350	72	2123	72	4851	72	7537	72
16	3357	76	6238	76	9063	76	1836	77	4563	77	7249	77
17	3072	81	5952	81	8776	81	1548	81	4274	82	6960	82
18	2787	85	5667	86	8489	86	1261	86	3986	86	6671	87
19	2502	90	5381	91	8203	91	0973	91	3698	91	6382	91
20	2218	95	5095	95	7916	96	0686	96	3410	96	6093	96
21	201938	100	184809	100	167629	100	150398	101	133121	101	115804	101
22	1648	104	4523	105	7342	105	0111	106	2833	106	5515	106
23	1363	109	4237	110	7056	110	149823	111	2545	110	5226	111
24	1078	114	3951	115	6769	115	9535	116	2256	115	4937	116
25	0793	119	3665	119	6482	119	9248	120	1968	120	4648	120
26	0508	123	3380	124	6195	124	8960	125	1680	125	4359	125
27	0223	128	3094	129	5908	129	8672	130	1391	130	4070	130
28	199938	133	2808	134	5621	134	8385	135	1103	134	3781	135
29	9653	138	2522	138	5335	138	8097	140	0815	139	3492	140
30	9368	143	2236	143	5048	143	7809	144	0526	144	3203	144
31	199083	147	181950	148	164761	148	147522	149	130238	149	112914	149
32	8798	152	1664	153	4474	153	7234	153	129949	154	2625	154
33	8513	157	1377	157	4187	158	6946	158	9661	159	2336	159
34	8228	162	1091	162	3900	163	6659	163	9373	163	2047	164
35	7943	166	0805	167	3613	167	6371	168	9084	168	1758	169
36	7657	171	0519	172	3326	172	6083	172	8796	173	1469	174
37	7372	176	0233	176	3039	177	5795	177	8507	178	1180	179
38	7087	181	179947	181	2752	182	5508	182	8219	183	0891	184
39	6802	185	9661	186	2465	187	5220	187	7930	187	0602	189
40	6517	190	9375	191	2178	191	4932	192	7642	192	0313	193
41	196231	195	179088	195	161891	196	144644	196	127353	197	110023	198
42	5946	200	8802	200	1604	201	4356	201	7065	202	109734	203
43	5661	205	8516	205	1317	206	4068	206	6776	207	9445	208
44	5376	209	8230	210	1030	210	3781	211	6488	212	9156	212
45	5090	214	7944	214	0743	215	3493	215	6199	216	8867	217
46	4805	219	7657	219	0456	220	3205	220	5910	221	8578	222
47	4520	224	7371	224	0168	225	2917	225	5622	226	8289	227
48	4234	228	7085	229	159881	230	2629	230	5333	231	7999	231
49	3949	233	6798	234	9594	234	2341	235	5045	236	7710	236
50	3664	238	6512	238	9307	239	2053	240	4756	240	7421	241
51	193378	243	176226	243	159020	244	141765	244	124467	245	107132	246
52	3093	247	5940	248	8733	249	1477	249	4179	250	6843	250
53	2807	252	5653	253	8445	254	1189	254	3890	255	6553	255
54	2522	257	5367	257	8158	258	0901	259	3602	260	6264	260
55	2237	262	5080	262	7871	263	0613	264	3313	264	5975	265
56	1951	267	4794	267	7584	268	0325	268	3024	269	5686	270
57	1666	271	4508	272	7296	273	0037	273	2736	274	5396	275
58	1380	276	4221	276	7009	277	139749	278	2447	279	5107	279
59	1095	281	3935	281	6722	282	9461	283	2158	284	4818	284
60	0809	285	3648	286	6435	287	9173	287	1869	288	4529	289

′ / ″	84° Co-sine.	84° Parts for ″	85° Co-sine.	85° Parts for ″	86° Co-sine.	86° Parts for ″	87° Co-sine.	87° Parts for ″	88° Co-sine.	88° Parts for ″	89°. Co-sine.	89°. Parts for ″
0	104529	0	087156	0	069757	0	052336	0	034899	0	017452	0
1	4239	5	6866	5	9466	5	2046	5	4609	5	7162	5
2	3950	10	6576	10	9176	10	1755	10	4318	10	6871	10
3	3661	15	6286	15	8886	15	1465	15	4027	15	6580	15
4	3371	19	5997	19	8596	19	1174	19	3737	19	6289	19
5	3082	24	5707	24	8306	24	0884	24	3446	24	5998	24
6	2792	29	5417	29	8015	29	0593	29	3155	29	5707	29
7	2503	34	5127	34	7725	34	0302	34	2864	34	5417	34
8	2214	39	4837	39	7435	39	0012	39	2574	39	5126	39
9	1925	44	4547	44	7145	44	049721	44	2283	44	4835	44
10	1635	48	4258	48	6854	48	9431	48	1992	48	4544	49
11	101846	53	083968	53	066564	53	049140	53	031701	53	014253	53
12	1056	58	3678	58	6274	58	8850	58	1411	58	3962	58
13	0767	63	3388	63	5984	63	8559	63	1120	63	3671	63
14	0478	68	3098	68	5693	68	8269	68	0829	68	3381	68
15	0188	73	2808	73	5403	73	7978	73	0539	73	3090	73
16	099899	77	2518	77	5113	77	7688	77	0248	77	2799	78
17	9609	82	2228	82	4823	82	7397	82	029957	82	2508	83
18	9320	87	1939	87	4532	87	7107	87	9666	87	2217	87
19	9030	92	1649	92	4242	92	6816	92	9376	92	1926	92
20	8741	97	1359	97	3952	97	6525	97	9085	97	1635	97
21	098451	102	081069	102	063661	102	046235	102	028794	102	011344	102
22	8162	107	0779	107	3371	106	5944	106	8503	106	1054	107
23	7872	112	0489	112	3081	111	5654	111	8212	111	0763	112
24	7583	116	0199	116	2791	116	5363	116	7922	116	0472	116
25	7293	121	079909	121	2500	121	5072	121	7631	121	0181	121
26	7004	126	9619	126	2210	126	4782	126	7340	126	009890	126
27	6714	131	9329	131	1920	131	4491	131	7049	131	9599	131
28	6425	136	9039	136	1629	136	4201	136	6759	136	9308	136
29	6135	141	8749	141	1339	140	3910	140	6468	140	9017	141
30	5846	145	8459	145	1049	145	3619	145	6177	145	8726	145
31	095556	150	078169	150	060758	150	043329	150	025886	150	008436	150
32	5267	155	7879	155	0468	155	3038	155	5595	155	8145	155
33	4977	160	7589	160	0178	160	2748	160	5305	160	7854	160
34	4688	164	7299	164	059887	165	2457	165	5014	165	7563	165
35	4398	169	7009	169	9597	169	2166	169	4723	170	7272	170
36	4108	174	6719	174	9306	174	1876	174	4432	175	6981	175
37	3819	179	6429	179	9016	179	1585	179	4141	179	6690	179
38	3529	184	6139	184	8726	184	1294	184	3851	184	6400	184
39	3240	189	5849	189	8435	189	1004	189	3560	189	6109	189
40	2950	193	5559	193	8145	194	0713	194	3269	194	5818	194
41	092660	198	075269	198	057854	198	040422	198	022978	199	005527	199
42	2371	203	4979	203	7564	203	0132	203	2687	204	5236	204
43	2081	208	4689	208	7274	208	039841	208	2397	209	4945	209
44	1791	213	4399	213	6983	213	9551	213	2106	213	4654	213
45	1502	218	4109	218	6693	218	9260	218	1815	218	4363	218
46	1212	222	3818	222	6402	223	8969	223	1524	223	4072	223
47	0922	227	3528	227	6112	227	8679	227	1233	228	3782	228
48	0633	232	3238	232	5822	232	8388	232	0942	233	3491	233
49	0343	237	2948	237	5531	237	8097	237	0652	238	3200	238
50	0053	242	2658	242	5241	242	7807	242	0361	243	2909	243
51	089764	247	072368	247	054950	247	037516	247	020070	247	002618	247
52	9174	252	2078	252	4660	252	7225	252	019779	252	2327	252
53	9184	257	1788	257	4369	257	6934	257	9488	257	2036	257
54	8894	261	1497	261	4079	261	6644	261	9197	262	1745	262
55	8605	266	1207	266	3788	266	6353	266	8907	267	1454	267
56	8315	271	0917	271	3498	271	6062	271	8616	272	1164	272
57	8025	276	0627	276	3207	276	5772	276	8325	276	0873	276
58	7735	281	0837	281	2917	281	5481	281	8034	281	0582	281
59	7446	285	0047	285	2626	286	5190	286	7743	286	0291	286
60	7156	290	069757	290	2336	290	4899	290	7452	291	0000	291

TABLE 7.

s. "	h. m. 0° 0′	h. m. 0° 1′	h. m. 0° 2′	h. m. 0° 3′	h. m. 0° 4′	h. m. 0° 5′	h. m. 0° 6′	h. m. 0° 7′	h. m. 0° 8′	h. m. 0° 9′	h. m. 0° 10′
0		2·2553	1·9542	1·7782	1·6532	1·5563	1·4771	1·4102	1·3522	1·3010	1·2553
1	4·0334	481	506	757	514	549	759	091	513	002	545
2	3·7324	410	471	734	496	534	747	081	504	1·2994	538
3	5563	341	435	710	478	520	735	071	495	986	531
4	4314	272	400	686	460	506	723	061	486	978	524
5	3345	205	365	663	443	491	711	050	477	970	517
6	2553	139	331	639	425	477	699	040	468	962	510
7	1883	073	296	616	407	463	688	030	459	954	502
8	1303	009	262	593	390	449	676	020	450	946	1·2495
9	0792	2·1946	228	570	372	435	664	010	441	939	488
10	3·0334	2·1883	1·9195	1·7547	1·6355	1·5421	1·4652	1·4000	1·3432	1·2931	1·2481
11	2·9920	822	162	524	338	407	640	1·3989	423	923	474
12	9542	761	128	501	320	393	629	979	415	915	467
13	9195	701	096	479	303	379	617	969	406	907	460
14	8873	642	063	456	286	365	606	959	397	1·2899	453
15	8573	584	031	434	269	351	594	949	388	891	445
16	8293	526	1·8999	412	252	337	582	939	379	883	438
17	8030	469	967	390	235	324	571	929	371	876	431
18	7782	413	935	368	218	310	559	919	362	868	424
19	7547	358	904	346	201	296	548	910	353	860	417
20	2·7324	2·1303	1·8873	1·7324	1·6185	1·5283	1·4536	1·3900	1·3345	1·2852	1·2410
21	7112	249	842	302	168	269	525	890	336	845	403
22	6910	196	811	281	151	256	514	880	327	837	1·2396
23	6717	143	781	259	135	242	502	870	319	829	389
24	6532	091	751	238	118	229	491	860	310	821	382
25	6355	040	721	217	102	215	480	851	301	814	375
26	6185	2·0989	691	196	085	202	469	841	293	806	368
27	6021	939	661	175	069	189	457	831	284	1·2798	362
28	5863	889	632	154	053	175	446	821	276	791	355
29	5710	840	602	133	037	162	435	812	267	783	348
30	2·5563	2·0792	1·8573	1·7112	1·6021	1·5149	1·4424	1·3802	1·3259	1·2775	1·2341
31	5421	744	544	091	005	136	412	792	250	768	334
32	5283	696	516	071	1·5989	123	401	783	242	760	327
33	5149	649	487	050	973	110	390	773	233	753	320
34	5019	603	459	030	957	097	379	764	225	745	313
35	4894	557	431	010	941	084	368	754	216	738	307
36	4771	512	403	1·6990	925	071	357	745	208	730	300
37	4652	467	375	970	909	058	346	735	199	722	1·2293
38	4536	422	348	950	894	045	335	726	191	715	286
39	4424	378	320	930	878	032	325	716	183	707	279
40	2·4314	2·0334	1·8293	1·6910	1·5863	1·5019	1·4314	1·3707	1·3174	1·2700	1·2272
41	4206	291	266	890	847	007	303	697	166	1·2692	266
42	4102	248	239	871	832	1·4994	292	688	158	685	259
43	4000	206	212	851	816	981	281	678	149	678	252
44	3900	164	186	832	801	969	270	669	141	670	245
45	3802	122	159	812	786	956	260	660	133	663	239
46	3707	081	133	793	771	943	249	650	124	655	232
47	3613	040	107	774	755	931	238	641	116	648	225
48	3522	000	081	755	740	918	228	632	108	640	218
49	3432	1·9960	055	736	725	906	217	623	100	633	212
50	2·3345	1·9920	1·8030	1·6717	1·5710	1·4894	1·4206	1·3613	1·3091	1·2626	1·2205
51	3259	881	004	698	695	881	196	604	083	618	1·2198
52	3174	842	1·7979	679	680	869	185	595	075	611	192
53	3091	803	954	661	666	856	175	586	067	604	185
54	3010	765	929	642	651	844	164	576	059	1·2596	178
55	2931	727	904	624	636	832	154	567	051	589	172
56	2852	690	879	605	621	820	143	558	043	582	165
57	2775	652	855	587	607	808	133	549	034	574	159
58	2700	615	830	568	592	795	122	540	026	567	152
59	2626	579	806	550	578	783	112	531	018	560	145

s. ″	h. m. 0° 11′	h. m. 0° 12′	h. m. 0° 13′	h. m. 0° 14′	h. m. 0° 15′	h. m. 0° 16′	h. m. 0° 17′	h. m. 0° 18′	h. m. 0° 19′	h. m. 0° 20′	h. m. 0° 21′
0	1·2139	1·1761	1·1413	1·1091	1·0792	1·0512	1·0248	1·0000	9765	9542	9331
1	32	55	08	86	87	07	44	0·9996	61	39	27
2	26	49	02	81	82	02	40	92	58	35	24
3	19	43	1·1397	76	77	1·0498	35	88	54	32	20
4	13	37	91	71	73	93	31	84	50	28	17
5	06	31	86	66	68	89	27	80	46	24	13
6	1·2099	25	80	61	63	84	23	76	42	21	10
7	93	19	74	55	58	80	19	72	39	17	06
8	86	13	69	50	53	75	14	68	35	14	03
9	80	07	63	45	49	71	10	64	31	10	00
10	1·2073	1·1701	1·1358	1·1040	1·0744	1·0467	1·0206	9960	9727	9506	9296
11	67	1·1695	52	35	39	62	02	56	23	03	93
12	61	89	47	30	34	58	1·0197	52	20	9499	89
13	54	83	42	25	30	53	93	48	16	96	86
14	48	77	36	20	25	49	89	44	12	92	83
15	41	71	31	15	20	44	85	40	08	88	79
16	35	65	25	09	15	40	81	36	05	85	76
17	28	60	20	04	11	35	76	32	01	81	72
18	22	54	14	1·0999	06	31	72	28	9697	78	69
19	16	48	09	94	01	26	68	24	93	74	66
20	1·2009	1·1642	1·1303	1·0989	1·0696	1·0422	1·0164	9920	9690	9471	9262
21	03	36	1·1298	84	92	18	60	16	86	67	59
22	96	30	92	79	87	13	56	12	82	64	55
23	90	24	87	74	82	09	51	08	78	60	52
24	84	19	82	69	78	04	47	05	75	56	49
25	77	13	76	64	73	00	43	01	71	53	45
26	71	07	71	59	68	1·0395	39	9897	67	49	42
27	65	01	66	54	63	91	35	93	64	46	38
28	58	1·1595	60	49	59	87	31	89	60	42	35
29	52	89	55	44	54	82	26	85	56	39	32
30	1·1946	1·1584	1·1249	1·0939	1·0649	1·0378	1·0122	9881	9652	9435	9228
31	39	78	44	34	45	74	18	77	49	32	25
32	33	72	39	29	40	69	14	73	45	28	22
33	27	66	33	24	35	65	10	69	41	25	18
34	21	61	28	19	31	60	06	65	38	21	15
35	14	55	23	14	26	56	02	61	34	18	12
36	08	49	17	09	21	52	1·0098	58	30	14	08
37	02	43	12	04	17	47	93	54	26	11	05
38	1·1896	38	07	1·0899	12	43	89	50	23	07	01
39	89	32	01	94	08	39	85	46	19	04	9198
40	1·1883	1·1526	1·1196	1·0889	1·0603	1·0334	1·0081	9842	9615	9400	9195
41	77	20	91	84	1·0598	30	77	38	12	9397	91
42	71	15	86	80	94	26	73	34	08	93	88
43	65	09	80	75	89	21	69	30	04	90	85
44	58	03	75	70	85	17	65	27	01	86	81
45	52	1·1498	70	65	80	13	61	23	9597	83	78
46	46	92	64	60	75	08	57	19	93	79	75
47	40	86	59	55	71	04	53	15	90	76	72
48	34	81	54	50	66	00	49	11	86	72	69
49	28	75	49	45	62	1·0295	44	07	82	69	65
50	1·1822	1·1469	1·1143	1·0840	1·0557	1·0291	1·0040	9803	9579	9365	9162
51	16	64	38	35	52	87	36	00	75	62	58
52	09	58	33	31	48	82	32	9796	71	58	55
53	03	52	28	26	43	78	28	92	68	55	52
54	1·1797	47	23	21	39	74	24	88	64	51	48
55	91	41	17	16	34	70	20	84	61	48	45
56	85	36	12	11	30	65	16	80	57	44	42
57	79	30	07	06	25	61	12	77	53	41	38
58	73	24	02	01	21	57	08	73	50	37	35
59	67	19	1·1097	1·0797	16	52	04	69	46	34	32

s. ″	h. m. 0° 22′	h. m. 0° 23′	h. m. 0° 24′	h. m. 0° 25′	h. m. 0° 26′	h. m. 0° 27′	h. m. 0° 28′	h. m. 0° 29′	h. m. 0° 30′	h. m. 0° 31′	h. m. 0° 32′
0	9128	8935	8751	8573	8403	8239	8081	7929	7782	7639	7501
1	25	32	48	70	00	36	79	26	79	37	7499
2	22	29	45	68	8397	34	76	24	77	34	97
3	19	26	42	65	95	31	73	21	74	32	94
4	15	23	39	62	92	28	71	19	72	30	92
5	12	20	36	59	89	26	68	16	69	27	90
6	09	17	33	56	86	23	66	14	67	25	88
7	06	13	30	53	84	20	63	11	65	23	85
8	02	10	27	50	81	18	61	09	62	20	83
9	9099	07	24	47	78	15	58	06	60	18	81
10	9096	8904	8721	8544	8375	8212	8055	7904	7757	7616	7479
11	92	01	18	42	72	10	53	01	55	13	76
12	89	8898	15	39	70	07	50	7899	53	11	74
13	86	95	12	36	67	04	48	96	50	09	72
14	83	92	09	33	64	02	45	94	48	07	70
15	79	88	06	30	61	8199	43	91	45	04	67
16	76	85	03	27	59	96	40	89	43	02	65
17	73	82	00	24	56	94	37	87	41	00	63
18	70	79	8697	22	53	91	35	84	38	7597	61
19	66	76	94	19	50	88	32	82	36	95	58
20	9063	8873	8691	8516	8348	8186	8030	7879	7734	7593	7456
21	60	70	88	13	45	83	27	77	31	90	54
22	57	67	85	10	42	81	25	74	29	88	52
23	53	64	82	07	39	78	22	72	26	86	50
24	50	61	79	04	37	75	20	69	24	83	47
25	47	57	76	02	34	73	17	67	22	81	45
26	44	54	73	8499	31	70	14	64	19	79	43
27	41	51	70	96	28	67	12	62	17	77	41
28	37	48	67	93	26	65	09	59	14	74	38
29	34	45	64	90	23	62	07	57	12	72	36
30	9031	8842	8661	8487	8320	8159	8004	7855	7710	7570	7434
31	28	39	58	84	18	57	02	52	07	67	32
32	24	36	55	82	15	54	7999	50	05	65	29
33	21	33	52	79	12	52	97	47	03	63	27
34	18	30	49	76	09	49	94	45	00	60	25
35	15	27	46	73	07	46	92	42	7698	58	23
36	12	24	43	70	04	44	89	40	96	56	21
37	08	21	40	67	01	41	87	37	93	54	18
38	05	17	37	65	8298	38	84	35	91	51	16
39	02	14	35	62	96	36	81	32	88	49	14
40	8999	8811	8632	8459	8293	8133	7979	7830	7686	7547	7412
41	96	08	29	56	90	31	76	28	84	44	09
42	92	05	26	53	88	28	74	25	81	42	07
43	89	02	23	51	85	25	71	23	79	40	05
44	86	8799	20	48	82	23	69	20	77	38	03
45	83	96	17	45	79	20	66	18	74	35	01
46	80	93	14	42	77	17	64	15	72	33	7398
47	77	90	12	39	74	15	61	13	70	31	96
48	73	87	08	37	71	12	59	11	67	28	94
49	70	84	05	34	69	10	56	08	65	26	92
50	8967	8781	8602	8431	8266	8107	7954	7806	7663	7524	7390
51	64	78	8599	28	63	04	51	03	60	22	87
52	61	75	97	25	61	02	49	01	58	19	85
53	58	72	94	23	58	8099	46	7798	55	17	83
54	54	69	91	20	55	97	44	96	53	15	81
55	51	66	88	17	53	94	41	94	51	13	79
56	48	63	85	14	50	91	39	91	48	10	76
57	45	60	82	11	47	89	36	89	46	08	74
58	42	57	79	09	44	86	34	86	44	06	72
59	39	54	76	06	42	84	31	84	41	03	70

s. "	h. m. 0° 33′	h. m. 0° 34′	h. m. 0° 35′	h. m. 0° 36′	h. m. 0° 37′	h. m. 0° 38′	h. m. 0° 39′	h. m. 0° 40′	h. m. 0° 41′	h. m. 0° 42′	h. m. 0° 43′
0	7368	7238	7112	6990	6871	6755	6642	6532	6425	6320	6218
1	65	36	10	88	69	53	40	30	23	19	16
2	63	34	08	86	67	51	38	29	21	17	15
3	61	32	06	84	65	49	37	27	20	15	13
4	59	29	04	82	63	47	35	25	18	13	11
5	57	27	02	80	61	45	33	23	16	12	10
6	54	25	00	78	59	43	31	21	14	10	08
7	52	23	7098	76	57	42	29	19	13	08	06
8	50	21	96	74	55	40	27	18	11	06	05
9	48	19	93	72	53	38	25	16	09	05	03
10	7346	7217	7091	6970	6851	6736	6624	6514	6407	6303	6201
11	44	15	89	68	49	34	22	12	06	01	00
12	41	12	87	66	47	32	20	10	04	00	6198
13	39	10	85	64	45	30	18	09	02	6298	96
14	37	08	83	62	43	28	16	07	00	96	95
15	35	06	81	60	41	26	14	05	6398	94	93
16	33	04	79	58	40	25	12	03	97	93	91
17	30	02	77	56	38	23	11	01	95	91	90
18	28	00	75	54	36	21	09	00	93	89	88
19	26	7198	73	52	34	19	07	6498	91	88	86
20	7324	7196	7071	6950	6832	6717	6605	6496	6390	6286	6185
21	22	93	69	48	30	15	03	94	88	84	83
22	20	91	67	46	28	13	01	92	86	82	81
23	17	89	65	44	26	11	00	91	84	81	79
24	15	87	63	42	24	09	6598	89	83	79	78
25	13	85	61	40	22	08	96	87	81	77	76
26	11	83	59	38	20	06	94	85	79	76	74
27	09	81	57	36	18	04	92	84	77	74	73
28	07	79	55	34	16	02	90	82	76	72	71
29	04	77	52	32	14	00	89	80	74	71	69
30	7302	7175	7050	6930	6812	6698	6587	6478	6372	6269	6168
31	00	72	48	28	10	96	85	76	71	67	66
32	7298	70	46	26	09	94	83	75	69	65	65
33	96	68	44	24	07	92	81	73	67	64	63
34	94	66	42	22	05	91	79	71	65	62	61
35	91	64	40	20	03	89	78	69	64	60	60
36	89	62	38	18	01	87	76	67	62	59	58
37	87	60	36	16	6799	85	74	66	60	57	56
38	85	58	34	14	97	83	72	64	58	55	55
39	83	56	32	12	95	81	70	62	57	54	53
40	7281	7154	7030	6910	6793	6679	6568	6460	6355	6252	6151
41	79	52	28	08	91	77	67	59	53	50	50
42	76	49	26	06	89	76	65	57	51	48	48
43	74	47	24	04	87	74	63	55	50	47	46
44	72	45	22	02	85	72	61	53	48	45	45
45	70	43	20	00	84	70	59	51	46	43	43
46	68	41	18	6898	82	68	58	50	44	42	41
47	66	39	16	96	80	66	56	48	43	40	40
48	64	37	14	94	78	64	54	46	41	38	38
49	61	35	12	92	76	63	52	44	39	37	36
50	7259	7133	7010	6890	6774	6661	6550	6443	6338	6235	6135
51	57	31	08	88	72	59	48	41	36	33	33
52	55	29	06	86	70	57	47	39	34	32	31
53	53	27	04	84	68	55	45	37	32	30	30
54	51	24	02	82	66	53	43	35	31	28	28
55	49	22	00	81	64	51	41	34	29	26	26
56	46	20	6998	79	63	50	39	32	27	25	25
57	44	18	96	77	61	48	38	30	25	23	23
58	42	16	94	75	59	46	36	28	24	21	21
59	40	14	92	73	57	44	34	27	22	20	20

s. "	h. m. 0° 44′	h. m. 0° 45′	h. m. 0° 46′	h. m. 0° 47′	h. m. 0° 48′	h. m. 0° 49′	h. m. 0° 50′	h. m. 0° 51′	h. m. 0° 52′	h. m. 0° 53′	h. m. 0° 54′
0	6118	6021	5925	5832	5740	5651	5563	5477	5393	5310	5229
1	17	19	24	30	39	49	62	76	91	09	27
2	15	17	22	29	37	48	60	74	90	07	26
3	13	16	20	27	36	46	59	73	89	06	25
4	12	14	19	26	34	45	57	71	87	05	23
5	10	13	17	24	33	43	56	70	86	03	22
6	08	11	16	23	31	42	54	69	84	02	21
7	07	09	14	21	30	40	53	67	83	00	19
8	05	08	13	19	28	39	51	66	82	5299	18
9	03	06	11	18	27	37	50	64	80	98	17
10	6102	6005	5909	5816	5725	5636	5549	5463	5379	5296	5215
11	00	03	08	15	24	35	47	61	77	95	14
12	6099	01	06	13	22	33	46	60	76	94	13
13	97	00	05	12	21	32	44	59	75	92	11
14	95	5998	03	10	19	30	43	57	73	91	10
15	94	97	02	09	18	29	41	56	72	90	09
16	92	95	00	07	16	27	40	54	70	88	07
17	90	93	5898	06	15	26	38	53	69	87	06
18	89	92	97	04	13	24	37	52	68	85	05
19	87	90	95	03	12	23	36	50	66	84	03
20	6085	5989	5894	5801	5710	5621	5534	5449	5365	5283	5202
21	84	87	92	00	09	20	33	47	64	81	01
22	82	85	91	5798	07	18	31	46	62	80	5199
23	81	84	89	96	06	17	30	45	61	79	98
24	79	82	88	95	04	15	28	43	59	77	97
25	77	81	86	93	03	14	27	42	58	76	96
26	76	79	84	92	01	13	26	40	57	75	94
27	74	77	83	90	00	11	24	39	55	73	93
28	72	76	81	89	5698	10	22	37	54	72	91
29	71	74	80	87	97	08	21	36	53	71	90
30	6069	5973	5878	5786	5695	5607	5520	5435	5351	5269	5189
31	67	71	77	84	94	05	18	33	50	68	87
32	66	69	75	83	92	04	17	32	48	66	86
33	64	68	74	81	91	02	16	30	47	65	85
34	63	66	72	80	89	01	14	29	46	64	83
35	61	65	70	78	88	5599	13	28	44	62	82
36	59	63	69	77	86	98	11	26	43	61	81
37	58	61	67	75	85	96	10	25	41	60	79
38	56	60	66	74	83	95	08	23	40	58	78
39	55	58	64	72	82	94	07	22	39	57	77
40	6053	5957	5863	5771	5680	5592	5506	5421	5337	5256	5175
41	51	55	61	69	79	91	04	19	36	54	74
42	50	54	60	68	77	89	03	18	35	53	73
43	48	52	58	66	76	88	01	16	33	52	72
44	46	50	56	65	74	86	00	15	32	50	70
45	45	49	55	63	73	85	5498	14	31	49	69
46	43	47	53	61	71	83	97	12	29	48	68
47	42	46	52	60	70	82	96	11	28	46	66
48	40	44	50	58	69	80	94	09	26	45	65
49	38	42	49	57	67	79	93	08	25	44	64
50	6037	5941	5847	5755	5666	5578	5491	5407	5324	5242	5162
51	35	39	46	54	64	76	90	05	22	41	61
52	33	38	44	52	63	75	88	04	21	40	60
53	32	36	43	51	61	73	87	02	20	38	58
54	30	35	41	49	60	72	86	01	18	37	57
55	29	33	39			70	84	00	17	35	56
56			38			69	83	5398	15	34	54
57			36			67	81	97	14	33	53
58	27	31	35	43	58	66	80	95	13	31	52
59	[illegible]	[illegible]	33	46	[illegible]	64	78	94	11	30	50

s. ″	h. m. 0° 55′	h. m. 0° 56′	h. m. 0° 57′	h. m. 0° 58′	h. m. 0° 59′	h. m. 1° 0′	h. m. 1° 1′	h. m. 1° 2′	h. m. 1° 3′	h. m. 1° 4′	h. m. 1° 5′
0	5149	5071	4994	4918	4844	4771	4699	4629	4559	4491	4424
1	48	70	93	17	43	70	98	28	58	90	22
2	46	68	91	16	42	69	97	26	57	89	21
3	45	67	90	15	41	68	96	25	56	88	20
4	44	66	89	13	39	66	95	24	55	86	19
5	43	64	88	12	38	65	93	23	54	85	18
6	41	63	86	11	37	64	92	22	52	84	17
7	40	62	85	10	36	63	91	21	51	83	16
8	39	61	84	08	34	62	90	19	50	82	15
9	37	59	83	07	33	60	89	18	49	81	14
10	5136	5058	4981	4906	4832	4759	4688	4617	4548	4480	4412
11	35	57	80	05	31	58	86	16	47	79	11
12	33	55	79	03	30	57	85	15	46	77	10
13	32	54	77	02	28	56	84	14	44	76	09
14	31	53	76	01	27	54	83	12	43	75	08
15	29	51	75	00	26	53	82	11	42	74	07
16	28	50	74	4899	25	52	80	10	41	73	06
17	27	49	72	97	23	51	79	09	40	72	05
18	25	48	71	96	22	50	78	08	39	71	04
19	24	46	70	95	21	48	77	07	38	69	02
20	5123	5045	4969	4894	4820	4747	4676	4606	4536	4468	4401
21	22	44	67	92	19	46	75	04	35	67	00
22	20	43	66	91	17	45	73	03	34	66	4399
23	19	41	65	90	16	44	72	02	33	65	98
24	18	40	64	89	15	42	71	01	32	64	97
25	16	39	62	87	14	41	70	00	31	63	96
26	15	37	61	86	12	40	69	4599	30	62	95
27	14	36	60	85	11	39	68	97	28	60	94
28	12	35	59	84	10	38	66	96	27	59	93
29	11	34	57	82	09	36	65	95	26	58	91
30	5110	5032	4956	4881	4808	4735	4664	4594	4525	4457	4390
31	08	31	55	80	06	34	63	93	24	56	89
32	07	30	54	79	05	33	62	92	23	55	88
33	06	28	52	77	04	32	60	90	22	54	87
34	05	27	51	76	03	30	59	89	20	53	86
35	03	26	50	75	01	29	58	88	19	52	85
36	02	25	49	74	00	28	57	87	18	50	84
37	01	23	47	73	4799	27	56	86	17	49	83
38	5099	22	46	71	98	26	55	85	16	48	81
39	98	21	45	70	97	24	53	84	15	47	80
40	5097	5019	4943	4869	4795	4723	4652	4582	4514	4446	4379
41	95	18	42	68	94	22	51	81	12	45	78
42	94	17	41	66	93	21	50	80	11	44	77
43	93	16	40	65	92	20	49	79	10	43	76
44	92	14	38	64	91	18	48	78	09	41	75
45	90	13	37	63	89	17	46	77	08	40	74
46	89	12	36	61	88	16	45	75	07	39	73
47	88	11	35	60	87	15	44	74	06	38	72
48	86	09	33	59	86	14	43	73	05	37	70
49	85	08	32	58	85	12	42	72	03	36	69
50	5084	5007	4931	4856	4783	4711	4640	4571	4502	4435	4368
51	82	05	30	55	82	10	39	70	01	34	67
52	81	04	28	54	81	09	38	69	00	33	66
53	80	03	27	53	80	08	37	67	4499	31	65
54	79	02	26	52	78	07	36	66	98	30	64
55	77	00	25	50	77	05	35	65	97	29	63
56	76	4999	23	49	76	04	33	64	95	28	62
57	75	98	22	48	75	03	32	63	94	27	61
58	73	97	21	47	74	02	31	62	93	26	59
59	72	95	20	45	72	01	30	60	92	25	58

s. "	h. m. 1° 6′	h. m. 1° 7′	h. m. 1° 8′	h. m. 1° 9′	h. m. 1° 10′	h. m. 1° 11′	h. m. 1° 12′	h. m. 1° 13′	h. m. 1° 14′	h. m. 1° 15′	h. m. 1° 16′
0	4357	4292	4228	4164	4102	4040	3979	3919	3860	3802	3745
1	56	91	27	63	01	39	78	19	59	01	44
2	55	90	26	62	00	38	77	18	58	00	43
3	54	89	24	61	4099	37	76	17	57	3799	42
4	53	88	23	60	98	36	75	16	56	98	41
5	52	87	22	59	97	35	74	15	56	97	40
6	51	85	21	58	96	34	73	14	55	96	39
7	50	84	20	57	95	33	72	13	54	95	38
8	49	83	19	56	93	32	71	12	53	94	37
9	47	82	18	55	92	31	70	11	52	93	36
10	4346	4281	4217	4154	4091	4030	3969	3910	3851	3792	3735
11	45	80	16	53	90	29	68	09	50	92	34
12	44	79	15	52	89	28	67	08	49	91	33
13	43	78	14	51	88	27	66	07	48	90	32
14	42	77	13	50	87	26	65	06	47	89	31
15	41	76	12	49	86	25	64	05	46	88	30
16	40	75	11	47	85	24	63	04	45	87	29
17	39	74	10	46	84	23	62	03	44	86	28
18	38	73	09	45	83	22	61	02	43	85	27
19	36	71	07	44	82	21	60	01	42	84	27
20	4335	4270	4206	4143	4081	4020	3959	3900	3841	3783	3726
21	34	69	05	42	80	19	58	3899	40	82	25
22	33	68	04	41	79	18	57	98	39	81	24
23	32	67	03	40	78	17	56	97	38	80	23
24	31	66	02	39	77	16	55	96	37	79	22
25	30	65	01	38	76	15	54	95	36	78	21
26	29	64	00	37	75	14	53	94	35	77	20
27	28	63	4199	36	74	13	52	93	34	76	19
28	27	62	98	35	73	12	51	92	33	75	18
29	26	61	97	34	72	11	50	91	32	74	17
30	4325	4260	4196	4133	4071	4010	3949	3890	3831	3773	3716
31	23	59	95	32	70	09	48	89	30	72	15
32	22	58	94	31	69	08	47	88	29	71	14
33	21	56	93	30	68	07	46	87	28	70	13
34	20	55	92	29	67	06	45	86	27	69	12
35	19	54	91	28	66	05	44	85	26	68	11
36	18	53	89	27	65	04	43	84	25	68	10
37	17	52	88	26	64	03	42	83	24	67	09
38	16	51	87	25	63	02	41	82	23	66	09
39	15	50	86	24	62	01	40	81	22	65	08
40	4314	4249	4185	4122	4061	4000	3939	3880	3821	3764	3707
41	13	48	84	21	60	3999	38	79	20	63	06
42	11	47	83	20	59	98	37	78	20	62	05
43	10	46	82	19	58	97	36	77	19	61	04
44	09	45	81	18	56	96	35	76	18	60	03
45	08	44	80	17	55	95	34	75	17	59	02
46	07	43	79	16	54	93	33	74	16	58	01
47	06	41	78	15	53	92	32	73	15	57	00
48	05	40	77	14	52	91	31	72	14	56	3699
49	04	39	76	13	51	90	30	71	13	55	98
50	4303	4238	4175	4112	4050	3989	3929	3870	3812	3754	3697
51	02	37	74	11	49	88	28	69	11	53	96
52	01	36	73	10	48	87	27	68	10	52	95
53	00	35	72	09	47	86	26	67	09	51	94
54	4298	34	71	08	46	85	25	66	08	50	93
55	97	33	69	07	45	84	24	65	07	49	93
56	96	32	68	06	44	83	23	64	06	48	92
57	95	31	67	05	43	82	22	63	05	47	91
58	94	30	66	04	42	81	21	62	04	46	90
59	93	29	65	03	41	80	20	61	03	46	89

s. ″	h. m. 1° 17′	h. m. 1° 18′	h. m. 1° 19′	h. m. 1° 20′	h. m. 1° 21′	h. m. 1° 22′	h. m. 1° 23′	h. m. 1° 24′	h. m. 1° 25′	h. m. 1° 26′	h. m. 1° 27′
0	3688	3632	3576	3522	3468	3415	3362	3310	3259	3208	3158
1	87	31	76	21	67	14	61	09	58	07	57
2	86	30	75	20	66	13	60	08	57	06	56
3	85	29	74	19	65	12	59	07	56	05	55
4	84	28	73	18	64	11	58	06	55	04	54
5	83	27	72	17	63	10	58	06	54	04	53
6	82	26	71	16	63	09	57	05	53	03	53
7	81	25	70	15	62	08	56	04	53	02	52
8	80	24	69	14	61	08	55	03	52	01	51
9	79	23	68	14	60	07	54	02	51	00	50
10	3678	3623	3567	3513	3459	3406	3353	3301	3250	3199	3149
11	77	22	66	12	58	05	52	00	49	98	48
12	77	21	65	11	57	04	51	00	48	98	48
13	76	20	65	10	56	03	51	3299	47	97	47
14	75	19	64	09	55	02	50	98	47	96	46
15	74	18	63	08	54	01	49	97	46	95	45
16	73	17	62	07	54	00	48	96	45	94	44
17	72	16	61	06	53	00	47	95	44	93	43
18	71	15	60	06	52	3399	46	94	43	93	43
19	70	14	59	05	51	98	45	94	42	92	42
20	3669	3613	3558	3504	3450	3397	3345	3293	3242	3191	3141
21	68	12	57	03	49	96	44	92	41	90	40
22	67	11	56	02	48	95	43	91	40	89	39
23	66	10	55	01	47	94	42	90	39	88	38
24	65	10	55	00	46	93	41	89	38	88	38
25	64	09	54	3499	46	93	40	88	37	87	37
26	63	08	53	98	45	92	39	88	36	86	36
27	63	07	52	97	44	91	38	87	36	85	35
28	62	06	51	97	43	90	38	86	35	84	34
29	61	05	50	96	42	89	37	85	34	83	33
30	3660	3604	3549	3495	3441	3388	3336	3284	3233	3183	3133
31	59	03	48	94	40	87	35	83	32	82	32
32	58	02	47	93	39	86	34	82	31	81	31
33	57	01	46	92	38	86	33	82	31	80	30
34	56	00	45	91	38	85	32	81	30	79	29
35	55	3599	45	90	37	84	32	80	29	78	29
36	54	98	44	89	36	83	31	79	28	78	28
37	53	98	43	88	35	82	30	78	27	77	27
38	52	97	42	88	34	81	29	77	26	76	26
39	51	96	41	87	33	80	28	76	25	75	25
40	3650	3595	3540	3486	3432	3379	3327	3276	3225	3174	3124
41	49	94	39	85	31	79	26	75	24	73	24
42	49	93	38	84	31	78	25	74	23	73	23
43	48	92	37	83	30	77	25	73	22	72	22
44	47	91	36	82	29	76	24	72	21	71	21
45	46	90	35	81	28	75	23	71	20	70	20
46	45	89	35	80	27	74	22	70	20	69	19
47	44	88	34	80	26	73	21	70	19	68	19
48	43	87	33	79	25	72	20	69	18	68	18
49	42	87	32	78	24	72	19	68	17	67	17
50	3641	3586	3531	3477	3423	3371	3319	3267	3216	3166	3116
51	40	85	30	76	23	70	18	66	15	65	15
52	39	84	29	75	22	69	17	65	14	64	14
53	38	83	28	74	21	68	16	65	14	63	14
54	37	82	27	73	20	67	15	64	13	63	13
55	36	81	26	72	19	66	14	63	12		12
56	35	80	25	71	18	65	13	62	11		11
57	35	79	25	71	17	65	13	61	10		10
58	34	78	24	70	16	64	12	60	09	62	10
59	33	77	23	69	15	63	11	59	09	[illegible]	09

s. "	h. m. 1° 28′	h. m. 1° 29′	h. m. 1° 30′	h. m. 1° 31′	h. m. 1° 32′	h. m. 1° 33′	h. m. 1° 34′	h. m. 1° 35′	h. m. 1° 36′	h. m. 1° 37′	h. m. 1° 38′
0	3108	3059	3010	2962	2915	2868	2821	2775	2730	2685	2640
1	07	58	09	62	14	67	21	75	29	84	40
2	06	57	09	61	13	66	20	74	29	84	39
3	05	56	08	60	12	66	19	73	28	83	38
4	05	56	07	59	12	65	18	72	27	82	38
5	04	55	06	58	11	64	18	72	26	81	37
6	03	54	05	58	10	63	17	71	25	81	36
7	02	53	05	57	09	62	16	70	25	80	35
8	01	52	04	56	09	62	15	69	24	79	35
9	01	52	03	55	08	61	15	69	23	78	34
10	3100	3051	3002	2954	2907	2860	2814	2768	2722	2678	2633
11	3099	50	01	54	06	59	13	67	22	77	32
12	98	49	01	53	05	59	12	66	21	76	32
13	97	48	00	52	05	58	11	66	20	75	31
14	96	47	2999	51	04	57	11	65	19	75	30
15	96	47	98	50	03	56	10	64	19	74	29
16	95	46	97	50	02	55	09	63	18	73	29
17	94	45	97	49	01	55	08	63	17	72	28
18	93	44	96	48	01	54	08	62	16	72	27
19	92	43	95	47	00	53	07	61	16	71	26
20	3091	3043	2994	2946	2899	2852	2806	2760	2715	2670	2626
21	91	42	93	46	98	52	05	60	14	69	25
22	90	41	93	45	98	51	05	59	13	69	24
23	89	40	92	44	97	50	04	58	13	68	24
24	88	39	91	43	96	49	03	57	12	67	23
25	87	39	90	42	95	48	02	56	11	66	22
26	87	38	89	42	94	48	01	56	10	66	21
27	86	37	89	41	94	47	01	55	10	65	21
28	85	36	88	40	93	46	00	54	09	64	20
29	84	35	87	39	92	45	2799	53	08	63	19
30	3083	3034	2986	2939	2891	2845	2798	2753	2707	2663	2618
31	82	34	85	38	91	44	98	52	07	62	18
32	82	33	85	37	90	43	97	51	06	61	17
33	81	32	84	36	89	42	96	50	05	60	16
34	80	31	83	35	88	42	95	50	04	60	15
35	79	30	82	35	87	41	95	49	04	59	15
36	78	30	81	34	87	40	94	48	03	58	14
37	78	29	81	33	86	39	93	47	02	57	13
38	77	28	80	32	85	38	92	47	01	57	12
39	76	27	79	31	84	38	92	46	01	56	12
40	3075	3026	2978	2931	2883	2837	2791	2745	2700	2655	2611
41	74	26	77	30	83	36	90	44	2699	55	10
42	73	25	77	29	82	35	89	44	98	54	10
43	73	24	76	28	81	35	88	43	98	53	09
44	72	23	75	27	80	34	88	42	97	52	08
45	71	22	74	27	80	33	87	41	96	52	07
46	70	22	73	26	79	32	86	41	95	51	07
47	69	21	73	25	78	31	85	40	95	50	06
48	69	20	72	24	77	31	85	39	94	49	05
49	68	19	71	24	76	30	84	38	93	49	04
50	3067	3018	2970	2923	2876	2829	2783	2738	2692	2648	2604
51	66	18	69	22	75	28	82	37	92	47	03
52	65	17	69	21	74	28	82	36	91	46	02
53	65	16	68	20	73	27	81	35	90	46	01
54	64	15	67	20	73	26	80	35	89	45	01
55		14		19	72	25	79	34	89	44	00
56		14		18	71	25	79	33	88	43	2599
57		13		17	70	24	78	32	87	43	
58	63	12	66	16	69	23	77	32	87	42	99
59	[illegible]	11	[illegible]	16	69	22	76	31	86	41	98

s.	h. m. 1° 39′	h. m. 1° 40′	h. m. 1° 41′	h. m. 1° 42′	h. m. 1° 43′	h. m. 1° 44′	h. m. 1° 45′	h. m. 1° 46′	h. m. 1° 47′	h. m. 1° 48′	h. m. 1° 49′
0	2596	2553	2510	2467	2424	2382	2341	2300	2259	2218	2178
1	96	52	09	66	24	82	40	2299	58	18	78
2	95	51	08	65	23	81	39	98	58	17	77
3	94	51	07	65	22	80	39	98	57	16	76
4	93	50	07	64	22	80	38	97	56	16	76
5	93	49	06	63	21	79	37	96	56	15	75
6	92	48	05	62	20	78	37	96	55	14	74
7	91	48	04	62	19	78	36	95	54	14	74
8	91	47	04	61	19	77	35	94	53	13	73
9	90	46	03	60	18	76	35	94	53	12	72
10	2589	2545	2502	2460	2417	2375	2334	2293	2252	2212	2172
11	88	45	02	59	17	75	33	92	51	11	71
12	88	44	01	58	16	74	33	91	51	10	70
13	87	43	00	58	15	73	32	91	50	10	70
14	86	43	2499	57	15	73	31	90	49	09	69
15	85	42	99	56	14	72	31	89	49	08	69
16	85	41	98	55	13	71	30	89	48	08	68
17	84	40	97	55	12	71	29	88	47	07	67
18	83	40	97	54	12	70	28	87	47	06	67
19	83	39	96	53	11	69	28	87	46	06	66
20	2582	2538	2495	2453	2410	2368	2327	2286	2245	2205	2165
21	81	38	94	52	10	68	26	85	45	04	65
22	80	37	94	51	09	67	26	85	44	04	64
23	80	36	93	50	08	66	25	84	43	03	63
24	79	35	92	50	08	66	24	83	43	02	63
25	78	35	92	49	07	65	24	83	42	02	62
26	77	34	91	48	06	64	23	82	41	01	61
27	77	33	90	48	05	64	22	81	41	00	61
28	76	33	89	47	05	63	22	81	40	00	60
29	75	32	89	46	04	62	21	80	39	2199	59
30	2574	2531	2488	2445	2403	2362	2320	2279	2239	2198	2159
31	74	30	87	45	03	61	20	79	38	98	58
32	73	30	87	44	02	60	19	78	37	97	57
33	72	29	86	43	01	59	18	77	37	96	57
34	72	28	85	43	01	59	17	77	36	96	56
35	71	27	85	42	00	58	17	76	35	95	55
36	70	27	84	41	2399	57	16	75	35	94	55
37	69	26	83	41	98	57	15	74	34	94	54
38	69	25	82	40	98	56	15	74	33	93	53
39	68	25	82	39	97	55	14	73	33	92	53
40	2567	2524	2481	2438	2396	2355	2313	2272	2232	2192	2152
41	66	23	80	38	96	54	13	72	31	91	51
42	66	22	80	37	95	53	12	71	31	90	51
43	65	22	79	36	94	53	11	70	30	90	50
44	64	21	78	36	94	52	11	70	29	89	49
45	64	20	77	35	93	51	10	69	29	88	49
46	63	20	77	34	92	50	09	68	28	88	48
47	62	19	76	33	91	50	09	68	27	87	47
48	61	18	75	33	91	49	08	67	27	86	47
49	61	17	75	32	90	48	07	66	26	86	46
50	2560	2517	2474	2431	2389	2348	2307	2266	2225	2185	2145
51	59	16	73	31	89	47	06	65	25	84	45
52	59	15	72	30	88	46	05	64	24	84	44
53	58	15	72	29	87	46	04	64	23	83	43
54	57	14	71	29	87	45	04	63	23	82	43
55	56	13	70	28	86	44	03	62	22	82	42
56	56	12	70	27	85	44	02	62	21	81	41
57	55	12	69	26	84	43	02	61	20	80	41
58	54	11	68	26	84	42	01	60	20	80	40
59	53	10	67	25	83	42	00	60	19	79	39

s.	h. m. 1° 50′	h. m. 1° 51′	h. m. 1° 52′	h. m. 1° 53′	h. m. 1° 54′	h. m. 1° 55′	h. m. 1° 56′	h. m. 1° 57′	h. m. 1° 58′	h. m. 1° 59′	h. m. 2° 0′
0	2139	2099	2061	2022	1984	1946	1908	1871	1834	1797	1761
1	38	99	60	21	83	45	08	70	33	97	60
2	37	98	59	21	82	44	07	70	33	96	60
3	37	98	59	20	82	44	06	69	32	95	59
4	36	97	58	19	81	43	06	68	31	95	59
5	36	96	57	19	81	43	05	68	31	94	58
6	35	96	57	18	80	42	04	67	30	94	57
7	34	95	56	17	79	41	04	67	30	93	57
8	34	94	55	17	79	41	03	66	29	92	56
9	33	94	55	16	78	40	03	65	28	92	55
10	2132	2093	2054	2016	1977	1939	1902	1865	1828	1791	1755
11	32	92	53	15	77	39	01	64	27	91	54
12	31	92	53	14	76	38	01	63	27	90	54
13	30	91	52	14	75	38	00	63	26	89	53
14	30	90	52	13	75	37	1899	62	25	89	52
15	29	90	51	12	74	36	99	62	25	88	52
16	28	89	50	12	74	36	98	61	24	88	51
17	28	88	50	11	73	35	98	60	23	87	51
18	27	88	49	10	72	34	97	60	23	86	50
19	26	87	48	10	72	34	96	59	22	86	49
20	2126	2086	2048	2009	1971	1933	1896	1859	1822	1785	1749
21	25	86	47	09	70	33	95	58	21	85	48
22	24	85	46	08	70	32	94	57	20	84	48
23	24	85	46	07	69	31	94	57	20	83	47
24	23	84	45	07	68	31	93	56	19	83	46
25	22	83	44	06	68	30	93	55	19	82	46
26	22	83	44	05	67	29	92	55	18	81	45
27	21	82	43	05	67	29	91	54	17	81	45
28	20	81	42	04	66	28	91	54	17	80	44
29	20	81	42	03	65	28	90	53	16	80	43
30	2119	2080	2041	2003	1965	1927	1889	1852	1816	1779	1743
31	18	79	41	02	64	26	89	52	15	78	42
32	18	79	40	01	63	26	88	51	14	78	42
33	17	78	39	01	63	25	88	50	14	77	41
34	16	77	39	00	62	24	87	50	13	77	40
35	16	77	38	00	62	24	86	49	12	76	40
36	15	76	37	1999	61	23	86	49	12	75	39
37	15	75	37	98	60	23	85	48	11	75	39
38	14	75	36	98	60	22	84	47	11	74	38
39	13	74	35	97	59	21	84	47	10	74	37
40	2113	2073	2035	1996	1958	1921	1883	1846	1809	1773	1737
41	12	73	34	96	58	20	83	46	09	72	36
42	11	72	33	95	57	19	82	45	08	72	36
43	11	72	33	94	56	19	81	44	08	71	35
44	10	71	32	94	56	18	81	44	07	71	34
45	09	70	32	93	55	18	80	43	06	70	34
46	09	70	31	93	55	17	80	43	06	69	33
47	08	69	30	92	54	16	79	42	05	69	33
48	07	68	30	91	53	16	78	41	05	68	32
49	07	68	29	91	53	15	78	41	04	68	31
50	2106	2067	2028	1990	1952	1914	1877	1840	1803	1767	1731
51	05	66	28	89	51	14	76	39	03	66	30
52	05	66	27	89	51	13	76	39	02	66	30
53	04	65	26	88	50	13	75	38	02	65	29
54	03	64	26	87	50	12	75	38	01	65	28
55	03	64	25	87	49	11	74	37	00	64	28
56	02	63	25	86	48	11	73	36	00	63	27
57	01	62	24	86	48	10	73	36	1799	63	27
58	01	62	23	85	47	09	72	35	98	62	26
59	00	61	23	84	46	09	71	35	98	62	25

s. "	h. m. 2° 1′	h. m. 2° 2′	h. m. 2° 3′	h. m. 2° 4′	h. m. 2° 5′	h. m. 2° 6′	h. m. 2° 7′	h. m. 2° 8′	h. m. 2° 9′	h. m. 2° 10′	h. m. 2° 11′
0	1725	1689	1654	1619	1584	1549	1515	1481	1447	1413	1380
1	24	89	53	18	83	48	14	80	46	13	79
2	24	88	52	17	82	48	14	79	46	12	79
3	23	87	52	17	82	47	13	79	45	12	78
4	22	87	51	16	81	47	12	78	45	11	78
5	22	86	51	16	81	46	12	78	44	11	77
6	21	86	50	15	80	46	11	77	43	10	77
7	21	85	50	14	80	45	11	77	43	09	76
8	20	84	49	14	79	44	10	76	42	09	76
9	19	84	48	13	78	44	10	76	42	08	75
10	1719	1683	1648	1613	1578	1543	1509	1475	1441	1408	1374
11	18	83	47	12	77	43	08	74	41	07	74
12	18	82	47	12	77	42	08	74	40	07	73
13	17	81	46	11	76	42	07	73	40	06	73
14	17	81	45	10	76	41	07	73	39	06	72
15	16	80	45	10	75	40	06	72	38	05	72
16	15	80	44	09	74	40	06	72	38	04	71
17	15	79	44	09	74	39	05	71	37	04	71
18	14	78	43	08	73	39	04	70	37	03	70
19	14	78	43	07	73	38	04	70	36	03	70
20	1713	1677	1642	1607	1572	1538	1503	1469	1436	1402	1369
21	12	77	41	06	71	37	03	69	35	02	68
22	12	76	41	06	71	36	02	68	35	01	68
23	11	76		05	70	36	02	68	34	01	67
24	11	75			70	35	01	67	33	00	67
25	10	74			69	35	00	67	33	1399	66
26	09	74			69	34	00	66	32	99	66
27	09	73			68	34	1499	65	32	98	65
28	08	73	40	05	67	33	99	65	31	98	65
29	08	72	[illegible]	04	67	32	98	64	31	97	64
30	1707	1671	1636	1601	1566	1532	1498	1464	1430	1397	1363
31	06	71		00	66	31	97	63	29	96	63
32	06	70		00	65	31	96	63	29	96	62
33	05	70		1599	65	30	96	62	28	95	62
34	05	69		99	64	30	95	61	28	94	61
35	04	68		98	63	29	95	61	27	94	61
36	03	68		98	63	28	94	60	27	93	60
37	03	67		97	62	28	94	60	26	93	60
38	02	67		96	62	27	93	59	26	92	59
39	02	66	35	96	61	27	93	59	25	92	59
40	1701	1665	1630	1595	1561	1526	1492	1458	1424	1391	1358
41	00	65	30	95	60	26	91	58	24	91	57
42	00	64	29	94	59	25	91	57	23	90	57
43	1699	64	28	93	59	24	90	56	23	89	56
44	99	63	28	93	58	24	90	56	22	89	56
45	98	63	27	92	58	23	89	55	22	88	55
46	97	62	27	92	57	23	89	55	21	88	55
47	97	61	26	91	56	22	88	54	21	87	54
48	96	61	26	91	56	22	87	54	20	87	54
49	96	60	25	90	55	21	87	53	19	86	53
50	1695	1660	1624	1589	1555	1520	1486	1452	1419	1386	1352
51	94	59	24	89	54	20	86	52	18	85	52
52	94	58	23	88	54	19	85	51	18	84	51
53	93	58	23	88	53	19	85	51	17	84	51
54	93	57	22	87	52	18	84		17	83	50
55	92	57	21	87	52	18	83		16	83	50
56	92	56	21	86	51	17	83		16	82	49
57	91	55	20	85	51	16	82		15	82	49
58	90	55	20	85	50	16	82	50	14	81	48
59	90	54	19	84	50	15	81	[illegible]	14	81	48

s. "	h. m. 2° 12′	h. m. 2° 13′	h. m. 2° 14′	h. m. 2° 15′	h. m. 2° 16′	h. m. 2° 17′	h. m. 2° 18′	h. m. 2° 19′	h. m. 2° 20′	h. m. 2° 21′	h. m. 2° 22′
0	1347	1314	1282	1249	1217	1186	1154	1123	1091	1061	1030
1	46	14	81	49	17	85	53	22	91	60	29
2	46	13	81	48	16	84	53	22	90	60	29
3	45	13	80	48	16	84	52	21	90	59	28
4	45	12	80	47	15	83	52	20	89	58	28
5	44	11	79	47	15	83	51	20	89	58	27
6	44	11	78	46	14	82	51	19	88	57	27
7	43	10	78	46	14	82	50	19	88	57	26
8	43	10	77	45	13	81	50	18	87	56	26
9	42	09	77	45	13	81	49	18	87	56	25
10	1342	1309	1276	1244	1212	1180	1149	1117	1086	1055	1025
11	41	08	76	43	11	80	48	17	86	55	24
12	40	08	75	43	11	79	48	16	85	54	24
13	40	07	75	42	10	79	47	16	85	54	23
14	39	07	74	42	10	78	47	15	84	53	23
15	39	06	74	41	09	78	46	15	84	53	22
16	38	06	73	41	09	77	46	14	83	52	22
17	38	05	73	40	08	77	45	14	83	52	21
18	37	04	72	40	08	76	45	13	82	51	21
19	37	04	71	39	07	75	44	13	82	51	20
20	1336	1303	1271	1239	1207	1175	1143	1112	1081	1050	1020
21	35	03	70	38	06	74	43	12	81	50	19
22	35	02	70	38	06	74	42	11	80	49	19
23	34	02	69	37	05	73	42	11	80	49	18
24	34	01	69	37	05	73	41	10	79	48	18
25	33	01	68	36	04	72	41	10	79	48	17
26	33	00	68	35	04	72	40	09	78	47	17
27	32	00	67	35	03	71	40	09	78	47	16
28	32	1299	67	34	02	71	39	08	77	46	16
29	31	98	66	34	02	70	39	08	76	46	15
30	1331	1298	1266	1233	1201	1170	1138	1107	1076	1045	1015
31	30	97	65	33	01	69	38	06	75	45	14
32	29	97	64	32	00	69	37	06	75	44	14
33	29	96	64	32	00	68	37	05	74	44	13
34	28	96	63	31	1199	68	36	05	74	43	13
35	28	95	63	31	99	67	36	04	73	43	12
36	27	95	62	30	98	67	35	04	73	42	12
37	27	94	62	30	98	66	35	03	72	42	11
38	26	94	61	29	97	65	34	03	72	41	11
39	26	93	61	29	97	65	34	02	71	41	10
40	1325	1292	1260	1228	1196	1164	1133	1102	1071	1040	1009
41	25	92	60	27	96	64	32	01	70	40	09
42	24	91	59	27	95	63	32	01	70	39	08
43	23	91	59	26	95	63	31	00	69	39	08
44	23	90	58	26	94	62	31	00	69	38	07
45	22	90	57	25	93	62	30	1099	68	37	07
46	22	89	57	25	93	61	30	99	68	37	06
47	21	89	56	24	92	61	29	98	67	36	06
48	21	88	56	24	92	60	29	98	67	36	05
49	20	88	55	23	91	60	28	97	66	35	05
50	1320	1287	1255	1223	1191	1159	1128	1097	1066	1035	1004
51	19	87	54	22	90	59	27	96	65	34	04
52	19	86	54	22	90	58	27	96	65	34	03
53	18	85	53	21	89	58	26	95	64	33	03
54	17	85	53	21	89	57	26	95	64	33	02
55	17	84	52	20	88	57	25	94	63	32	02
56	16	84	52	19	88	56	25	94	63	32	01
57	16	83	51	19	87	56	24	93	62	31	01
58	15	83	50	18	87	55	24	92	62	31	00
59	15	82	50	18	86	54	23	92	61	30	00

s. ″	h. m. 2° 23′	h. m. 2° 24′	h. m. 2° 25′	h. m. 2° 26′	h. m. 2° 27′	h. m. 2° 28′	h. m. 2° 29′	h. m. 2° 30′	h. m. 2° 31′	h. m. 2° 32′	h. m. 2° 33′
0	0999	0969	0939	0909	0880	0850	0821	0792	0763	0734	0706
1	99	69	39	09	79	50	20	91	62	34	05
2	98	68	38	08	79	49	20	91	62	33	05
3	98	68	38	08	78	49	19	90	62	33	04
4	97	67	37	07	78	48	19	90	61	32	04
5	97	67	37	07	77	48	18	89	61	32	03
6	96	66	36	06	77	47	18	89	60	31	03
7	96	66	36	06	76	47	17	88	60	31	03
8	95	65	35	05	76	46	17	88	59	30	02
9	95	65	35	05	75	46	16	87	59	30	02
10	0994	0964	0934	0904	0875	0845	0816	0787	0758	0730	0701
11	94	64	34	04	74	45	16	87	58	29	01
12	93	63	33	03	74	44	15	86	57	29	00
13	93	63	33	03	73	44	15	86	57	28	00
14	92	62	32	02	73	43	14	85	56	28	0699
15	92	62	32	02	72	43	14	85	56	27	99
16	91	61	31	01	72	42	13	84	55	27	98
17	91	61	31	01	71	42	13	84	55	26	98
18	90	60	30	00	71	41	12	83	54	26	97
19	90	60	30	00	70	41	12	83	54	25	97
20	0989	0959	0929	0899	0870	0840	0811	0782	0753	0725	0696
21	89	59	29	99	69	40	11	82	53	24	96
22	88	58	28	98	69	39	10	81	52	24	95
23	88	58	28	98	68	39	10	81	52	23	95
24	87	57	27	97	68	38	09	80	51	23	95
25	87	57	27	97	67	38	09	80	51	22	94
26	86	56	26	96	67	37	08	79	51	22	94
27	86	56	26	96	66	37	08	79	50	21	93
28	85	55	25	95	66	36	07	78	50	21	93
29	85	55	25	95	65	36	07	78	49	21	92
30	0984	0954	0924	0894	0865	0835	0806	0777	0749	0720	0692
31	84	54	24	94	64	35	06	77	48	20	91
32	83	53	23	93	64	34	05	76	48	19	91
33	83	53	23	93	63	34	05	76	47	19	90
34	82	52	22	92	63	34	04	75	47	18	90
35	82	52	22	92	62	33	04	75	46	18	89
36	81	51	21	91	62	33	03	74	46	17	89
37	81	51	21	91	61	32	03	74	45	17	88
38	80	50	20	90	61	32	02	74	45	16	88
39	80	50	20	90	60	31	02	73	44	16	87
40	0979	0949	0919	0889	0860	0831	0801	0773	0744	0715	0687
41	79	49	19	89	59	30	01	72	43	15	86
42	78	48	18	88	59	30	01	72	43	14	86
43	78	48	18	88	58	29	00	71	42	14	86
44	77	47	17	87	58	29	00	71	42	13	85
45	77	47	17	87	57	28	0799	70	41	13	85
46	76	46	16	86	57	28	99	70	41	12	84
47	76	46	16	86	56	27	98	69	40	12	84
48	75	45	15	85	56	27	98	69	40	11	83
49	75	45	15	85	55	26	97	68	40	11	83
50	0974	0944	0914	0884	0855	0826	0797	0768	0739	0711	0682
51	74	44	14	84	55	25	96	67	39	10	82
52	73	43	13	83	54	25	96	67	38	10	81
53	73	43	13	83	54	24	95	66	38	09	81
54	72	42	12	83	53	24	95	66	37	09	80
55	72	42	12	82	53	23	94	65	37	08	80
56	71	41	11	82	52	23	94	65	36	08	79
57	71	41	11	81	52	22	93	64	36	07	79
58	70	40	10	81	51	22	93	64	35	07	78
59	70	40	10	80	51	21	92	63	35	06	78

s. ″	h. m. 2° 34′	h. m. 2° 35′	h. m. 2° 36′	h. m. 2° 37′	h. m. 2° 38′	h. m. 2° 39′	h. m. 2° 40′	h. m. 2° 41′	h. m. 2° 42′	h. m. 2° 43′	h. m. 2° 44′
0	0678	0649	0621	0594	0566	0539	0512	0484	0458	0431	0404
1	77	49	21	93	66	38	11	84	57	30	04
2	77	48	21	93	65	38	11	84	57	30	03
3	76	48	20	92	65	37	10	83	56	30	03
4	76	48	20	92	64	37	10	83	56	29	03
5	75	47	19	91	64	36	09	82	55	29	02
6	75	47	19	91	63	36	09	82	55	28	02
7	74	46	18	91	63	36	08	81	54	28	01
8	74	46	18	90	62	35	08	81	54	27	01
9	73	45	17	90	62	35	07	80	54	27	00
10	0673	0645	0617	0589	0562	0534	0507	0480	0453	0426	0400
11	72	44	16	89	61	34	07	80	53	26	0399
12	72	44	16	88	61	33	06	79	52	26	99
13	71	43	15	88	60	33	06	79	52	25	99
14	71	43	15	87	60	32	05	78	51	25	98
15	70	42	15	87	59	32	05	78	51	24	98
16	70	42	14	86	59	31	04	77	50	24	97
17	70	41	14	86	58	31	04	77	50	23	97
18	69	41	13	85	58	31	03	76	50	23	96
19	69	41	13	85	57	30	03	76	49	22	96
20	0668	0640	0612	0585	0557	0530	0502	0475	0449	0422	0395
21	68	40	12	84	57	29	02	75	48	22	95
22	67	39	11	84	56	29	02	75	48	21	95
23	67	39	11	83	56	28	01	74	47	21	94
24	66	38	10	83	55	28	01	74	47	20	94
25	66	38	10	82	55	27	00	73	46	20	93
26	65	37	09	82	54	27	00	73	46	19	93
27	65	37	09	81	54	26	0499	72	46	19	92
28	64	36	09	81	53	26	99	72	45	18	92
29	64	36	08	80	53	26	98	71	45	18	91
30	0663	0635	0608	0580	0552	0525	0498	0471	0444	0418	0391
31	63	35	07	79	52	25	98	71	44	17	91
32	63	34	07	79	52	24	97	70	43	17	90
33	62	34	06	79	51	24	97	70	43	16	90
34	62	34	06	78	51	23	96	69	42	16	89
35	61	33	05	78	50	23	96	69	42	15	89
36	61	33	05	77	50	22	95	68	42	15	88
37	60	32	04	77	49	22	95	68	41	14	88
38	60	32	04	76	49	21	94	67	41	14	88
39	59	31	03	76	48	21	94	67	40	14	87
40	0659	0631	0603	0575	0548	0521	0493	0467	0440	0413	0387
41	58	30	02	75	47	20	93	66	39	13	86
42	58	30	02	74	47	20	93	66	39	12	86
43	57	29	02	74	46	19	92	65	38	12	85
44	57	29	01	73	46	19	92	65	38	11	85
45	56	28	01	73	46	18	91	64	38	11	84
46	56	28	00	73	45	18	91	64	37	10	84
47	55	28	00	72	45	17	90	63	37	10	84
48	55	27	0599	72	44	17	90	63	36	10	83
49	55	27	99	71	44	17	89	62	36	09	83
50	0654	0626	0598	0571	0543	0516	0489	0462	0435	0409	0382
51	54	26	98	70	43	16	89	62	35	08	82
52	53	25	97	70	42	15	88	61	34	08	81
53	53	25	97	69	42	15	88	61	34	07	81
54	52	24	96	69	41	14	87	60	34	07	81
55	52	24	96	68	41	14	87	60	33		80
56	51	23	96	68	41	13	86	59	33		80
57	51	23	95	68	40	13	86	59	32		79
58	50	22	95	67	40	12	85	58	32	06	79
59	50	22	94	67	39	12	85	58	31	06	78

s. "	h. m. 2° 45′	h. m. 2° 46′	h. m. 2° 47′	h. m. 2° 48′	h. m. 2° 49′	h. m. 2° 50′	h. m. 2° 51′	h. m. 2° 52′	h. m. 2° 53′	h. m. 2° 54′	h. m. 2° 55′
0	0378	0352	0326	0300	0274	0248	0223	0197	0172	0147	0122
1	77	51	25	0299	73	48	22	97	72	47	22
2	77	51	25	99	73	47	22	97	71	46	22
3	77	50	24	98	73	47	21	96	71	46	21
4	76	50	24	98	72	47	21	96	71	46	21
5	76	49	23	97	72	46	21	95	70	45	20
6	75	49	23	97	71	46	20	95	70	45	20
7	75	49	23	97	71	45	20	94	69	44	19
8	74	48	22	96	70	45	19	94	69	44	19
9	74	48	22	96	70	44	19	94	69	43	19
10	0374	0347	0321	0295	0270	0244	0219	0193	0168	0143	0118
11	73	47	21	95	69	44	18	93	68	43	18
12	73	46	20	94	69	43	18	92	67	42	17
13	72	46	20	94	68	43	17	92	67	42	17
14	72	46	19	94	68	42	17	92	66	41	17
15	71	45	19	93	67	42	16	91	66	41	16
16	71	45	19	93	67	41	16	91	66	41	16
17	70	44	18	92	67	41	16	90	65	40	15
18	70	44	18	92	66	41	15	90	65	40	15
19	70	43	17	91	66	40	15	89	64	39	14
20	0369	0343	0317	0291	0265	0240	0214	0189	0164	0139	0114
21	69	42	16	91	65	39	14	89	63	39	14
22	68	42	16	90	64	39	13	88	63	38	13
23	68	42	16	90	64	38	13	88	63	38	13
24	67	41	15	89	64	38	13	87	62	37	12
25	67	41	15	89	63	38	12	87	62	37	12
26	66	40	14	88	63	37	12	87	61	36	12
27	66	40	14	88	62	37	11	86	61	36	11
28	66	39	13	88	62	36	11	86	61	36	11
29	65	39	13	87	61	36	11	85	60	35	10
30	0365	0339	0313	0287	0261	0235	0210	0185	0160	0135	0110
31	64	38	12	86	61	35	10	84	59		10
32	64	38	12	86	60	35	09	84	59		09
33	63	37	11	85	60	34	09	84	58		09
34	63	37	11	85	59	34	08	83	58		08
35	63	36	10	85	59	33	08	83	58		08
36	62	36	10	84	58	33	08	82	57		07
37	62	36	10	84	58	33	07	82	57		07
38	61	35	09	83	58	32	07	81	56	34	07
39	61	35	09	83	57	32	06	81	56	33	06
40	0360	0334	0308	0282	0257	0231	0206	0181	0156	01	0106
41	60	34	08	82	56	31	05	80	55		05
42	59	33	07	82	56	30	05	80	55		05
43	59	33	07	81	55	30	05	79	54		05
44	59	33	07	81	55	30	04	79	54		04
45	58	32	06	80	55	29	04	79	53		04
46	58	32	06	80	54	29	03	78	53		03
47	57	31	05	79	54	28	03	78	53		03
48	57	31	05	79	53	28	02	77	52		03
49	56	30	04	79	53	27	02	77	52	31	02
50	0356	0330	0304	0278	0252	0227	0202	0176	0151	0126	0102
51	56	29	04	78	52	27	01	76	51		01
52	55	29	03	77	52	26	01	76	51		01
53	55	29	03	77	51	26	00	75	50	26	00
54	54	28	02	76	51	25	00	75	50	25	00
55		28	02	76	50	25	00	74	49	24	00
56		27	01	76	50	24	0199	74	49	24	0099
57		27	01	75	50	24	99	74	48	24	99
58	54	26	00	75	49	24	98	73	48	23	98
59	53	26	00	74	49	23	98	73	48	23	98

s. "	h. m. 2° 56′	h. m. 2° 57′	h. m. 2° 58′	h. m. 2° 59′	s. "	h. m. 2° 56′	h. m. 2° 57′	h. m. 2° 58′	h. m. 2° 59′
0	0098	0073	0049	0024	30	0085	0061	0036	0012
1	97	73	48	24	31	85	60	36	12
2	97	72	48	23	32	84	60	36	11
3	96	72	47	23	33	84	60	35	11
4	96	71	47	23	34	84	59	35	10
5	96	71	46	22	35	83	59	34	10
6	95	71	46	22	36	83	58	34	10
7	95	70	46	21	37	82	58	34	09
8	94	70	45	21	38	82	57	33	09
9	94	69	45	21	39	82	57	33	08
10	0093	0069	0044	0020	40	0081	0057	0032	0008
11	93	68	44	20	41	81	56	32	08
12	93	68	44	19	42	80	56	31	07
13	92	68	43	19	43	80	55	31	07
14	92	67	43	19	44	80	55	31	06
15	91	67	42	18	45	79	55	30	06
16	91	66	42	18	46	79	54	30	06
17	91	66	42	17	47	78	54	29	05
18	90	66	41	17	48	78	53	29	05
19	90	65	41	17	49	77	53	29	04
20	0089	0065	0040	0016	50	0077	0053	0028	0004
21	89	64	40	16	51	77	52	28	04
22	89	64	40	15	52	76	52	27	03
23	88	64	39	15	53	76	51	27	03
24	88	63	39	15	54	75	51	27	02
25	87	63	38	14	55	75	51	26	02
26	87	62	38	14	56	75	50	26	02
27	87	62	38	13	57	74	50	25	01
28	86	62	37	13	58	74	49	25	01
29	86	61	37	12	59	73	49	25	00

TABLE 8.

FOR FINDING THE DISTANCE OF OBJECTS AT SEA.

Ht. of eye in feet.	Dist. in miles.	Ht. of eye in feet.	Dist. in miles.	Ht. of eye in feet.	Dist. in miles.	Ht. of eye in feet.	Dist. in miles.	Ht. of eye in feet.	Dist. in miles.	Height of eye in feet.	Dist. in miles.	Height of eye in feet.	Dist. in miles
1	1·15	25	5·74	49	8·0	180	15·4	420	23·5	820	32·9	2500	57·4
2	1·62	26	5·86	50	8·1	190	15·8	430	23·8	840	33·3	2600	58·6
3	1·99	27	5·97	55	8·5	200	16·2	440	24·1	860	33·7	2700	59·7
4	2·30	28	6·08	60	8·9	210	16·6	450	24·4	880	34·1	2800	60·8
5	2·57	29	6·18	65	9·3	220	17·0	460	24·6	900	34·5	2900	61·8
6	2·81	30	6·30	70	9·6	230	17·4	470	24·9	920	34·8	3000	63·0
7	3·04	31	6·40	75	9·9	240	17·8	480	25·2	940	35·2	3100	64·0
8	3·25	32	6·50	80	10·3	250	18·2	490	25·4	960	35·6	3200	65·0
9	3·45	33	6·60	85	10·6	260	18·5	500	25·7	980	36·0	3300	66·0
10	3·63	34	6·70	90	10·9	270	18·9	520	26·2	1000	36·3	3400	67·0
11	3·81	35	6·80	95	11·2	280	19·2	540	26·7	1100	38·1	3500	68·0
12	3·98	36	6·90	100	11·5	290	19·6	560	27·2	1200	39·8	3600	69·0
13	4·14	37	6·99	105	11·8	300	19·9	580	27·7	1300	41·4	3700	69·9
14	4·30	38	7·09	110	12·1	310	20·2	600	28·1	1400	43·0	3800	70·9
15	4·45	39	7·17	115	12·3	320	20·6	620	28·6	1500	44·5	3900	71·7
16	4·60	40	7·27	120	12·6	330	20·9	640	29·1	1600	46·0	4000	72·7
17	4·73	41	7·36	125	12·8	340	21·2	660	29·5	1700	47·3	4100	73·6
18	4·87	42	7·44	130	13·1	350	21·5	680	30·0	1800	48·7	4200	74·4
19	5·01	43	7·54	135	13·3	360	21·8	700	30·4	1900	50·1	4300	75·4
20	5·14	44	7·62	140	13·6	370	22·1	720	30·8	2000	51·4	4400	76·2
21	5·26	45	7·70	145	13·8	380	22·4	740	31·2	2100	52·6	4500	77·0
22	5·39	46	7·79	150	14·1	390	22·7	760	31·7	2200	53·9	4700	78·8
23	5·51	47	7·88	160	14·5	400	23·0	780	32·1	2300	55·1	5000	81·2
24	5·62	48	7·96	170	15·0	410	23·3	800	32·5	2400	56·2	1 mile	83·5

Table 9.

Declination.		JANUARY. 1.	JANUARY. 11.	JANUARY. 21.	FEBRUARY. 1.	FEBRUARY. 11.	FEBRUARY. 21.
° ′		h. m.	h. m.	h. m.	h. m.	h. m.	h. m.
28 19 N.	α Androm.	5 16	4 37	3 58	3 15	2 35	1 56
14 24 N.	γ Pegasi	21	42	4 3	20	40	2 1
78 3 S.	β Hydri	34	55	16	33	53	14
55 46 N.	α Cassiop.	48	5 9	29	46	3 6	28
18 46 S.	β Ceti	52	13	33	51	11	32
88 33 N.	Pole Star	6 19	5 40	5 0	4 17	3 38	2 59
8 55 S.	θ^1 Ceti	32	53	13	30	50	3 12
57 58 S.	α Eridani	48	6 9	29	46	4 6	28
22 48 N.	α Arietis	7 14	35	55	5 12	32	54
2 38 N.	γ Ceti	51	7 12	6 32	50	5 10	4 31
3 32 N.	α Ceti	8 10	7 31	6 51	6 8	5 28	4 49
49 21 N.	α Persei	29	50	7 10	27	47	5 8
23 40 N.	η Tauri	54	8 15	35	53	6 13	34
13 55 S.	γ^1 Eridani	9 7	28	48	7 6	26	47
16 13 N.	α Tauri	43	9 4	8 24	41	7 1	6 22
45 51 N.	α Aurigæ	10 20	9 42	9 2	8 19	7 39	7 0
8 22 S.	β Orionis	22	44	4	21	41	2
28 29 N.	β Tauri	32	52	14	31	51	12
0 24 S.	δ Orionis	39	10 0	21	38	58	19
17 56 S.	α Leporis	41	2	23	40	8 0	21
1 18 S.	ι Orionis	10 43	10 4	9 25	8 42	8 2	7 23
34 9 S.	α Columbæ	49	10	31	48	8	29
7 23 N.	α Orionis	11 2	23	44	9 1	21	42
22 35 N.	μ Geminor.	29	50	10 10	28	48	8 9
52 37 S.	α Argus	36	57	17	35	55	16
87 15 N.	51 Cephei	11 42	11 3	10 23	9 41	9 1	8 22
16 31 S.	α Canis Maj.	53	14	34	52	12	33
28 47 S.	ε Canis Maj.	12 8	29	49	10 6	27	48
22 14 N.	δ Geminor.	26	47	11 7	24	45	9 6
32 12 N.	α^2 Geminor.	40	12 1	21	38	58	20
5 35 N.	α Canis Min.	12 46	12 7	11 27	10 44	10 4	9 26
28 22 N.	β Geminor.	51	12	32	49	10	31
23 54 S.	15 Argus	13 16	37	57	11 14	34	55
6 56 N.	ε Hydræ	54	13 15	12 35	52	11 12	10 23
48 36 N.	ι Ursæ Maj.	14 4	25	45	12 2	22	43
58 41 S.	ι Argus	14 28	13 49	13 9	12 26	11 46	11 7
8 3 S.	α Hydræ	35	56	16	33	53	14
52 19 N.	θ Ursæ Maj.	38	59	19	36	56	17
24 26 N.	ε Leonis	52	14 13	33	50	12 10	31
12 40 N.	α Leonis	15 15	36	56	13 13	33	54
20 33 N.	γ^1 Leonis	15 26	14 47	14 7	13 24	12 44	12 5
58 56 S.	η Argus	52	15 15	35	52	13 12	33
62 31 N.	α Ursæ Maj.	16 8	30	50	14 7	27	48
21 18 N.	δ Leonis	20	42	15 2	19	39	13 0
14 1 S.	δ Hydræ	26	48	8	25	45	6
15 22 N.	β Leonis	16 55	16 16	15 37	14 54	14 14	13 35
54 29 N.	γ Ursæ Maj.	17 0	21	42	59	19	40
78 31 S.	β Chamæl.	23	44	16 4	15 22	42	14 3
62 19 S.	α^1 Crucis	32	53	13	31	51	12
22 37 S.	β Corvi	40	17 1	21	39	59	20

Declination.		JANUARY.			FEBRUARY.		
		1.	11.	21.	1.	11.	21.
° ′		h. m.	h. m.	h. m.	h. m.	h. m.	h. m.
5 N.	12 Can. Ven.	18 3	17 24	16 44	16 1	15 22	14 43
25 S.	α Virginis ...	31	52	17 12	29	50	15 11
1 N.	η Ursæ Maj.	55	18 16	36	53	16 13	35
39 6 N.	η Bootis	19 1	22	42	59	19	41
[illegible] 41 S.	β Centauri ...	7	28	48	17 5	25	47
N.	α Bootis	19 23	18 44	18 4	17 21	16 41	16 2
S.	α² Centauri ...	43	19 4	24	41	17 1	22
N.	ι Bootis	52	13	33	50	10	31
19 55 S.	α² Libræ	56	17	37	54	14	35
[illegible] N.	β Ursæ Min.	20 5	26	46	18 3	17 23	44
8 52 S.	β Libræ	20 23	19 44	19 4	18 21	17 41	17 2
27 12 N.	α Coro. Bor.	42	20 3	23	40	18 0	21
6 52 N.	α Serpentis...	51	12	32	49	9	30
78 14 N.	ζ Ursæ Min.	21 4	25	45	19 2	22	43
19 25 S.	β¹ Scorpii	10	31	51	8	28	49
3 20 S.	δ Ophiuchi ...	21 20	20 41	20 1	19 18	18 38	17 59
26 7 S.	α Scorpii	34	55	15	32	52	18 13
61 50 N.	η Draconis ...	36	57	17	34	54	15
68 46 S.	α Tria. Aust.	46	21 7	27	44	19 4	25
82 16 N.	ε Ursæ Min.	22 16	37	57	20 14	34	55
14 33 N.	α Herculis ...	22 22	21 43	21 3	20 20	19 40	19 1
52 24 N.	β Draconis ...	44	22 5	25	42	20 2	23
12 40 N.	α Ophiuchi ...	45	6	26	43	3	24
89 17 S.	σ Octantis ...	23 0	20	40	57	18	39
51 30 N.	γ Draconis ...	6	27	48	21 5	25	46
21 6 S.	μ¹ Sagittarii...	23 18	22 39	22 0	21 17	20 37	19 58
86 36 N.	δ Ursæ Min.	35	56	16	34	54	20 15
38 39 N.	α Lyræ	45	23 6	26	44	21 4	25
33 12 N.	β Lyræ	57	18	38	56	16	37
13 39 N.	ζ Aquilæ	0 15	36	56	22 13	34	54
2 50 N.	δ Aquilæ	0 35	23 56	23 16	22 33	21 54	21 14
10 16 N.	γ Aquilæ	56	0 17	37	54	22 14	34
8 30 N.	α Aquilæ	1 0	21	41	58	18	38
6 3 N.	β Aquilæ	5	26	46	23 3	23	43
88 53 N.	λ Ursæ Min.	21	42	0 2	19	39	59
59 S.	α² Capricor. ...	1 27	0 48	0 8	23 25	22 45	22 6
11 S.	α Pavonis ...	31	52	12	28	48	9
46 N.	α Cygni	53	1 14	34	47	23 7	28
12 3 N.	61¹ Cygni	2 17	38	58	0 15	31	52
[illegible] 39 N.	ζ Cygni	23	44	1 4	21	37	58
61 59 N.	α Cephei	2 32	1 53	1 13	0 30	23 46	23 7
6 12 S.	β Aquarii ...	40	2 1	21	33	54	15
69 56 N.	β Cephei	44	5	25	42	0 2	19
9 14 N.	ι Pegasi	54	15	35	52	12	29
1 1 S.	α Aquarii ...	3 15	36	56	1 13	33	50
39 S.	α Gruis	3 15	2 36	1 56	1 13	0 33	23 50
47 5 N.	ζ Pegasi	51	3 12	2 32	49	1 9	0 30
30 22 S.	α Pisc. Aust.	4 5	27	47	2 4	24	45
14 27 N.	α Pegasi	13	35	53	12	32	53
4 51 N.	ι Piscium ...	48	4 10	3 30	47	2 7	1 28

Declination.		MARCH. 1.	MARCH. 11.	MARCH. 21.	APRIL. 1.	APRIL. 11.	APRIL. 21.
° ′		h. m.	h. m.	h. m.	h. m.	h. m.	h. m.
28 19 N.	α Androm. ...	1 24	0 45	0 6	23 18	22 39	21 59
14 24 N.	γ Pegasi	29	50	11	23	44	22 4
78 3 S.	β Hydri	42	1 3	24	36	57	17
55 46 N.	α Cassiop. ...	56	17	38	50	23 11	31
18 46 S.	β Ceti	2 0	21	42	54	15	35
88 38 N.	Pole Star	2 27	1 48	1 9	0 25	23 42	23 2
8 55 S.	θ¹ Ceti	40	2 1	22	38	55	15
57 58 S.	α Eridani ...	56	17	38	54	0 15	31
22 48 N.	α Arietis	3 22	43	2 4	1 20	41	0 1
2 38 N.	γ Ceti	59	3 20	41	57	1 18	38
3 32 N.	α Ceti	4 17	3 39	3 0	2 16	1 37	0 57
49 21 N.	α Persei	36	58	19	35	56	1 16
23 40 N.	η Tauri	5 2	4 23	44	3 0	2 21	41
13 55 S.	γ¹ Eridani ...	15	36	57	13	34	54
16 13 N.	α Tauri	50	5 11	4 32	49	3 10	2 30
45 N.	α Aurigæ ...	6 28	5 49	5 10	4 26	3 48	3 8
8 S.	β Orionis......	30	51	12	28	50	10
28 N.	β Tauri	40	6 1	22	38	59	20
0 51 S.	δ Orionis......	47	8	29	45	4 6	27
17 [illegible] S.	α Leporis......	49	10	31	47	8	29
1 18	ι Orionis......	6 51	6 12	5 33	4 49	4 10	3 31
34 9	α Columbæ ...	57	18	39	55	16	37
7 23	α Orionis......	7 10	31	52	5 8	29	50
22 35 S.	μ Geminor....	37	58	6 19	35	56	4 16
52 37 [illegible]	α Argus	44	7 5	26	42	5 3	23
N.	51 Cephei	7 50	7 11	6 32	5 48	5 9	4 29
S.	α Canis Maj.	8 1	22	48	59	20	40
S.	ε Canis Maj.	16	37	58	6 14	35	55
37 15 N.	δ Geminor....	34	55	7 16	32	53	5 13
[illegible] [illegible] N.	α² Geminor....	48	8 9	30	46	6 7	27
5 N.	α Canis Min.	8 54	8 15	7 36	6 52	6 13	5 33
28 N.	β Geminor....	59	20	41	57	18	38
23 S.	15 Argus	9 24	45	8 6	7 22	43	6 3
6 35 N.	ε Hydræ	10 2	9 23	44	8 0	7 21	41
48 [illegible] N.	ι Ursæ Maj.	12	33	54	10	31	51
58 S.	ι Argus	10 35	9 57	9 18	8 34	7 55	7 15
8 S.	α Hydræ	42	10 3	25	41	8 2	22
52	θ Ursæ Maj.	45	6	28	44	5	25
24 41 N.	ε Leonis	59	20	42	58	19	39
12 [illegible] N.	α Leonis	11 22	43	10 4	9 21	42	8 2
20 33 N.	γ¹ Leonis	11 33	10 54	10 15	9 32	8 53	8 13
58 56 S.	η Argus ..	12 1	11 22	43	10 0	9 21	41
62 31 N.	α Ursæ Maj.	16	37	58	14	36	56
21 18 N.	δ Leonis	28	49	11 10	26	48	9 8
14 1 S.	δ Hydræ	34	55	16	32	54	14
15 22 N.	β Leonis	13 3	12 24	11 45	11 1	10 22	9 43
54 29 N.	γ Ursæ Maj.	8	29	50	6	27	48
78 31 S.	β Chamæl. ...	31	52	12 13	29	50	10 10
62 19 S.	α Crucis	40	13 1	22	38	59	19
22 37 S.	β Corvi	48	9	30	46	11 7	27

Declination.		MARCH.			APRIL.		
		1.	11.	21.	1.	11.	21.
° ′		h. m.	h. m.	h. m.	h. m.	h. m.	h. m.
5 N.	12 Can. Ven.	14 11	13 32	12 53	12 9	11 30	10 50
25 S.	α Virginis ...	39	14 0	13 21	37	58	11 18
1 N.	η Ursæ Maj.	15 3	24	45	13 1	12 22	42
39 6 N.	η Bootis	9	30	51	7	28	48
59 41 S.	β Centauri ...	15	36	57	13	34	54
19 55 N.	α Bootis	15 31	14 52	14 13	13 29	12 50	12 10
60 15 S.	$α^2$ Centauri ...	51	15 12	33	49	13 10	30
27 40 N.	ι Bootis	16 0	21	42	58	19	39
15 27 S.	$α^2$ Libræ	4	25	46	14 2	23	43
74 44 N.	β Ursæ Min.	13	34	55	11	32	52
8 52 S.	β Libræ	16 30	15 52	15 13	14 29	13 50	13 10
27 12 N.	α Coro. Bor.	49	16 10	32	48	14 9	29
6 52 N.	α Serpentis...	58	19	41	57	18	38
78 14 N.	ζ Ursæ Min.	17 11	32	53	15 10	31	51
19 25 S.	$β^1$ Scorpii	17	38	58	16	37	57
3 20 S.	δ Ophiuchi ...	17 27	16 48	16 9	15 26	14 47	14 7
26 7 S.	α Scorpii	41	17 2	23	40	15 1	21
61 50 N.	η Draconis ...	43	4	25	42	3	23
68 46 S.	α Tria. Aust.	53	14	35	52	13	33
82 16 N.	ε Ursæ Min.	18 23	44	17 5	16 22	43	15 3
14 33 N.	α Herculis ...	18 29	17 50	17 11	16 28	15 49	15 9
52 24 N.	β Draconis ...	51	18 11	32	48	16 8	28
12 40 N.	α Ophiuchi ...	52	12	33	49	9	29
89 17 S.	σ Octantis ...	19 8	28	49	17 5	25	45
51 30 N.	γ Draconis ...	14	35	56	12	33	54
21 6 S.	$μ^1$ Sagittarii...	19 26	18 47	18 8	17 24	16 45	16 6
86 36 N.	δ Ursæ Min.		19 4	25	41	17 2	22
38 39 N.	α Lyræ.........		14	35	51	12	32
33 12 N.	β Lyræ.........	20 43	26	47	18 3	24	44
13 39 N.	ζ Aquilæ......	53	44	19 15	21	42	16 2
2 50 N.	δ Aquilæ	20 42	20 3	19 24	18 40	18 1	17 21
10 16 N.	γ Aquilæ	21 2	22	42	58	18 19	39
8 30 N.	α Aquilæ......	6	26	46	19 2	23	43
6 3 N.	β Aquilæ	9	30	51	7	28	48
88 53 N.	λ Ursæ Min.	28	49	20 9	25	46	18 6
12 59 S.	$α^2$ Capricor. ...	21 34	20 55	20 15	19 31	18 52	18 12
57 11 S.	α Pavonis ...	38	59	20	36	57	17
44 46 N.	α Cygni	57	21 18	39	55	19 16	36
38 3 N.	61^1 Cygni	22 21	42	21 3	20 19	40	19 0
29 39 N.	ζ Cygni	27	47	9	25	46	6
61 59 N.	α Cephei	22 35	21 56	21 18	20 35	19 54	19 15
6 12 S.	β Aquarii ...	43	22 4	26	42	20 3	23
69 56 N.	β Cephei	47	8	30	46	7	27
9 14 N.	ε Pegasi	57	18	40	56	17	37
1 1 S.	α Aquarii ...	23 18	39	22 0	21 17	38	58
47 39 S.	α Gruis	23 18	22	22 0	21 17	20 38	19 58
10 5 N.	ζ Pegasi	54	23	36	53	21 14	20 34
30 22 S.	α Pisc. Aust.	0 13		51	22 7	29	49
14 27 N.	α Pegasi	21	39	59	15	37	57
4 51 N.	ι Piscium ..	56	0 [illegible]	23 34	50	22 12	21 32

Declination.		MAY.			JUNE.		
		1.	11.	21.	1.	11.	21.
° ′		h. m.	h. m.	h. m.	h. m.	h. m.	h. m.
28 N.	α Androm.	21 21	20 41	20 2	19 19	18 39	18 0
14 N.	γ Pegasi	26	46	7	24	44	5
78 S.	β Hydri	39	59	20	37	57	18
53 19 N.	α Cassiop. ...	53	21 13	34	51	19 11	32
18 [illegible] S.	β Ceti	56	17	38	55	15	36
88 33 N.	Pole Star	22 23	21 44	21 5	20 22	19 42	19 3
8 53 S.	θ^1 Ceti	36	57	18	35	55	16
57 58 S.	α Eridani ...	52	22 12	34	51	20 11	32
22 48 N.	α Arietis......	23 18	38	22 0	21 17	37	58
2 38 N.	γ Ceti	53	23 15	37	54	21 14	20 35
3 N.	α Ceti	0 18	23 34	22 55	22 12	21 33	20 54
49 N.	α Persei	37	53	23 14	31	52	21 13
23 N.	η Tauri	1 2	0 22	39	56	22 17	38
13 32 S.	γ^1 Eridani ...	15	35	52	23 9	30	51
16 [illegible] N.	α Tauri	51	1 11	0 32	45	23 5	22 26
45 51 N.	α Aurigæ ...	2 29	1 49	1 10	0 27	23 43	23 4
8 22 S.	β Orionis......	31	51	12	29	45	6
28 29 N.	β Tauri	41	2 1	22	39	55	16
0 24 S.	δ Orionis......	48	8	29	46	0 6	23
17 56 S.	α Leporis ...	50	10	31	48	8	25
1 18 S.	ε Orionis......	2 52	2 12	1 33	0 50	0 10	23 27
34 9 S.	α Columbæ....	58	18	39	56	16	33
7 23 N.	α Orionis......	3 11	31	52	1 9	29	46
22 35 N.	μ Geminor....	38	58	2 19	36	58	0 17
52 37 S.	α Argus	45	3 5	26	43	1 3	23
87 15 N.	51 Cephei	3 51	3 11	2 32	1 49	1 9	0 29
16 31 S.	α Canis Maj.	4 1	22	43	2 0	20.	41
28 47 S.	ε Canis Maj.	16	37	58	15	35	56
22 14 N.	δ Geminor....	34	55	3 16	33	53	1 14
32 12 N.	α^2 Geminor....	48	4 8	30	47	2 7	28
5 35 N.	α Canis Min.	4	4 14	3 36	2 53	2 13	1 31
28 22 N.	β Geminor. ..		19	41	58	18	39
23 51 S.	15 Argus	5	44	4 5	3 23	43	2 4
6 56 N.	ε Hydræ	6 54	5 22	43	4 1	3 21	42
48 36 N.	ι Ursæ Maj.	[illegible]	32	53	11	31	52
58 41 S.	ι Argus	6 36	5 56	5 17	4 34	3 55	3 16
8 8 S.	α Hydræ	43	6 3	24	41	4 1	23
52 19 N.	θ Ursæ Maj.	46	6	27	44	4	26
24 26 N.	ε Leonis	7 0	20	41	58	18	30
12 40 N.	α Leonis	22	42	6 4	5 21	41	4 2
20 33 N.	γ^1 Leonis	7 34	6 54	6 15	5 32	4 52	4 13
58 56 S.	η Argus	8 2	7 22	43	6 0	5 20	41
62 31 N.	α Ursæ Maj.	17	37	58	15	35	56
21 18 N.	δ Leonis	29	49	7 10	27	47	5 8
14 1 S.	δ Hydræ	35	55	16	33	53	14
15 22 N.	β Leonis	9 4	8 24	7 45	7 2	6 22	5 43
54 29 N.	γ Ursæ Maj.	8	29	50	7	27	48
78 31 S.	β Chamæl. ...	32	52	8 13	30	50	6 11
62 19 S.	α^1 Crucis	41	9 1	22	39	59	20
22 37 S.	β Corvi	49	9	30	47	7 7	28

Declination.				MAY.			JUNE.		
				1.	11.	21.	1.	11.	21.
°	′			h. m.	h. m.	h. m.	h. m.	h. m.	h. m.
	5	.	12 Can. Ven...	10 11	9 32	8 53	8 10	7 30	6 51
	25		α Virginis ...	30	10 0	9 21	38	58	7 19
	1		η Ursæ Maj..	11 3	23	45	9 2	8 22	43
39	6	N	η Bootis	9	29	51	8	28	49
[illegible]	41	[illegible]	β Centauri ...	15	35	57	14	34	55
		N.	α Bootis	11 31	10 51	10 12	9 30	8 50	8 11
		S.	α² Centauri ...	51	11 11	32	50	9 10	31
		N.	ε Bootis	12 0	20	41	50	19	40
19	55	S.	α² Libræ	4	24	45	10 8	23	44
[illegible]	[illegible]	N.	β Ursæ Min..	13	33	54	12	32	53
8	52	S.	β. Libræ	12 31	11 51	11 12	10 30	9 50	9 11
27	12	N.	α Coro. Bor...	50	12 10	31	48	10 8	30
6	52	N.	α Serpentis...	59	19	40	57	17	39
78	14	N.	ζ Ursæ Min..	13 12	32	53	11 10	30	52
19	25	S.	β¹ Scorpii	18	38	59	16	36	58
		S.	δ Ophiuchi ...	13 28	12 48	12 9	11 26	10 46	10 8
		S.	α Scorpii	42	13 2	23	40	11 0	21
		N.	η Draconis ...	44	4	25	42	2	23
3	20	S.	α Tria. Aust.	54	14	35	52	12	33
[illegible]	[illegible]	N.	ε Ursæ Min..	14 24	44	13 5	12 22	42	11 4
14	33	N.	α Herculis ...	14 30	13 50	13 11	12 28	11 48	11 10
52	24	N.	β Draconis ...	49	14 9	30	47	12 7	28
12	40	N.	α Ophiuchi...	50	10	31	48	8	29
89	17	S.	σ Octantis ...	15 10	30	51	13 8	28	49
51	30	N.	γ Draconis ...	15	35	56	13	33	54
21	6	S.	μ¹ Sagittarii...	15 27	14 47	14 8	13 25	12 45	12 6
86	36	N.	δ Ursæ Min..	43	15 4	25	42	13 2	23
38	39	N.	α Lyræ.........	53	14	35	52	12	33
33	12	N.	β Lyræ.........	16 5	26	47	14 4	24	45
13	39	N.	ζ Aquilæ	22	43	15 4	21	41	13 2
2	50	N.	δ Aquilæ......	16 42	16 3	15 4	14 41	14 1	13 22
10	16	N.	γ Aquilæ......	17 0	20	42	59	19	40
8	30	N.	α Aquilæ	4	24	46	15 3	23	44
6	3	N.	β Aquilæ......	9	29	51	8	28	49
88	53	N.	λ Ursæ Min..	28	48	16 9	26	47	14 7
12	59	S.	α² Capricor....	17 33	16 53	16 14	15 31	14 51	14 12
57	11	S.	α Pavonis ...	38	58	19	37	57	18
44	46	N.	α Cygni	57	17 17	38	56	15 16	37
38	3	N.	61¹ Cygni	18 21	41	17 2	16 19	40	15 0
29	39	N.	ζ Cygni	27	47	8	25	45	6
61	59	N.	α Cephei	18 36	17 56	17 17	16 34	15 54	15 15
6	12	S.	β Aquarii ...	44	18 4	25	42	16 2	23
69	56	N.	β Cephei	48	8	29	46	6	27
9	14	N.	ε Pegasi	58	18	39	56	16	38
1	1	S.	α Aquarii ...	19 19	39	18 0	17 17	37	58
47	39	S.	α Gruis	19 19	18 39	18 0	17 17	16 37	15 58
10	5	N.	ζ Pegasi	55	19 15	36	53	17 13	16 34
30	22	S.	α Pisc. Aust..	20 10	30	51	18 8	28	49
14	27	N.	α Pegasi	18	38	59	16	36	57
4	51	N.	ι Piscium ...	53	20 13	19 34	51	18 11	17 32

Declination.		JULY.			AUGUST.		
		1.	11.	21.	1.	11.	21.
° ′		h. m.	h. m.	h. m.	h. m.	h. m.	h. m.
28 19 N.	α Androm. ...	17 20	16 41	16 2	15 19	14 40	14 1
14 24 N.	γ Pegasi	25	46	7	24	45	6
78 3 S.	β Hydri	38	59	20	37	58	19
55 46 N.	α Cassiop. ...	52	17 13	34	51	15 12	33
18 46 S.	β Ceti	56	17	38	55	16	37
88 33 N.	Pole Star	18 23	17 44	17 5	16 22	15 43	15 4
8 55 S.	θ^1 Ceti	36	57	18	34	56	17
57 58 S.	α Eridani ...	52	18 13	34	50	16 11	33
22 48 N.	α Arietis	19 18	39	18 0	17 16	37	59
2 38 N.	γ Ceti	55	19 16	37	53	17 14	16 36
3 32 N.	α Ceti	20 14	19 35	18 56	18 12	17 33	16 54
49 21 N.	α Persei	33	54	19 15	31	52	17 13
23 40 N.	η Tauri	58	20 19	40	56	18 17	38
13 55 S.	γ^1 Eridani ...	21 11	32	53	19 9	30	51
16 13 N.	α Tauri	47	21 8	20 29	45	19 6	18 27
45 51 N.	α Aurigæ......	22 24	21 45	21 6	20 23	19 44	19 5
8 22 S.	β Orionis......	26	48	9	25	46	7
28 29 N.	β Tauri	36	57	19	35	56	17
0 24 S.	δ Orionis......	43	22 4	26	42	20 3	24
17 56 S.	α Leporis......	45	6	28	44	5	26
1 18 S.	ι Orionis......	22 47	22 8	21 31	20 46	20 7	19 28
34 9 S.	α Columbæ ...	53	14	36	52	13	34
7 23 N.	α Orionis......	23 6	27	49	21 5	26	47
22 35 N.	μ Geminor. ...	33	54	22 10	32	53	20 14
52 37 S.	α Argus	40	23 1	22	39	21 0	21
87 15 N.	51 Cephei	23 49	23 9	22 30	21 47	21 8	20 29
16 31 S.	α Canis Maj.	0 2	22	43	22 0	21	42
28 47 S.	ε Canis Maj.	16	36	57	4	35	56
22 14 N.	ζ Geminor. ...	34	54	23 15	32	53	21 14
32 12 N.	α^2 Geminor. ...	48	0 8	29	46	22 7	29
5 35 N.	α Canis Min.	0 54	0 15	23 36	22 52	22 13	21 34
28 22 N.	β Geminor. ...	59	20	41	57	18	39
23 54 S.	15 Argus	1 24	45	0 6	23 22	43	22 4
6 56 N.	ι Hydræ	2 2	1 23	44	0 0	23 21	42
48 36 N.	ι Ursæ Maj. .	12	33	54	10	31	52
53 41 S.	ι Argus	2 36	1 57	1 18	0 34	23 54	23 15
8 3 S.	α Hydræ	43	2 4	25	41	0 2	23
52 19 N.	θ Ursæ Maj. .	46	7	28	44	5	26
24 26 N.	ε Leonis	3 0	21	42	58	19	40
12 40 N.	α Leonis	23	44	2 5	1 21	42	0 3
20 33 N.	γ^1 Leonis	3 34	2 55	2 10	1 32	0 53	0 14
58 56 S.	η Argus	4 1	3 23	44	2 0	1 21	42
62 31 N.	α Ursæ Maj. .	16	38	59	15	36	57
21 18 N.	δ Leonis	28	50	3 11	27	48	1 9
14 1 S.	δ Hydræ	34	56	17	33	54	15
15 22 N.	β Leonis	5 3	4 24	3 45	3 2	2 23	1 44
54 29 N.	γ Ursæ Maj. .	8	29	51	7	28	49
78 31 S.	β Chamæl. ...	31	52	4 14	30	51	2 12
62 19 S.	α Crucis	40	5 1	22	39	3 0	21
22 37 S.	β Corvi	48	9	30	47	8	29

Declination.		JULY.			AUGUST.		
		1.	11.	21.	1.	11.	21.
° ′		h. m.	h. m.	h. m.	h. m.	h. m.	h. m.
39 5 N.	12 Can. Ven...	6 11	5 32	4 53	4 9	3 31	2 52
10 25 S.	α Virginis ..	39	6 0	5 21	37	59	3 20
50 1 N.	η Ursæ Maj..	7 3	24	45	5 1	4 22	44
19 6 N.	η Bootis	9	30	51	7	28	50
59 41 S.	β Centauri ...	15	36	57	13	34	56
19 55 N.	α Bootis	7 31	6 52	6 13	5 29	4 50	4 11
60 15 S.	α^2 Centauri ...	51	7 12	33	49	5 10	31
27 40 N.	ε Bootis	8 0	21	42	58	19	40
15 27 S.	α^2 Libræ	4	25	46	6 2	23	44
74 44 N.	β Ursæ Min.	13	34	55	11	32	53
8 52 S.	β Libræ	8 31	7	7 13	6 29	5 50	5 11
27 12 N.	α Coro. Bor...	50	8	32	48	6 9	30
6 52 N.	α Serpentis ..	59		41	57	18	39
78 14 N.	ζ Ursæ Min.	9 12	52	54	7 10	31	52
19 25 S.	β^1 Scorpii	18	[illegible]	8 0	16	37	58
3 20 S.	δ Ophiuchi ...	9 23	8 49	8 10	7 26	6 47	6 8
26 7 S.	α Scorpii	42	9 3	24	40	7 1	22
61 50 N.	η Draconis ...	44	5	26	42	3	24
68 46 S.	α Tria. Aust.	54	15	36	52	13	34
82 16 N.	ε Ursæ Min.	10 23	45	9 6	8 22	43	7 4
14 33 N.	α Herculis ...	10 29	9	9 12	8 28	7 49	7 10
52 24 N.	β Draconis ...	48	10	31	47	8 8	29
12 40 N.	α Ophiuchi ...	49		32	48	9	30
89 17 S.	σ Octantis ...	11 10	51	51	9 7	28	49
51 30 N.	γ Draconis ...	16	36	57	13	34	55
21 6 S.	μ^1 Sagittarii...	11 26	10 47	10 9	9 25	8 46	8 7
86 36 N.	δ Ursæ Min.	43	11 4	25	42	9 3	24
38 39 N.	α Lyræ	53	14	35	52	13	34
33 12 N.	β Lyræ	12 5	26	47	10 4	25	46
13 39 N.	ζ Aquilæ......	22	43	11 4	21	42	9 3
2 50 N.	δ Aquilæ......	12 42	12 3	11 24	10 41	10 2	9 23
10 16 N.	γ Aquilæ......	13 0	21	42	58	19	41
8 30 N.	α Aquilæ......	4	25	46	11 2	23	45
6 3 N.	β Aquilæ......	9	30	51	7	28	50
85 53 N.	λ Ursæ Min.	29	49	12 10	26	47	10 8
12 59 S.	α^2 Capricor....	13 33	12 53	12 14	11 30	10 51	10 12
57 11 S.	α Pavonis ...	38	58	19	35	56	17
44 46 N.	α Cygni	57	13 18	39	55	11 16	37
38 3 N.	61^1 Cygni	14 21	42	13 3	12 19	40	11 1
29 39 N.	ζ Cygni	27	48	9	25	46	7
61 59 N.	α Cephei	14 36	13 57	13 18	12 34	11 55	11 16
6 12 S.	β Aquarii ...	44	14 5	26	42	12 3	24
69 56 N.	β Cephei	48	9	30	46	7	28
9 14 N.	ε Pegasi	58	19	40	56	17	38
1 1 S.	α Aquarii ...	15 19	40	14 1	13 17	38	59
47 39 S.	α Gruis	15 19	14 40	14 1	13 17	12 38	11 59
10 5 N.	ζ Pegasi	58	15 20	41	57	13 18	12 39
30 22 S.	α Pisc. Aust.	16 9	31	52	14 8	29	50
14 27 N.	α Pegasi	17	39	15 0	16	37	58
4 51 N.	ι Piscium ...	52	16 14	35	51	14 12	13 33

Declination.		September 1.	September 11.	September 21.	October 1.	October 11.	October 21.
° ′		h. m.	h. m.	h. m.	h. m.	h. m.	h. m.
N.	α Androm. ...	13 17	12 37	11 58	11 19	10 39	10 0
N.	γ Pegasi	22	42	12 3	24	44	5
S.	β Hydri	35	55	16	37	57	18
28 19 N.	α Cassiop. ...	49	13 9	30	51	11 11	32
[illegible] [illegible] S.	β Ceti	53	13	34	55	15	36
88 N.	Pole Star	14 20	13 40	13 1	12 22	11 42	11 3
8 S.	θ^1 Ceti	33	53	14	35	55	16
57 S.	α Eridani ...	49	14 9	30	51	12 11	32
22 33 N.	α Arietis......	15 15	35	56	13 17	37	58
2 [illegible] N.	γ Ceti	52	15 12	14 33	54	13 14	12 35
3 N.	α Ceti	16 10	15 31	14 52	14 13	13 33	12 54
49 N.	α Persei	29	50	15 11	32	52	13 13
23 N.	η Tauri	54	16 15	36	57	14 17	38
13 32 S.	γ^1 Eridani ...	17 7	28	49	15 10	30	51
16 [illegible] N.	α Tauri	43	17 3	16 24	46	15 6	14 27
45 51 N.	α Aurigæ ...	18 21	17 41	17 2	16 23	15 44	15 5
8 22 S.	β Orionis......	23	43	4	25	46	7
28 29 N.	β Tauri	33	53	14	35	56	17
0 24 S.	δ Orionis......	40	18 0	21	42	16 2	34
17 56 S.	α Leporis......	42	2	23	44	4	26
1 18 S.	ι Orionis......	18 44	18 4	17 25	16 46	16 6	15 28
34 S.	α Columbæ ...	50	10	31	52	12	34
7 N.	α Orionis......	19 3	23	44	17 5	25	47
22 9 N.	μ Geminor....	30	50	18 11	32	52	16 13
52 [illegible] S.	α Argus	37	57	18	39	59	20
87 N.	51 Cephei	19 45	19 5	18 26	17 47	17 7	16 28
16 15 S.	α Canis Maj.	58	18	39	18 0	20	41
28 [illegible] S.	ε Canis Maj.	20 12	32	53	14	34	55
22 14 N.	δ Geminor....	30	50	19 10	31	51	17 12
32 12 N.	α^2 Geminor....	45	20 5	26	46	18 7	28
5 N.	α Canis Min.	20 51	20 11	19 32	18 53	18 13	17 34
28 N.	β Geminor....	55	15	36	57	17	38
28 S.	15 Argus	21 20	30	20 1	19 22	42	18 3
6 35 N.	ε Hydræ	59	21 19	40	20 0	19 21	42
48 [illegible] N.	ι Ursæ Maj. .	22 9	29	50	10	31	52
58 41 S.	ι Argus	22 32	21 53	21 14	20 35	19 55	19 15
8 3 S.	α Hydræ	39	22 0	21	42	20 2	22
52 19 .	θ Ursæ Maj. .	42	3	24	45	5	25
24 26 N.	ε Leonis	56	17	38	59	19	40
12 40 N.	α Leonis	23 19	39	22 0	21 22	42	20 3
N.	γ^1 Leonis	23 26	22 46	22 7	21 29	20 49	20 10
S.	η Argus	54	23 14	35	56	21 17	38
N.	α Ursæ Maj. .	0 13	29	50	22 11	32	53
20 33 N.	δ Leonis	25	41	23 2	23	44	21 5
[illegible] [illegible] S.	δ Hydræ	31	47	8	29	50	11
15 22 N.	β Leonis	1 0	0 20	23 37	22 58	22 18	21 40
54 29 N.	γ Ursæ Maj. .	5	25	42	23 3	23	45
78 31 S.	β Chamæl. ...	28	48	0 9	26	46	22 8
62 19 S.	α^1 Crucis	37	57	18	35	55	16
22 37 S.	β Corvi	45	1 5	26	43	23 3	24

Declination.		SEPTEMBER.			OCTOBER.		
		1.	11.	21.	1.	11.	21.
° ′		h. m.	h. m.	h. m.	h. m.	h. m.	h m.
5 N.	12 Can. Ven...	2 8	1 28	0 49	0 10	23 26	22 47
25 S.	α Virginis ...	36	56	1 17	38	54	23 15
1 N.	η Ursæ Maj. .	3 0	2 20	41	1 2	0 22	39
39 6 N.	η Bootis	6	26	47	8	28	45
[illegible] 41 S.	β Centauri ...	12	32	53	14	34	51
19 N.	α Bootis	3 28	2 48	2 9	1 30	0 50	0 11
60 S.	α^2 Centauri ...	48	3 8	29	50	1 10	31
27 N.	ε Bootis	57	17	38	59	19	40
15 55 S.	α^2 Libræ	4 1	21	42	2 3	23	44
74 [illegible] N.	β Ursæ Min. .	10	30	51	12	32	53
8	β Libræ	4 27	3 48	3 9	2 30	1 50	1 11
27 .	α Coro. Bor...	46	4 6	28	49	2 9	30
6 .	α Serpentis...	55	15	37	58	18	39
78 52 S.	ζ Ursæ Min. .	5 8	28	50	3 11	31	52
19 [illegible] N.	β^1 Scorpii	14	34	56	17	37	58
20 S.	δ Ophiuchi ...	5 24	4 44	4 6	3 27	2 47	2 8
7 S.	α Scorpii	38	58	19	41	3 1	22
50 N.	η Draconis ...	40	5 0	21	43	3	24
3 46 S.	α Tria. Aust.	50	10	31	53	13	34
[illegible] 16 N.	ε Ursæ Min. .	6 20	40	5 1	4 22	43	3 4
N.	α Herculis ...	6 26	5 46	5 7	4 28	3 49	3 10
N.	β Draconis ...	45	6 5	26	47	4 7	29
N.	α Ophiuchi ...	46	6	27	48	8	30
14 33 S.	σ Octantis ...	7 5	25	46	5 6	27	47
[illegible] [illegible] N.	γ Draconis ...	11	31	52	13	33	55
6 S.	μ^1 Sagittarii...	7 23	6 43	6 4	5 25	4 45	4 7
36 N.	δ Ursæ Min. .	40	7 0	21	42	5 2	23
39 N.	α Lyræ	50	10	31	52	12	33
21 12 N.	β Lyræ	8 2	22	43	6 4	24	45
[illegible] 39 N.	ζ Aquilæ	19	39	7 0	21	41	5 2
2 50 N.	δ Aquilæ	8 39	7	7 20	6 41	6 1	5 22
10 16 N.	γ Aquilæ	57	8	38	59	19	40
8 30 N.	α Aquilæ	9 1		42	7 3	23	44
6 3 N.	β Aquilæ	6	59	47	8	28	49
85 53 N.	λ Ursæ Min. .	25	[illegible]	8 6	27	47	6 7
S.	α^2 Capricor. ...	9 29	8 49	8 10	7 31	6 51	6 11
S.	α Pavonis ...	33	53	14	35	53	15
N.	α Cygni	54	9 14	35	56	7 16	37
12 59 N.	61^1 Cygni	10 17	37	59	8 20	40	7 1
[illegible] [illegible] N.	ζ Cygni	23	43	9 5	26	46	7
61 59 N.	α Cephei	10 32	9 52	9 14	8 35	7 55	7 16
6 12 S.	β Aquarii ...	40	10 0	22	43	8 3	24
69 56	β Cephei	44	4	26	47	7	28
9 14 N.	ε Pegasi	54	14	36	57	17	38
1 1 [illegible]	α Aquarii ...	11 15	35	57	9 18	38	59
47 S.	α Gruis	11 15	10 35	9 57	9 18	8 38	7 59
10 N.	ζ Pegasi	51	11 11	10 32	54	9 14	8 35
30 S.	α Pisc. Aust.	12 6	26	47	10 9	29	50
1 89 N.	α Pegasi	14	34	55	16	37	58
4 [illegible] N.	ι Piscium ...	49	12 9	11 30	51	10 12	9 33

Declination.	Star.	NOVEMBER. 1.	NOVEMBER. 11.	NOVEMBER. 21.	DECEMBER. 1.	DECEMBER. 11.	DECEMBER. 21.
° ′		h. m.	h. m.	h. m.	h. m.	h. m.	h. m.
N.	α Androm. ...	9 16	8 37	7 58	7 18	6 39	5 59
N.	γ Pegasi	21	42	8 3	23	44	6 4
S.	β Hydri	34	55	16	36	57	17
28 19 N.	α Cassiop. ...	48	9 9	30	50	7 11	31
[illegible] S.	β Ceti	52	13	34	54	15	35
88 N.	Pole Star	10 19	9 40	9 1	8 21	7 42	7 2
8 S.	θ^1 Ceti	32	53	14	34	55	15
57 S.	α Eridani ...	48	10 9	30	50	8 11	31
22 33 N.	α Arietis......	11 14	35	56	9 16	9 37	57
2 [illegible] N.	γ Ceti	51	11 12	10 33	53	14	8 34
3 N.	α Ceti	12 10	11 31	10 52	10 12	9 33	8 53
49 N.	α Persei	29	50	11 11	31	52	9 12
23 N.	η Tauri	54	12 15	36	54	10 17	37
13 32 S.	γ^1 Eridani ...	13 7	28	49	11 9	30	50
16 [illegible] N.	α Tauri	43	13 4	12 25	45	11 6	10 26
45 N.	α Aurigæ......	14 21	13 42	13 3	12 23	11 44	11 4
8 S.	β Orionis......	23	44	5	25	46	6
28 N.	β Tauri	33	54	15	35	56	16
0 51	δ Orionis......	40	14 1	22	42	12 3	23
17 [illegible] S.	α Leporis......	42	3	24	44	5	25
1 18 S.	ι Orionis......	14 44	14 5	13 25	12 46	12 7	11 27
34 9 S.	α Columbæ ...	50	11	32	53	13	33
7 23 N.	α Orionis......	15 3	24	45	13 5	26	40
22 35 N.	μ Geminor. ...	30	51	14 12	32	53	12 13
52 37 S.	α Argus	37	58	19	39	13 0	20
87 15 N.	51 Cephei	15 45	15 6	14 27	13 47	13 8	12 28
16 31 S.	α Canis Maj..	58	19	40	14 0	21	41
28 47 S.	ε Canis Maj..	16 11	33	54	14	35	55
22 14 N.	δ Geminor. ...	28	50	15 11	31	52	13 12
32 12 N.	α^2 Geminor. ...	44	16 5	26	46	14 7	27
5 35 N.	α Canis Min.	16 50	16 11	15 31	14 51	14 12	13 32
28 22 N.	β Geminor. ...	54	15	35	55	16	36
23 54 S.	15 Argus	17 19	40	16 0	15 20	40	14 0
6 56 N.	ι Hydræ	58	17 19	39	59	15 19	39
48 36 N.	ι Ursæ Maj. .	18 8	29	49	16 9	29	49
58 41 S.	ι Argus	18 31	17 52	17 13	16 33	15 54	15 14
8 3 S.	α Hydræ	33	59	20	40	16 1	21
52 19 .	θ Ursæ Maj. .	41	18 2	23	43	4	24
24 26 N.	ε Leonis	56	17	38	58	19	39
12 40 N.	α Leonis	19 19	40	18 1	17 21	42	16 2
20 33 N.	γ^1 Leonis	19 26	18 47	18 8	17 28	16 49	16 9
58 56 S.	η Argus	54	19 15	36	56	17 17	37
62 31 N.	α Ursæ Maj. .	20 9	30	51	18 11	32	52
21 18 N.	δ Leonis	21	42	19 3	23	44	17 4
14 1 S.	δ Hydræ	27	48	9	29	50	10
15 22 N.	β Leonis	20 56	20 17	19 38	18 58	18 19	17 39
54 29 N.	γ Ursæ Maj. .	21 1	22	43	19 3	24	44
78 31 S.	β Chamæl. ...	24	45	20 6	26	47	18 7
62 19 S.	α^1 Crucis	33	54	15	35	56	16
22 37 S.	β Corvi.........	41	21 2	23	43	19 4	24

Declination.		NOVEMBER. 1.	NOVEMBER. 11.	NOVEMBER. 21.	DECEMBER. 1.	DECEMBER. 11.	DECEMBER. 21.
° ′		h. m.	h. m.	h. m.	h. m.	h. m.	h. m.
39 5 N.	12 Can. Ven...	22 3	21 25	20 46	20 6	19 27	18 47
10 25 S.	α Virginis ...	31	53	21 14	34	55	19 15
50 1 N.	η Ursæ Maj..	55	22 16	38	58	20 19	39
19 6 N.	η Bootis	23 1	22	44	21 4	25	45
59 41 S.	β Centauri ...	7	28	50	10	31	51
19 55 N.	α Bootis	23 23	22 44	22 5	21 26	20 47	20 7
60 15 S.	α² Centauri ...	43	23 4	25	46	21 7	27
27 40 N.	ε Bootis	52	13	34	55	16	36
15 27 S.	α² Libræ	56	17	38	59	20	40
74 44 N.	β Ursæ Min..	0 9	26	47	22 8	29	49
8 52 S.	β Libræ	0 27	23 44	23 5	22 25	21 47	21 7
27 12 N.	α Coro. Bor...	46	0 7	24	44	22 5	26
6 52 N.	α Serpentis...	55	16	33	53	14	35
78 14 N.	ζ Ursæ Min..	1 9	29	46	23 6	27	48
19 25 S.	β¹ Scorpii	15	35	52	12	33	54
3 20 S.	δ Ophiuchi ...	1 25	0 45	0 6	23 22	22 43	22 4
26 7 S.	α Scorpii	38	59	20	36	57	17
61 50 N.	η Draconis ...	40	1 1	22	38	59	19
68 46 S.	α Tria. Aust.	50	11	32	48	23 9	29
82 16 N.	ε Ursæ Min..	2 20	41	1 2	0 22	39	59
14 33 N.	α Herculis ...	2 26	1 47	1 8	0 28	23 45	23 5
52 24 N.	β Draconis ...	45	2 6	27	47	0 8	28
12 40 N.	α Ophiuchi ...	46	7	28	48	9	29
89 17 S.	σ Octantis ...	3 3	24	45	1 5	26	46
51 30 N.	γ Draconis ...	11	32	53	13	34	54
21 6 S.	μ¹ Sagittarii...	3 23	2 44	2 5	1 25	0 46	0 6
86 36 N.	δ Ursæ Min..	40	3 1	22	42	1 3	23
38 39 N.	α Lyræ.........	50	11	32	52	13	33
33 12 N.	β Lyræ.........	4 2	23	44	2 4	25	45
13 39 N.	ζ Aquilæ......	18	39	3 0	20	41	1 1
2 50 N.	δ Aquilæ......	4 38	3 59	3 20	2 40	2 1	1 21
10 16 N.	γ Aquilæ......	56	4 17	39	59	19	40
8 30 N.	α Aquilæ......	5 0	21	43	3 3	24	44
6 3 N.	β Aquilæ......	5	26	48	8	29	49
85 53 N.	λ Ursæ Min..	23	44	4 5	25	46	2 7
12 59 S.	α² Capricor....	5 28	4 49	4 10	3 30	2 51	2 12
57 11 S.	α Pavonis ...	32	53	14	34	55	16
44 46 N.	α Cygni	53	5 14	35	56	3 17	37
38 3 N.	61¹ Cygni	6 17	38	59	4 20	41	3 1
29 39 N.	ζ Cygni	33	44	5 5	26	47	7
61 59 N.	α Cephei......	6 32	5 53	5 14	4 35	3 56	3 16
6 12 S.	β Aquarii......	40	6 1	22	43	4 4	24
69 56 N.	β Cephei	44	5	26	46	7	28
9 14 N.	ε Pegasi	54	15	36	56	17	38
1 1 S.	α Aquarii ...	7 15	36	57	5 17	38	59
47 39 S.	α Gruis	7 15	6 36	5 57	5 17	4 38	3 59
10 5 N.	ζ Pegasi	51	7 12	6 33	53	5 14	4 35
30 22 S.	α Pisc. Aust..	8 6	27	48	6 8	29	49
14 27 N.	α Pegasi	14	35	56	16	37	57
4 51 N.	ι Piscium ...	49	8 10	7 31	51	6 12	5 32

Table 10.

Lat.	Declination of the same name as the Latitude.												
	0°	2°	4°	6°	8°	10°	12°	14°	16°	18°	20°	22°	24°
°	h.m.	h.m.	h.m.	h.m.	h.m.	h.m.	h.m.	h.m.	h.m.	h.m.	h.m.	h.m.	h.m.
1	6 0	4 0	5 2	5 22	5 31	5 37	5 41	5 44	5 46	5 48	5 49	5 50	5 51
2	0	0 0	4 0	4 42	2	14	22	28	32	35	38	40	42
3	0	3 13	2 46	0	4 32	4 51	3	11	18	23	27	30	33
4	0	4 0	0 0	3 13	1	27	4 43	4 55	4	10	16	20	24
5	0	26	2 28	2 15	3 26	1	23	38	4 49	4 58	4	10	15
6	0	42	3 13	0 0	2 46	3 34	1	20	34	45	4 53	0	5
7	0	54	41	2 5	1 56	3	3 39	2	19	31	41	4 49	4 56
8	0	5 2	4 1	46	0 0	2 29	14	3 43	3	17	29	39	46
9	0	9	15	3 14	1 50	1 44	2 47	22	3 46	3	17	28	37
10	6 0	5 14	4 27	3 34	2 29	0 0	2 16	3 0	3 28	3 49	4 4	4 16	4 27
11	0	16	36	49	55	1 40	1 35	2 35	9	33	3 51	5	16
12	0	22	43	4 1	3 14	2 16	0 0	6	2 49	17	37	3 53	6
13	0	25	49	12	30	41	1 32	1 29	26	2 59	23	41	3 55
14	0	28	55	20	43	3 0	2 6	0 0	1 58	40	7	28	44
15	0	30	59	28	53	15	30	1 26	23	18	2 50	14	32
16	0	32	5 4	34	4 3	28	49	58	0 0	1 52	32	2 59	20
17	0	34	7	40	11	39	3 4	2 21	1 21	19	11	43	7
18	0	35	10	45	17	49	17	40	52	0 0	1 47	26	2 53
19	0	37	13	49	24	57	28	54	2 14	1 17	16	6	37
20	6 0	5 38	5 16	4 53	4 29	4 4	3 37	3 7	2 32	1 47	0 0	1 43	2 21
21	0	39	18	56	34	11	46	18	47	2 9	1 14	13	2
22	0	40	20	5 0	39	16	53	28	59	26	43	0 0	1 39
23	0	41	22	3	43	22	4 0	36	3 10	40	2 4	1 11	10
24	0	42	24	5	46	27	6	44	20	53	21	39	0 0
25	0	43	26	8	50	31	12	51	28	3 3	35	2 0	1 9
26	0	44	27	10	53	35	17	57	36	13	47	16	36
27	0	44	28	12	56	39	21	4 3	43	22	3 0	30	56
28	0	45	30	14	59	43	26	8	49	29	7	42	2 18
29	0	46	31	16	5 1	46	30	13	55	36	16	53	26
30	6 0	5 46	5 32	5 18	5 4	4 49	4 34	4 18	4 1	3 43	3 24	3 2	2 38
31	0	47	33	20	6	52	37	22	6	49	31	11	49
32	0	47	34	21	8	54	40	26	11	55	38	19	58
33	0	48	35	23	10	57	44	30	15	4 0	44	26	3 7
34	0	48	36	24	12	59	47	33	19	5	49	33	15
35	0	49	37	25	14	5 2	49	37	23	9	55	39	22
36	0	49	38	27	15	4	52	40	27	14	4 0	45	29
37	0	49	39	28	17	6	54	43	31	18	4	50	35
38	0	50	39	29	19	8	57	46	34	22	9	55	41
39	0	50	40	30	20	10	59	48	37	25	13	4 0	47
40	6 0	5 50	5 41	5 31	5 21	5 11	5 1	4 51	4 40	4 29	4 17	4 5	3 52
41	0	51	42	32	23	13	3	53	43	32	21	9	57
42	0	51	42	33	24	15	5	56	46	35	25	13	4 1
43	0	51	43	34	25	16	7	58	48	38	28	17	6
44	0	52	43	35	27	18	9	5 0	51	41	31	21	10
45	0	52	44	36	28	19	11	2	53	44	35	25	14
46	0	52	45	37	29	21	13	4	56	47	38	28	18
47	0	53	45	38	30	22	14	6	58	49	41	31	22
48	0	53	46	38	31	23	16	8	5 0	52	43	35	25
49	0	53	46	39	32	25	17	10	2	54	46	38	29
50	6 0	5 53	5 47	5 40	5 33	5 26	5 19	5 12	5 4	4 57	4 49	4 41	4 32
52	0	54	47	41	35	28	22	15	8	5 1	54	46	39
54	0	54	48	42	37	31	24	18	12	5	59	52	45
56	0	55	49	44	38	33	27	21	15	9	5 3	57	50
58	0	55	50	45	40	35	29	24	19	13	7	5 2	55
60	0	55	51	46	41	37	32	27	22	17	11	6	5 0
62	0	56	51	47	43	38	34	30	25	20	15	10	5
64	0	56	52	48	44	40	36	32	28	24	19	15	10
66	0	56	53	49	46	42	38	35	31	27	23	19	14
68	0	57	54	50	47	44	40	37	33	30	26	22	19
70	0	57	54	51	48	45	42	39	36	33	30	26	23

Lat.	Declination of the same name as the Latitude.												
	26°	28°	30°	32°	34°	36°	38°	40°	42°	44°	46°	48°	50°
°	h.m.	h.m.	h.m.	h.m.	h.m.	h.m.	h.m.	h.m.	h.m.	h.m.	h.m.	h.m.	h.m.
1	5 52	5 52	5 53	5 54	5 54	5 54	5 55	5 55	5 56	5 56	5 56	5 56	5 57
2	44	45	46	47	48	49	50	50	51	52	52	53	53
3	35	37	39	41	42	43	45	46	47	48	48	49	50
4	27	30	32	34	36	38	39	41	42	43	45	46	47
5	19	22	25	28	30	32	34	36	38	39	41	42	43
6	10	14	18	21	24	27	29	31	33	35	37	38	40
7	2	7	11	15	18	21	24	26	29	31	33	35	36
8	4 53	4 59	4	8	12	15	19	21	24	27	29	31	33
9	44	51	4 56	1	6	10	13	16	19	22	25	27	29
10	4 35	4 43	4 49	4 54	4 59	5 4	5 8	5 11	5 15	5 18	5 21	5 23	5 26
11	26	34	41	48	53	58	2	6	10	14	17	20	22
12	17	26	34	40	47	52	4 57	1	5	9	13	16	19
13	7	17	26	33	40	46	51	4 56	1	5	8	12	15
14	3 57	8	18	26	33	40	46	51	4 56	0	4	8	12
15	47	3 59	9	18	26	33	40	46	51	4 56	0	4	8
16	36	49	1	11	19	27	34	40	46	51	4 56	0	4
17	25	40	3 52	3	12	20	28	35	41	46	51	4 56	1
18	13	29	43	3 55	5	14	22	29	35	41	47	52	4 57
19	0	19	34	46	3 57	7	15	23	30	36	42	48	53
20	2 47	3 7	3 24	3 38	3 49	4 0	4 9	4 17	4 25	4 31	4 38	4 43	4 49
21	32	2 55	13	28	41	3 52	2	11	19	26	33	39	45
22	16	42	2	19	33	45	3 55	5	13	21	28	35	41
23	1 58	28	2 51	9	24	37	48	3 58	7	16	23	30	37
24	36	13	38	2 58	15	29	41	52	1	10	18	25	32
25	8	1 55	25	47	5	20	33	45	3 55	5	13	21	28
26	0 0	34	9	35	2 55	11	25	38	49	3 59	8	16	23
27	1 7	6	1 52	21	44	2	17	30	42	53	2	11	19
28	34	0 0	32	7	32	2 52	8	23	35	46	3 56	6	14
29	53	1 6	5	1 50	19	41	2 59	15	28	40	50	0	9
30	2 9	1 32	0 0	1 30	2 5	2 30	2 49	3 6	3 20	3 33	3 44	3 55	4 4
31	23	51	1 4	4	1 48	17	39	2 57	13	26	38	49	3 59
32	35	2 7	30	0 0	28	3	28	47	4	19	32	43	54
33	45	20	49	1 3	3	1 47	15	37	2 55	11	25	37	48
34	55	32	2 5	28	0 0	27	1	26	46	3	17	30	42
35	3 3	42	18	47	1 2	2	1 45	14	36	2 54	10	24	36
36	11	52	30	2 3	27	0 0	26	0	25	45	2	17	30
37	19	3 0	40	16	46	1 2	1	1 44	13	35	2 53	9	23
38	25	8	49	28	2 1	26	0 0	26	1 59	24	44	1	16
39	32	16	58	38	14	45	1 1	1	44	12	34	2 53	9
40	3 38	3 23	3 6	2 47	2 26	2 0	1 26	0 0	1 25	1 59	2 23	2 44	3 1
41	43	29	14	56	36	13	44	1 1	0	43	12	34	2 53
42	49	35	20	3 4	46	25	59	25	0 0	25	1 58	23	44
43	54	41	27	12	55	35	2 12	48	1 0	0	43	12	34
44	59	46	33	19	3 4	45	24	59	25	0 0	25	1 58	23
45	4 3	52	39	25	10	54	34	2 12	43	1 0	0	43	12
46	8	56	44	32	17	3 2	44	23	58	25	0 0	25	1 59
47	12	4 1	50	37	24	9	53	34	2 12	43	1 0	0	43
48	16	6	55	43	30	17	3 1	44	23	58	25	0 0	25
49	20	10	4 0	48	36	23	9	53	34	2 12	43	1 0	0
50	4 23	4 14	4 4	3 54	3 42	3 30	3 16	3 1	2 44	2 23	1 59	1 25	0 0
52	30	22	13	4 3	53	42	30	16	3 1	44	2 24	59	1 26
54	37	29	21	12	4 3	53	42	30	17	3 2	45	2 25	2 0
56	43	36	28	20	12	4 3	53	42	30	17	3 3	46	26
58	49	42	35	28	20	12	4 3	54	43	32	19	3 4	47
60	55	48	42	35	28	21	13	4 4	55	44	33	20	3 6
62	5 0	54	48	42	36	29	22	14	4 6	56	46	35	23
64	5	5 0	55	49	43	37	30	23	16	4 8	59	49	38
66	10	5	5 0	55	50	45	39	32	25	18	4 10	4 1	52
68	15	10	6	5 2	57	52	46	41	35	28	21	13	4 5
70	19	15	11	7	5 3	59	54	49	43	38	31	22	17

Table 11.

Declination of the same name as the Latitude.

Lat.	0°		2°		4°		6°		8°		10°		12°		14°		16°		18°		20°		22°		24°	
°	°	'	°	'	°	'	°	'	°	'	°	'	°	'	°	'	°	'	°	'	°	'	°	'	°	'
1	0	0	30	0	14	29	9	37	7	12	5	46	4	49	4	8	3	38	3	14	2	55	2	40	2	28
[illegible]		0	90	0	30	1	19	30	14	31	11	36	9	40	[illegible]	18	7	16	6	29	5	51	5	21	4	55
		0	41	49	48	37	30	3	22	5	17	32	14	35	1	30	10	57	9	45	8	48	8	2	7	24
		0	30	1	90	0	41	52	30	5	23	41	19	36	1	45	14	40	13	[illegible]	11	46	10	44	9	53
		0	23	36	53	10	56	30	38	46	30	8	24	47	2	7	18	26	16	2	14	46	13	27	12	22
		0	19	30	41	52	90	0	48	41	37	1	30		25	36	22	17	19	[illegible]	17	48	16	12	14	53
		0	16	38	34	55	59	4	61	7	44	34	35	53	30	15	26		23		20	52	18	59	17	26
8		0	14	31	30	5	48	41	90	0	53	16	42	1	35	7	30	20	26	46	24	1	21	49	20	1
9		0	12	53	26	29	41	56	62	50	64	16	48	48	40	17	34	35	30	25	27	13	24	41	22	37
10	0	0	11	36	23	41	37	1	53	16	90	0	56	38	45	52	39	3	34	11	30	31	27	37	25	16
11		0	10	32	21	27	33	13	46	50	65	31	[illegible]	36	52	4	[illegible]	[illegible]	38	8	33	55	30	37	27	59
12		0	9	40	19	36	30	11	42	1	56	38		0	59	15			42	17	37	26	33	43	30	45
13		0	8	56	18	4	27	41	38	13	50	32		33	68	25			46	43	41	8	36	54	33	35
14		0		18	16	45	25	36	35	7	45	52		15	90	0			51	32	45	1	40	14	36	30
15		0	7	45	15	38	23	49	32	32	42	8		27	69	11	69		56	53	49	11	43	42	39	31
16		0		16	14	40	22	17	30	20	39	3		58	61	22	90		63	7	53	42	47	22	42	40
17		0	6	51	13	48	20	57	28	26	36	26	45	20	53	50	70		71	7	58	45	51	18	45	57
18		0		29	13	3	19	46	26	46	34	11	42	17	51	32	63		90	0	64	37	55	35	49	27
19		0		9	12	22	18	44	25	18	32	14	39	41	48	0	57		71	39	72	9	60	21	53	10
20	0	0	5	51	11	46	17	48	24	1	30	31	37	26	45	1	53	42	64	37	90	0	65	55	57	14
21		0		35		13	16	57	22	51	28	59	35	28	42	28	50	17	59	34	72	38	73	4	61	47
22				21	10	44		12	21	49	27	37	33	43	40	14	47	22	55	35	65	55	90	0	67	5
23				7		17	15	31	20	52	26	23	32	9	38	15	44	52	52	16	61	5	73	29	73	53
24			4	55	9	53	14	53		1	25	16	30	45	36	30	42	40	49	27	57	14	67	5	90	0
25				44		30		19	19	14	24	16	29	28	34	55	40	43	46	59	54	1	62	25	74	16
26				31		9	13	48	18	31	23	20	28	19	33	30	38	58	44	49	51	17	58	43	68	5
27				25	8	50		19	17	51	22	29	27	15	32	12	37	23	42	54	48	53	55	36	63	37
28				16		33	12	52		15	21	42	26	17	31	1	35	57	41	10	46	46	52	56	60	3
29				8		16		27	16	41	20	59	25	24	29	56	34	39	39	36	44	52	50	36	57	2
30	0	0	4	0	8	1	12	4	16	10	20	19	24	34	28	56	33	27	38	10	43	10	48	31	54	26
31		0	3	53	7	47	11	43	15	41	19	42	23	49		1	32	21	36	52	41	37	46	40	52	10
32				47		34		23		14		8		6	27	10	31	21	35	40	40	12	44	59	50	9
33				40		22		4	14	48	18	36	22	26	2[illegible]	22	30	24	34	34	38	54	43	27	48	18
34				35		10	10	46		[illegible]		5	21	50	2	38	29	32	33	33	37	42	42	4	46	39
35				29	6	59		30			17	37		15	2	57	28	43	32	36	36	36	40	47	45	11
36				24		49		16	13			11	20	43		18	27	58	31	43	35	35	39	36	43	46
37				19		39		0		22	16	46		13	23	42		16	30	54	34	38	38	30	42	31
38				15		31	9	47		4		23	19	44		8	26	36		8	33	45	37	29	41	20
39				11		22		34	12	47		1		18	22	36	25	59	29	25	32	55	36	32	40	16
40	0	0	3	7	6	14	9	22	12	30	15	40	18	52	22	7	25	24	28	44	32	9	35	39	39	15
41		0		3		6		10		15		21		29	21	38	24	51		6	31	25	34	49	38	18
42		0	2	59	5	59	8	59		0		2		6		12		20	27	30	30	44		3	37	25
43		0		56		52		49	11	47	14	45	17	45	20	47	23	50	26	57		6	33	19	36	36
44		0		53		46		39		33		29		25		23		23		25	29	30	32	38	35	49
45				50		40		30		21		13		6		0	22	57	25	55	28	56	31	59		6
46				47		34		21		9	13	58	16	48	19	39		32		26		28		25	34	25
47				44		28		13	10	58		41		31		19		8		0	27	53	30	49	33	47
48				42		23		5		48		31		15		0	21	46	24	34		24		16		12
49		0		39		18	7	58		38		18	15	59	18	42		25		10	26	57	29	46	32	35
50	0	0	2	37	5	13	7	51	10	28	13	6	15	45	18	25	21	5	23	47	26	31	29	17	32	4
52		0		32		5		37		10	12	44		18	17	53	20	28		5	25	43	28	23	31	7
54				28	4	57		25	9	54		24	14	54		24	19	55	22	27		1	27	35	30	10
56				5		50		15		40		5		31	16	58		25	21	53	24	22	26	52	29	22
58				2		43		5		27	11	49		11		35	18	58		22	23	47		13	28	39
60				9		37	6	56		15		34	13	53		13		34	20	54		16	25	38		0
62				16		31		47		8		19		37	15	54		12		29	22	48		6	27	26
64				14		27		41	8	54		8		23		37	17	52		7		22	24	38	26	54
66				11		23		34		46	10	57		9		21		34	19	46	21	59		13		26
68				9		19		28		28		48	12	57		7		18		28		39	23	50		1
70				8		15		23		31		39		47	14	55		3		12		21		30	25	39

Lat.	Declination of the same name as the Latitude.																									
	26°		28°		30°		32°		34°		36°		38°		40°		42°		44°		46°		48°		50°	
°	°	'	°	'	°	'	°	'	°	'	°	'	°	'	°	'	°	'	°	'	°	'	°	'	°	'
1	2	17	2	8	2	0	1	53	1	47	1	42	1	37	1	33	1	30	1	26	1	23	1	21	1	[illegible]
2	4	34	4	16	4	0	3	47	3	35	3	24	3	15	3	7	2	59	2	53	2	47	2	42	[illegible]	
3	6	51	6	24	6	0	5	[illegible]	5	22	5	6	4	53	4	40	4	29	4	19	4	10	4	2		
4	9	9	8	33	8	1	7		7	10	6	49	6	31	6	14	5	59	5	46	5	34	5	23		
5	11	28	10	42	10	2	9	28	8	58	8	32	8	8	7	48	7	29	7	12	6	58	6	44		
6	13	48	[illegible]	52	12	4	11	23	10	46	10	16	9	47	9	22	8	59	8	39	8	21	8	5		
7	16	8		3	14	6	13	18	12	35	11	58	11	25	10	56	10	30	10	6	9	45	9	26		
8	18	31		15	16	10	15	14	14	25	13	42	13	4	12	30	12	0	11	33	11	9	10	48	1	
9	20	54		28	18	14	17	10	16	15	15	26	14	43	14	5	13	31	13	1	12	34	12	9	1	
10	23	20	2	42	20	19	19	8	18	5	17	11	16	23	15	40	15	2	14	29	13	58	13	31	13	6
11	25	48		59	22	26	23	6	19	57	18	57	[illegible]	3	17	16	16	31	15	57	28	23	14	53	14	25
12	28	19		17	24	34		6	21	50	20	43	9	41	18	52	18	6	17	25		48	16	15	15	45
13	30	52		38	26	44	25	7	23	43	22	30	2	26	20	29	19	39	18	54		15	17	37	17	5
14	33	30	1	1	28	56	27	10	25	38	24	18		8	22	7	21	12	20	23		39	19	6	18	25
15	36	11		27	31	10	29	14	27	34	26	8	2	52	23	45	22	45	21	53	1	5	20	23	19	45
16	38	58		57	33	27	33	21	29	32	27	58		36	25	24	24	20	23	23	22	32	21	46	21	5
17	41	50		31	35	47		29	31	31	29	50		21	27	3	25	55	24	53	23	59	23	10	22	26
18	44	49	1	10	38	10		40	33	33	31	43		58	28	44	27	30	26	35	25	26	24	34	23	47
19	47	58		54	40	38		54			33	38	1		30	26	29	7	27	7		55	25	59	25	9
20	51	17	46	46	43	10	40	12	37	42	35	35	33	45	32	9	30	44	29	30	28	23	27	24	26	31
21	54	50	52	[illegible]	45	47	42	33	39	51	37	34	[illegible]	36	33	53	[illegible]	23	[illegible]	3	[illegible]	[illegible]	[illegible]	50	27	54
22	58	48			[illegible]	31	44	59	42	4	39	36		29	35	39		3		38				16	29	17
23	63	2	56			24	[illegible]	30	44	20	41	40		24	37	26		44		14				43	30	40
24	68	6	60			26		8	[illegible]	40	43	47		21	39	15		26		50	34			11	32	4
25	74	3	64	1		42	2	54		6	45	58		21	41	6		10	7	28	35			40	33	29
26	90	0	69			15	5	49		37	48	14	45	24	43	0		56		8				9	34	54
27	74	56	[illegible]			14		57		17	50	34	47	31	44	56		44		49				39	36	21
28	69	2				52	62	22	7	6	53	0	[illegible]	41	46	55		38		31	40			11	37	48
29	64	48			5	50	66	11		7	55	34		57	48	57		26		16	42			43	39	16
30	61	15	69	52	90	0	70	39	63	24	58	17	54	18	51	4	48	21	46	2	44	2	42	17	40	45
31	58	20	68	43	[illegible]	[illegible]	76	23	67	5	61	11	[illegible]	47	53	15	50	20	47	[illegible]	45	43	45	52	42	15
32	55	49		22			90	0	71	23	64	22		24	55	32	2	22	49		47	27	4	29	43	46
33	53	36	59	32			76	39	76	54	67	55	6	12	57	55	54	29	[illegible]		49	13	47	8	45	19
34	51	37	57	6			71	23	90	0	72	3		16	60	27	56	41		37	51	1	[illegible]	48	46	53
35	49	51	[illegible]	56			67	30	77	8	77	22	68	42	63	10	59	0		40	52	53		31	48	29
36	48	14		0			64	[illegible]	72	3	90	0	72	[illegible]	66	8	61	27	7	48	54	48		16	50	7
37	46	45		16			[illegible]		68	18	77	36	77		69	26	64	5		2	56	47		5	51	47
38	45	24		41					65	16	72	42	90		73	18	66	56		25	58	51		56	53	29
39	44	9		15				4	62	42	69	4	78		78	15	7	8		57	61	2		52	55	14
40	43	0	46	55	51	4	55	32	60	27	66	8	73	[illegible]	90	0	73	52	67	43	63	20	59	53	37	3
41	41	56	[illegible]	42	48	39	[illegible]	53	58	28	63	38	[illegible]		78	28	78	39	70	49	65	47	[illegible]	59	[illegible]	55
42	40	56		33		21	2	22	56	41	61	27			73	52	90	0	74	25	68	28		13		52
43		0		30	4	9		59	55	5	59	32		1	70	29	78	51	79	3	71	27		36		55
44	39	8		31	46	2		43	53	37	57	48			67	43	74	25	90	0	74	57		11		4
45	38	19		36	[illegible]	0		32	52	16	[illegible]	14			65	22	71	8	79	14	79	25		5		23
46	37	33		45		2		27	51	1		[illegible]		51	63	20	68	28	74	57	90	0		28		53
47	36	50		56		8		[illegible]	49	52	53			20	61	31	66	12	71	46	79	36	79	47		42
48		9		11		17			48	48	52			56	59	53	64	13	69	11	75	28	90	0		57
49	35	31	38	28		29			47	49	51			40	58	24	62	27	66	59	72	23	79	57	80	8
50	34	54	37	48	40	45	43	46	46	53	50	7	53	29	57	3	60	52	65	4	69	53	75	57	90	0
52	33	48	[illegible]	[illegible]	30	[illegible]	[illegible]	[illegible]	45	12	48	[illegible]	[illegible]	23	54	40	58	7	61	50	65	51	70	34	76	25
54	32	49			33				43	43	46			33	52	37	55	48	59	10	62	46	66	43	71	14
56	31	55			37				[illegible]	25	45			57	50	50	53	49	56	55	60	11	63	41	67	31
58		8			36					15	43		46	33	49	17	52	6	55	0	58	1	61	12	64	[illegible]
60	30	25			35					13	42		45	19	47	55	[illegible]	36	53	20	56	10	59	6	62	
62	29	46		7	34			53		18	[illegible]		[illegible]	13	46	43		17	51	53	54	32	57	19	60	
64		11	31	30	33			8		28				14	45	40		7	50	37	53	10	55	47	58	
66	28	41	30	[illegible]			35	[illegible]	7	45				22	44	43		6	49	30	51	57	54	26	56	
68		13			32		34			6	39			36	43	53		12	48	31	50	53	53	16	55	
70	27	48	29						36	31	38			56		10		24	47	4	49	57	52	16	54	

PROPER NAMES OF CERTAIN OF THE PRINCIPAL FIXED STARS.

α Ursæ Minoris,	that is,	α of Ursa Minor (Little Bear)	Pole Star.
α Andromedæ	„	α of Andromeda	Alpheratz.
γ Pegasi	„	γ of Pegasus	Algenib.
α Cassiopeæ	„	α of Cassiopea	Schedar.
α Eridani	„	α of Eridanus	Achernar.
α Persei	„	α of Perseus	Mirfack.
β Persei	„	β of Perseus	Algol.
α Tauri	„	α of Taurus (Bull)	Aldebaran.
α Aurigæ	„	α of Auriga (Charioteer)	Capella.
α Orionis	„	α of Orion	Betelgeuse.
β Orionis	„	β of Orion	Rigel.
γ Orionis	„	γ of Orion	Bellatrix.
α Argus	„	α of Argo	Canopus.
α Canis Majoris,	„	α of Canis Major (Great Dog)	Sirius.
α Canis Minoris	„	α of Canis Minor (Little Dog)	Procyon.
α^2 Geminorum	„	α^2 of Gemini (Twins)	Castor.
β Geminorum	„	β of Gemini	Pollux.
α Leonis	„	α of Leo (Lion)	Regulus.
β Leonis	„	β of Leo	Denebola.
α Ursæ Majoris	„	α of Ursa Major (Great Bear)	Dubhe.
α Virginis	„	α of Virgo (Virgin)	Spica.
α Bootis	„	α of Bootes (Herdsman)	Arcturus.
α Scorpii	„	α of Scorpio (Scorpion)	Antares.
α Lyræ	„	α of Lyra (Harp)	Vega.
α Aquilæ	„	α of Aquila (Eagle)	Altair.
α Cygni	„	α of Cygnus (Swan)	Deneb.
α Piscis Australis	„	α of Piscis Aust. (Southern Fish)	Fomalhaut.
α Pegasi	„	α of Pegasus	Markab.

α (a) Alpha, β (b) Beta, γ (g) Gamma, δ (d) Delta, ε (ĕ) Epsilon, ζ or ϛ (z) Zeta, η (ē) Eta, θ (th) Theta, ι (i) Iota, κ (k) Kappa, λ (l) Lambda, μ (m) Mu, ν (n) Nu, ξ (x) Xi, ο (ŏ) Omicron, π, ϖ (p) Pi, ϱ (r) Ro, σ (s) Sigma, τ (t) Tau, υ (u) Upsilon, φ (ph) Phi, χ (ch) Chi, ψ (ps) Psi, ω (ō) Omega.

NOTE

ON USING THE BLANK FORMS FOR THE DIFFERENT COMPUTATIONS OF NAUTICAL ASTRONOMY.

In the Treatise to which the foregoing Tables are more especially adapted, a variety of *Blank Forms* are given to supply the place of verbal rules; but, as noticed in the Preface, the narrowness of the page rendered it sometimes necessary to depart from the arrangement of the several steps of the work as they appeared in the manuscript. We shall here exhibit a portion of one of these Forms, the Form at page 194, with the steps arranged as originally intended.

It may be remarked, however, that the relative positions on the paper, of the several distinct items of work, is a matter of little or no consequence. The main thing to be attended to, in filling up the blanks, is, first, to postpone reference to the Nautical Almanac and to the Book of Tables as long as possible, forwarding the work as much as we can without the aid of either; and, second, when the Nautical Almanac or the volume of Tables is once in hand, to make all the use of it we can before laying it down.

Thus, the partial Form on next page consists of four distinct items of computation; and, whether these be placed with respect to one another as here, or as in the Treatise referred to, we should proceed to fill up the blanks in the following manner:—

We should first insert the *Observed Altitude* in [1]; then passing to [2], we should fill up all the three blanks, and thus get the *Greenwich Date*. Taking now the first volume of the Tables, we should return to [1], and insert from them the corrections for *Index and Dip*, and for *Refraction—Parallax*.

We are now to take up the Nautical Almanac, and to extract from it the *Semidiameter*, the *Equation of Time*, and the *Noon Declination*, all of which particulars appear on the same page, inserting each as the blank form directs: and then, laying the Almanac aside, we are to complete this portion of the work, thus obtaining the True Altitude, the Equation of Time, and the Polar Distance.

Partial Blank Form. (Nautical Astronomy, page 194.)

For the True Altitude, Polar Distance, and Equation of Time

[1.]

		° ′ ″
Obs. alt.		. . ° . . ′ . . ″
Index and Dip	. . ′ . . ″	. . .
Semi-diam.	. . . .	
App. alt. of centre		. . . . . .
Ref.—Parallax		— . . . .
True Alt. of Centre		. . . . . .

[3.]

Equa. of Time	. . m . . s	Diff. ″ . .	
Cor. for the hrs.	. .	× . . hrs.	
Equa. of Time	. . . .	Cor. . . ″	

[2.]

Time, per Watch	. . h . . m . . s
Long. in time	. . . . . .
Mean Time at G.	. . . . . .

[4.]

Non Dec. at G	. . ° . . ′ . . ″	Diff. . . ″
Cor. for the hrs.	. . . . . .	× . . hrs.
Declination	. . . . . .	
	90	60). . . .
Polar Dist.	. . . . . .	Cor. . . ′ . . ″

BRADBURY AND EVANS, PRINTERS, WHITEFRIARS.

CPSIA information can be obtained
at www.ICGtesting.com
Printed in the USA
BVHW09*1515200818
525056BV00011B/678/P

9 780260 084231